Basic
Structural Analysis

CIVIL ENGINEERING AND ENGINEERING MECHANICS SERIES

N. M. Newmark and W. J. Hall, editors

Basic
Structural Analysis

KURT H. GERSTLE
University of Colorado

Prentice-Hall, Inc. Englewood Cliffs, New Jersey

Library of Congress Cataloging in Publication Data

GERSTLE, KURT H
    Basic structural analysis.

    (Civil engineering and engineering mechanics
series)

    Bibliography: p.
    1. Structures, Theory of. I. Title.
TA645.G44   624'.171   73–8912
ISBN 0–13–069393–6

© 1974 by
PRENTICE-HALL, INC.
Englewood Cliffs, New Jersey 07632

10 9 8 7 6 5 4 3 2 1

Printed in the United States of America

PRENTICE-HALL INTERNATIONAL, INC., *London*
PRENTICE-HALL OF AUSTRALIA, PTY LTD., *Sidney*
PRENTICE-HALL OF CANADA, LTD., *Toronto*
PRENTICE-HALL OF INDIA PRIVATE LIMITED, *New Delhi*
PRENTICE-HALL OF JAPAN, INC., *Tokyo*

In Memory of My Father
Siegfried Gerstle
1883–1961

Contents

Preface

No area of engineering has undergone greater changes within the last decades than that of structural analysis. The matrix formulation of structural analysis has not only shed a great deal of light on the basic unity of seemingly unrelated methods, but also has provided a bookkeeping scheme which makes it possible to treat structures of a size and complexity which were previously too great to contemplate. The use of the digital computer, whose development was intimately connected with that of matrix methods, finally enabled the structural engineer to carry such analyses to their logical conclusion in applying them to the design process to arrive at increasingly more functional, economic, and beautiful structures.

Many excellent texts have appeared in recent years which present the newer methods of analysis on a senior and graduate level; they usually presume at least one semester's exposure to the basics of structural analysis. This first course often follows classical treatment, emphasizing the physical behavior of structures under load, and presenting the different methods of analysis in order of their historical appearance. The transition from such a first course to the subsequent exposure to modern methods presents the structures student with unwarranted hardship; in all too many cases, he will have to start from scratch with seemingly new approaches.

The present book is intended for a first course in structural analysis, following the usual course in mechanics of deformable bodies (or strength of materials, as it is commonly misnamed). Its aim is to provide a smooth transition from the classical approaches through a development of a feel for the physical behavior of structures in terms of their deformed shape, to a formal treatment of a general class of structures by means of matrix formulation.

In particular, the presentation tries to avoid the "culture shock" which often affects the student when he is asked to reformulate previously well-understood relations in matrix form. To this end, these relations are presented in this book in matrix form at as early a stage as possible, immediately after their development in classical ways and before the going gets heavy. In this way, it is hoped to enable the student to acquire the necessary building blocks one at a time, for readiness when needed in the more powerful approaches. In keeping with the power of the modern methods, two- and three-dimensional structures are presented side by side. While the former are often used to explain the concept, no effort has been made to avoid more general forms in this text.

The background required of the student for this course conforms to most present-day engineering curricula. It is expected that elementary structural mechanics, including determination of reactions, calculation of internal forces in statically determinate beams and trusses, and the like, has already been touched upon in earlier courses in mechanics of rigid and deformable bodies; this material is therefore covered relatively lightly. It is, of course, up to the instructor whether to emphasize this material, or to treat it more by way of review.

Some background in the use of vector methods is also useful here. This approach, which is now commonly followed in courses of mechanics of rigid bodies, has unfortunately been largely neglected in structural analysis. Not only does it furnish a powerful tool for analyzing certain structures, but its use here would also provide purpose and motivation for its presentation in the earlier course. Similarly, the student is expected to be familiar with the elementary operations of matrix algebra, such as matrix multiplication, inversion, partitioning, and similar operations which are usually treated in previous mathematics courses. The use of the computer programs calls on some prior knowledge of FORTRAN programming, which most present-day students possess at this stage. Indeed, the use and development of this skill should be emphasized consistently in undergraduate courses.

The book is intended to be used for either a one-semester introductory course or a two-semester sequence. The topic coverage is arranged so as to provide freedom in the selection of material. The general presentation stresses the attainment of both a physical feel for structural action, as well as the skill to formulate problems in a general matrix way. If the former is to be emphasized, the advanced matrix-oriented topics, as well as the computer programs, can be deleted. If the orientation is to be toward the latter,

Chapters 6 and 7, as well as Chapters 12 and 14, can be left out. All sections dealing with trusses can also be skipped without affecting the development of the theory. Chapters 15 and 16 would probably only be covered in a two-semester sequence.

A thorough understanding of the methods of structural analysis, and the necessary skill in applying them to real-life analyses, is developed only through solving problems such as the ones provided at the ends of the chapters. No effort should be made to avoid some of the more demanding ones; it may well be that more is learned from one challenging problem than from a string of routine exercises. None of the problems intended for longhand calculation requires an undue amount of work, but care should be exercised in their selection, since some of them require the results of previous exercises. Problems intended for computer use should be assigned only if the computer programs provided in Appendix A, or similar routines, have been suitably prepared for student use. A number of subroutines is also listed in Appendix A to allow flexibility in preparing program variations.

A persistent problem in the teaching of structural analysis is that of motivation; therefore, it should throughout the course be held before the students' eyes that structural analysis is not an end in itself, but only a tool for good design. The constructive implications of the results of analyses, comparison between the behavior of statically determinate and indeterminate structures, use of models, and critical discussions of real structures and their idealizations could well be interlaced in the theoretical analyses by the instructor in order to reinforce the students' interest.

Much of the material in this text has been presented in the author's classes at the University of Colorado, and a good deal of the presentation reflects feedback from his students; this is acknowledged with thanks. The author also thanks the University of Colorado for a sabbatical, which, in conjunction with a delightful Fulbright year spent at the Technical University of Munich, Germany, made this work possible. Mrs. Elizabeth Stimmel did an exemplary job of typing, and, most importantly, my wife lent her constant support and encouragement to this effort.

Kurt H. Gerstle
Boulder, Colorado

I

Introduction

Before embarking on the study of structural analysis, we shall have to know the "big picture." What is the purpose of this topic, where does it fit into the field of engineering, and what are the rules of the game?

Accordingly, in Chapter 1 we discuss very briefly the scope of structural engineering, the types of structures encountered, and their purpose; a section on the loads that they may have to carry follows. In Sec. 1.4, we touch upon the all-important topic of the modeling of structures, that is, the way in which they can be simplified for analysis purposes so that a reasonable compromise is reached between reality and simplicity. Finally a short discussion of the underlying thought and methods covered in this book is presented.

CHAPTER 1

Structural Engineering
and Analysis

1.1. The Scope of Structural Analysis

Structural engineering is concerned with the design of the works of man for adequate strength and stiffness. These requirements are to be achieved without impairing the intended primary function of the structure, and at a minimum of cost. The structural conception thus must evolve as a result of deliberations between experts of all types and with due regard for functional, aesthetic, and economic considerations.

Thus, detailed planning for a building should be preceded by studies involving city planners, owners, users, architects, and the different types of engineers. In such cases, the architect is frequently the owner's delegate and the structural engineer will be responsible to him, although sometimes in primarily industrial projects their role is reversed. Recently, offices combining the services of architects and engineers, the A–E firms, have rendered the entire range of services to owners.

Similar combinations of experts will be involved in the planning of bridges, industrial plants, aerospace structures, and the like. In any case, it is clear that structural engineering is only one of a great number of inputs that go into the final design; conflicts can arise between competing demands and must be resolved by suitable compromises.

Once the structural configuration has been arrived at, load criteria must be established and the internal forces and deformations determined in a preliminary way. If unfavorable response is indicated for the structure, requiring undue member size or expense for proper functioning, then it may be necessary to modify or alter the original concept to arrive at a better design. This process should be continued until an optimal design results; only then should precise calculations of stresses and deformations he made in order to enable final design of the structural components for adequate strength and stiffness.

We see that the structural analyst's contribution consists of structural configurations, preliminary analysis, possible structural modifications, and final analysis. All results required for the design of the structure are presented for maximum convenience of the designer, who then sizes the members, designs the connections, and prepares the plans and specifications necessary for construction.

It should be clear that the processes of structural analysis and design are intimately related; for instance, the member weight and their stiffness, quantities that are known only after the members have been designed, are required for analysis. Thus, analysis and design are mutually interacting and should not properly be separated. Nevertheless, the convenience of considering one aspect at a time forces us to do just this. However, we should never lose sight of the fact that structural analysis is primarily a tool for good design, and should therefore never be an end in itself.

1.2. Types of Structures

The structural analyst should be prepared to cope with a great variety of different structures; we shall here enumerate a few types:

1. Buildings.
2. Bridges.
3. Underground structures, tunnels.
4. Industrial structures, pressure vessels, reactor containers.
5. Aerospace structures: planes, missiles.
6. Vehicle structures: automobiles, railcars, ships.
7. Machines, cranes, microwave aerials, and so forth.

We can also organize structures according to the nature of their components; they can be conveniently divided into three major classes:

1. Linear or uniaxial members, as shown in Fig. 1.1: Two-force members (as in trusses), beams, columns, arches, and their combinations. These

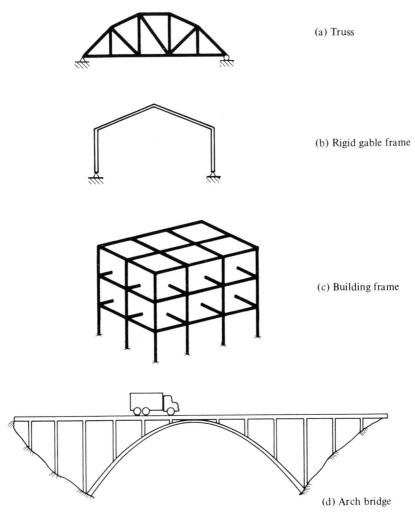

(a) Truss

(b) Rigid gable frame

(c) Building frame

(d) Arch bridge

Fig. 1.1   Framed structures.

are the simplest to analyze, and are, accordingly, particularly suitable for elementary presentation of structural theory. Often it is also possible to idealize more complex structures as assemblies of such members for purposes of analysis. The bulk of this presentation will be concerned with the analysis of such members.

2. Two-dimensional components, such as the plate, shell, and planar elements shown in Fig. 1.2: Although the analysis of such structures has traditionally been considered a branch of the theory of elasticity, the newer methods enable us to perform such analyses according to principles of structural theory to any desired degree of accuracy.

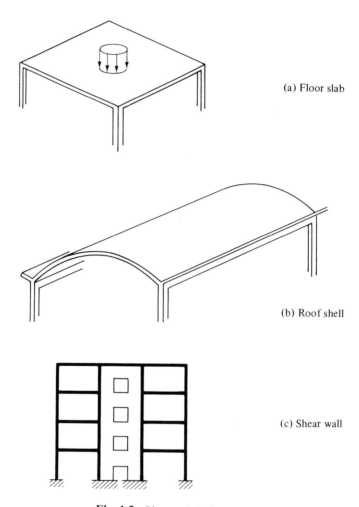

(a) Floor slab

(b) Roof shell

(c) Shear wall

Fig. 1.2   Plate and shell structures.

3. Three-dimensional bodies, such as complex machine parts, soil, and rock foundations: Detailed stress analyses of structural joints and members may also require three-dimensional treatment according to the theory of elasticity.

Whereas the classical approach taken in civil engineering books has been to concentrate attention on those structures which could be modeled as assemblies of linear members, such as beams, struts, or columns, the modern analyst is expected to be able to attack any structure, and the modern methods which use the power of matrix formulation and computer solutions are sufficiently general so that a much vaster class of structures can be calculated, including any combination of components. This approach will be stressed in this book.

1.3. Loads

Before a structure can be analyzed, the nature and magnitudes of the loads must be known. Our assumptions here are often only crude approximations. We shall here list and discuss some of the more important types of loads.

1. *Dead load.* The self-weight of the structures: This can be precisely known only after the structure has been designed. Obviously, the less the ratio of the dead load to the useful load, the more efficient is the structure. Vast structures such as long-span bridges may have to carry dead load many times their live load. For such structures, the use of lightweight materials and efficient structural shapes is indicated.

2. *Live load.* The useful load carried by the structure: If due to human occupancy and activities, it should be predicted by the use of probability theory. For building design, live-load values for different types of occupancy are usually taken from building code provisions and considered uniformly distributed over entire, or portions of, structures. Table 1.1, for instance, showing building live loads, is taken from the *Uniform Building Code.** In bridges, the live load is moving, consisting of trains of vehicles of specified weight traversing the structure. The job here is to determine that location of the load which will be critical in its influence on the structure; this matter is covered in Chapter 6. Standard bridge loadings are given in the appropriate codes.†

3. *Wind, earthquake, and aerodynamic forces.* All these effects should really be computed dynamically, since these loads are often of such short or cyclic variation so as to cause inertia forces in the structures. The field of structural dynamics, which will not be considered further here, is in an active stage of development. Although forces of this type acting on building frames have usually been assumed statically applied, there is an increasing trend toward dynamic analysis for tall, slender, or special structures. In aerospace structures the aerodynamic loads are computed from aerodynamic theory or obtained from wind-tunnel tests, and the resulting internal forces are computed from dynamic computer analysis. The results of such dynamic analyses show that there is a pronounced interaction between structural characteristics

*Uniform Building Code, International Conference of Building Officials, Pasadena, Calif., 1969.

†*Standard Specifications for Highway Bridges*, 10th ed., American Association of State Highway Officials, Washington, D.C., 1969.

TABLE 1.1. Building Live Loads (From *Uniform Building Code,* International Conference of Building Officials, Pasadena, Calif., 1967)

OCCUPANCY	LOAD IN POUNDS PER SQUARE FOOT OF HORIZONTAL PROJECTION
Apartments	40
Armories	150
Auditoriums—fixed seats	50
Movable seats	100
Balconies and galleries—fixed seats	50
Movable seats	100
Cornices	60
Corridors, public	100
Dance halls	100
Drill rooms	100
Dwellings	40
Exterior balconies	100
Fire escapes	100
Garages—storage or repair	100
Garages—storage private pleasure cars	50
Gymnasiums	100
Hospitals—wards and rooms	40
Hotels—guest rooms and private corridors	40
Libraries—reading rooms	60
Stack rooms	125
Loft buildings	100
Manufacturing—light	75
Heavy	125
Marquees	60
Offices	50
Printing plants—press rooms	150
Composing and linotype rooms	100
Public rooms	100
Rest rooms	50
Reviewing stands and bleachers	100
Roof loads	(See Section 2305)
Schools—classrooms	40
Sidewalks	250
Skating rinks	100
Stairways	100
Storage—light	125
Heavy (load to be determined from proposed use or occupancy, but never less than)	250
Stores—retail (light merchandise)	75
Wholesale (light merchandise)	100

and dynamic loads. Thus, the stiffer and heavier the structure, the higher may be the forces due to dynamic excitation. It may actually make the structure more resistant to this type of load to make it lighter and more flexible! This is quite opposite from its response to static loads, and should certainly be considered in design.

In usual practice, earthquake loads are simulated by horizontal (lateral) loads applied at each story level, of magnitude proportional to the mass of the story; the proportionality constant varies with the seismicity of the region, and can be found from maps that are part of building codes.* These lateral earthquake forces range up to 20 per cent of the building's dead weight in highly seismic zones.

Similarly, design wind pressures (which, under ideal conditions, depend on wind velocities) are specified for different areas on the basis of records of weather stations. Usually, boundary layer effects diminish wind velocities close to the ground, and therefore the wind pressure will increase with building height. The shape of the structure has an important influence, and suction may have to be considered on the leeward side or on roofs, as shown by many instances of flying roofs during gale winds. Data for such pressure distributions are usually obtained from wind-tunnel tests.†

4. *Gas, liquid, and earth pressures.* Gas and liquid containers are usually designed for a rated capacity. Vacuum and reactor containment vessels may have to withstand tremendous but well-defined applied pressures. The action of soil bearing against a retaining wall or against a tunnel lining is much less clear, because it varies between the extreme active and passive cases, depending on the soil–structure interaction. There are a number of theories which, though based on highly simplifying assumptions, have been used successfully to predict such loads.

5. *Self-strains due to support settlement, temperature changes, creep, or prestressing.* In addition to applied loads, there are effects that cause dimensional changes in the structure; if these changes are prevented by the support conditions of the structure, internal stresses arise that will have to be determined by the analyst. In such cases, the specification of the load is in terms of the expected support settlement, maximum temperature change, creep strains, prestressing force, or the like.

We observe that any or all of these effects may be acting simultaneously, and it is the analyst's job to superpose those cases leading to critical conditions. The probability of maximum loads due to several causes coinciding is

*Uniform Building Code, International Conference of Building Officials, Pasadena, Calif., 1969.

†"Wind Forces on Structures," A Symposium, *Proc. Am. Soc. Civil Engrs, J. Struct. Div.*, No. ST4, July 1958.

of course less the more different causes are considered. Probability theory can be used here to arrive at logical criteria. For instance, several building codes allow design for $\frac{3}{4}$(dead load + live load + lateral load) for joint action of gravity and lateral loads, because of the unlikelihood of occurrence of this combination of critical values. Similarly, in building design the design floor live load may be reduced when vast tributary floor areas are considered, because of the unlikelihood of the entire area being subjected to the full design load simultaneously. Sometimes the critical load combinations are not immediately obvious. For instance, if we consider a concrete column in bending in a building, the addition of an axial compressive force will increase the safety, as shown by the interaction curve of Fig. 1.3; in this case, the

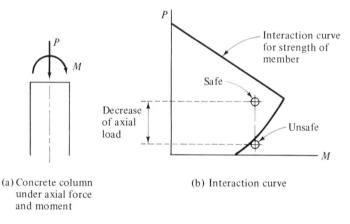

(a) Concrete column   (b) Interaction curve
under axial force
and moment

Fig. 1.3   Critical combination of forces on column.

critical condition will be due to loads that cause a maximum of bending in conjunction with a minimum of axial force, a fact of which the analyst must be aware.

The subject of loads on structures is no less important than the problems of analysis, but has received relatively little attention because of its complexity and because of the tedious work involved in gathering sufficient data for statistical evaluation. We should be aware of the futility of trying to improve our analysis methods while still being in the dark about loads. The need is therefore for more rational and realistic ways to predict loads on structures.

1.4. Idealizations and Assumptions of Structural Analysis

Real structures are usually much too complex for rational analysis; often they have to be reduced to simplified models prior to quantitative treatment. This modeling is one of the most important jobs of the analyst and requires experience and judgment so that the resulting model strikes a happy compromise between reality and simplicity. In this book, the numerous "structures" discussed and shown are really only models of the real thing.

We can consider the modeling according to various viewpoints:

1. *Geometry of the structure and interaction.* Structures, such as the portion of a building frame shown in Fig. 1.4(a), are in general three-

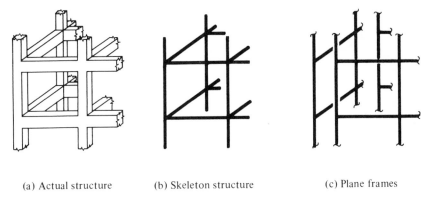

(a) Actual structure   (b) Skeleton structure   (c) Plane frames

Fig. 1.4   Modeling of building frame.

dimensional solids. For analysis we often try to consider them as an assembly of linear members (struts, beams, columns) with negligible lateral dimensions so that they may be assumed concentrated along their axes. The model then appears as a gridwork of members, as shown in Fig. 1.4(b), which can be analyzed. The class of such idealized structures is called *framed structures.* Since analysis in three dimensions is more involved than that of planar framed structures, the further decomposition into the individual plane frames is often made as shown in Fig. 1.4(c). Under this assumption, each frame is considered responsible for the portion of the load tributary to it, without considering the interaction of these individual components.

Whether such a decomposition does justice to reality depends on the particular case; for instance, the framed structure of Fig. 1.5 carries the applied torque as a closed thin-walled cell; if it were decomposed into individual walls, this behavior could not be modeled.

The concrete arch bridge shown in Fig. 1.6(a) consists of arch, piers, and deck all cast integrally; for preliminary analysis, the components are often considered disassembled and assumed acting independently, as shown in Fig. 1.6(b). Once the member sizes are estimated on the basis of these assumptions, the final analysis might be performed on the total structure, a formidable job but one that can be handled routinely by modern methods.

In tall buildings the resistance to lateral loads is often provided by shear walls of concrete (which may consist of elevator shafts or stair wells), as

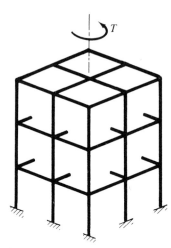

Fig. 1.5   Building frame under torsion.

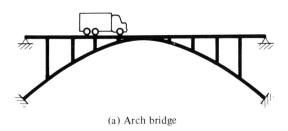

(a) Arch bridge

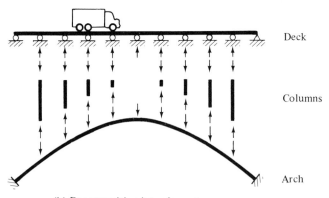

(b) Decomposition into elements

Fig. 1.6   Modeling of arch-deck bridge.

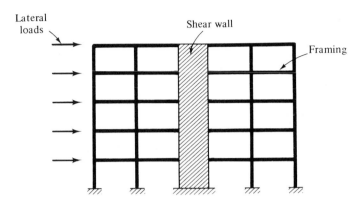

Fig. 1.7   Building frame with shear wall.

shown in Fig. 1.7. Such shear walls are often considered as beams for purposes of analysis; whether this assumption is justified depends on their proportions and should be decided from case to case.

A consequence of the idealization of members into line elements is that their joints are reduced to points, as shown in Fig. 1.4. Obviously, this prevents a close investigation of the stress distribution within these connections; if this is necessary, a more realistic picture, as shown in Fig. 1.8, is required, along with an accordingly more difficult analysis as a two- or three-dimensional body.

2. *Connections and support conditions.* Simplifications are also often effected by idealizing the joints and support conditions of structures. Thus, in truss-type structures the usual assumption is that of ideal frictionless pins connecting the members which allow full rotation, even if the joints are in fact welded. Standard beam connections for steel beams, as shown in Fig. 1.8(a), are also considered as simple supports, even though capable of resisting some moment.

Conversely, fully welded steel frames, or those cast monolithically in concrete, are assumed perfectly rigid, even though this may be far from reality. A better model of a "semi-rigid connection" of the joint action may be a spring of appropriate stiffness, as shown in Fig. 1.8. Since this assumption complicates the analysis, it is not customarily followed up—a typical case of simplicity taking precedence over accuracy; although the results in general bear out the validity of the procedure, the engineer should always be able to refine his analysis where indicated.

A similar situation prevails with respect to support conditions of structures. If we, for instance, consider the typical spread column footing of Fig. 1.9, the question arises as to the ability of this footing to resist rotation. Most probably, some intermediate rotation resistance between full fixity and com-

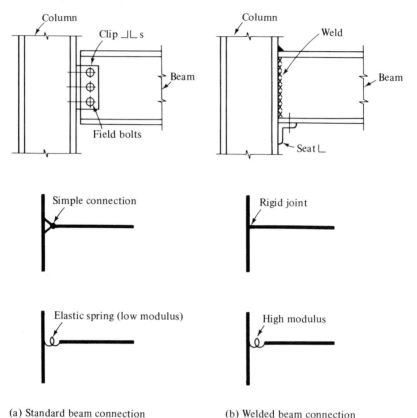

(a) Standard beam connection   (b) Welded beam connection

Fig. 1.8   Modeling of connections.

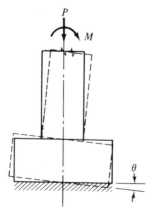

Fig. 1.9   Rotation of column footing.

plete freedom exists; the usual assumptions are either one extreme or the other, depending on the footing design and foundation conditions, whereas some intermediate assumption (which could again be modeled by an elastic spring) would do better justice to reality.

Another situation in which the complexity of the soil–structure interaction often forces us to simplify the matter is shown in Fig. 1.10. The possible deformed shape of the strip footing of Fig. 1.10(a) shows variable compres-

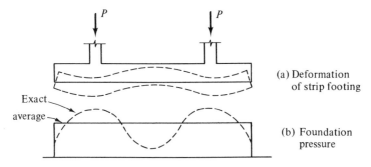

(a) Deformation of strip footing

(b) Foundation pressure

Fig. 1.10   Soil pressure under strip footing.

sion of the soil, with a maximum under the columns. The variability depends on the footing and soil stiffness, matters that are often unknown. To avoid the question, we assume that the footing is infinitely stiff and deforms only as a rigid body, leading to the linear soil pressure distribution shown in Fig. 1.10(b), which is easily calculated. An analysis of the moments in the footing due to the exact (curved) and the approximate (linear) soil pressures shows that the latter leads to higher moments and shears; thus, the assumption will lead to a conservative design. Considerations of this type are always useful to predict the consequences of an approximate analysis.

3. *Material behavior and stability.* Most engineering materials possess a linearly elastic range, for which Hooke's law holds, over only a limited range of stress. In usual cases, the allowable stresses under service loads are sufficiently low so that elastic action prevails, and the assumption of linear behavior is valid. In classical structural design, provision of a specified safety factor against inelastic action is considered adequate to ensure the required safety.

Beyond the elastic range, the material becomes progressively less stiff, as shown by Fig. 1.11. In ductile materials, such as mild steel, continued deformation under essentially constant yield stress will set in thereafter. In many other metallic materials, such as high-strength steels or aluminum, as well as in reinforced-concrete structures, an additional nonlinear range of behavior is available prior to failure. In such cases, the attainment of the elastic limit does not signify the limit of carrying capacity of the structure, and linearly elastic

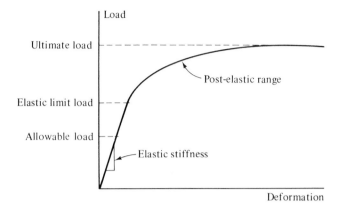

Fig. 1.11   Load–deformation relation.

theory cannot therefore give information about the safety against collapse. This "ultimate" range of behavior, shown in Fig. 1.11, is studied in the *inelastic theory of structures*, of which *theory of plasticity*, or *limit analysis*, is a special part. Although we shall restrict ourselves to the linearly elastic range of structures in this text, the reader should also be aware of the post-elastic response of structures to gain insight into the entire range of structural behavior.

We shall also neglect all stability aspects of structural behavior. It is well known that slender members under axial compression can fail by buckling. The usual procedure in structural design is to perform the analysis on the assumption that no buckling will take place, and then design the members so as to provide the specified safety factor against buckling failure. In some thin-walled components of structures, especially in aerospace design, buckling can be tolerated when it does not lead to overall failure, but such considerations are beyond the scope of the present treatment.

We see that elastic analyses of the type considered in this text can only give information about a portion of the total structural response. Nevertheless, such elastic analyses provide the cornerstone of structural design; refinement to account for the secondary effects cited can only be tackled with a thorough understanding of the primary response of structures to loads.

To summarize the vital aspects of the modeling of real structures for purposes of analysis, we consider the common case of a rectangular one-story building consisting of block, or panel, walls carrying a flat roof, as shown in Fig. 1.12(a). Such structures are often used for industrial or public purposes.

We can consider five basic components of this structure: four walls and the roof. The analyst must now visualize the role of these components in resisting the applied loads and their interaction.

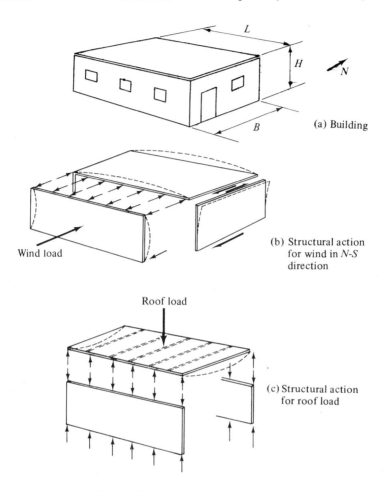

(a) Building

(b) Structural action for wind in *N-S* direction

Wind load

Roof load

(c) Structural action for roof load

Fig. 1.12 Modeling of one-story building.

For instance, the wind forces in the north–south direction will exert a pressure on the windward wall; this pressure is resisted primarily by bending of the wall in its short, or vertical, direction. It should be designed as a simple beam of length H spanning between wall foundation and roof; therefore, both foundation and roof must be designed to resist these lateral reactions. The roof will bend as a beam of length L and depth B in its horizontal plane under the influence of these forces, as shown in Fig. 1.12(b). Its reactions are furnished by the east and west walls acting as cantilever beams or shear panels of height H and depth B, transmitting these forces as shears to their foundations. All these components, as well as their connections, must be designed adequately to transmit these forces from windward wall to roof to side walls to ground.

Similar reasoning should serve to understand how roof loads are carried to earth, as shown in Fig. 1.12(c). If roof joists, or rafters, are spanning a distance *B* in the short direction, the roof loads will be primarily carried by the longitudinal (east–west) walls as axial forces in their plane. Under this assumption, the short (north–south) walls will carry only a minimal amount of the roof load, which is often disregarded.

We tried to avoid obscuring this discussion by commenting on structural details. The emphasis here was on the basic action of the structure in resisting loads. Irrespective of the complications of structural interaction or the obscurity of numerical computations, this basic structural action must always be kept in mind. Often a simple model of the structure can convey a good feel for the way in which the forces are carried to its supports.

1.5. Methods of Analysis

The field of structural analysis is currently in a state of flux; the classical methods were largely based on a physical understanding of the structural behavior, which could be summarized in the statement "if you cannot draw the deformed structure under load, you cannot analyze the structure." While this approach has its merits, it obviously also has its limitations; it restricts analysis to structures that are simple enough to be visualized, and all too often leads to crude "seat-of-the-pants" calculations, which are not in keeping with a scientific approach to the subject. Nevertheless, this plan of attack can furnish a good physical picture, which is especially useful for the beginner, and which is necessary for many of the approximate methods used in pre-liminary analysis.

The availability of high-speed computers has changed the picture in recent years. The approaches have become more and more tailored to a format that can be understood by computers: the matrix formulation of the calculations. Matrix methods are essentially a bookkeeping method that is suitable for keeping track of large quantities of numbers, and thus provides the means of analyzing ever-larger and more complex structures. In addition, the matrix method provides such an economical and elegant means of showing basic relationships that it has shed considerable light on the interrelationships between the many different methods which have been developed. The use of matrix formulation has thus led to a remarkable reorganization and unification of the field of structural analysis, resulting in an economy of basic ideas, which has sharpened our attack on structural problems.

The danger of such an approach is the loss of physical feel for the structural action, which is inherent in a highly formalized and abstract set of calculations. The results of such a powerful matrix–computer approach can be data in such profusion that it may be hard to tell whether or not they make sense. No matter how sophisticated the approach, the analyst should never lose sight of the connection between numerical results and reality.

In this text an attempt is made to combine the best features of the classical and the modern methods. The basic methods of structural mechanics are presented in the classical way, emphasizing their physical meaning. Following this, they are cast in discrete form and expressed in terms of matrices, starting with the simplest relationships in such a way that they can be gradually combined to yield formulations capable of analyzing general classes of structures. Since the appeal of matrix computations lies in their computer application, appropriate programs are furnished from time to time to demonstrate the computer use of the methods and to give the reader a tool for analysis. It is hoped that in this way a sound understanding of basic principles and their efficient application is developed, which will lead to the ability to determine the response of general structures under any loading.

II

Statics of Structures

Since all structures to be considered in this book are to be in static equilibrium, any free body constituting all or part of the structure must satisfy the equations of statics. It is thus the first, and most important, part of structural analysis to be able to express the statical relations between all forces acting on or in the body.

Chapter 2 contains a review of the methods of statical analysis by use of free bodies. In Chapter 3 we apply these methods to flexural members, including three-dimensional cases, and introduce matrix formulation for equilibrium relations. Chapter 4 continues the application of equilibrium conditions for the case of trusses, where we again aim at a general formulation by means of matrices tailored toward computer use.

Chapter 5 introduces the theorem of virtual work, and uses a form of this theorem for statical computations. In Chapter 6, finally, we consider the analysis of statically determinate structures under moving loads by use of influence lines.

The importance of this material cannot be overemphasized. No structure can possibly stand if equilibrium is violated, and all other considerations and calculations must be deficient if based on faulty statics.

CHAPTER 2

Force Systems

In Sec. 2.1 the all-important concept of free bodies is reviewed. Since a clear picture of the forces and force systems occurring in members and structures is necessary, these are classified in Secs. 2.2 and 2.3. The concepts of static determinacy and indeterminacy are explained in Sec. 2.4, and the calculation of reactions is touched upon in Sec. 2.5.

2.1. Free Bodies

The concept of free bodies, covered in courses in statics and mechanics of deformable bodies, should be thoroughly reviewed at this time. A free body is a portion of a structure with all forces external to it indicated in their proper position. Here, and throughout this book, a *force* (also called *action*) is used in its general sense to include both forces proper as well as moments.

Figures 2.1 and 2.2 demonstrate appropriately drawn free bodies. Figure 2.1(a) shows a complete truss under load with all reactions indicated. Note that all member forces are internal to the free body, and thus will not appear in the calculations. Figure 2.1(b) shows a free body of the same truss cut so that three member forces appear as external forces. Free bodies must always be selected in such a fashion that the desired forces are external to them.

In Fig. 2.2(a), a free body of a complete beam is drawn for the determination of the reactions. If the internal forces at a section of the beam are desired,

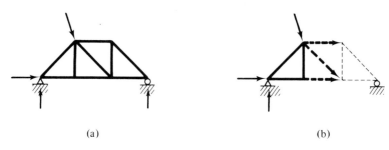

Fig. 2.1 Free bodies.

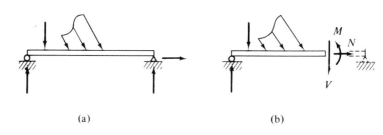

(a)   (b)

Fig. 2.2 Free bodies.

a partial free body which contains these as external forces is considered, as in Fig. 2.2(b).

Equilibrium calculations by means of free bodies are considered further in Sec. 2.5 and Chapters 3 and 4.

2.2. Forces

The forces occurring in structures can be considered as three different types: applied loads, reactions, and internal forces or stress resultants. These will be considered shortly.

1. *Loads.* Some types of loads applied to structures have already been discussed in Sec. 1.3. In most of the problems considered here, the loads are specified quantities applied to the structure in either concentrated or distributed manner. Schematically, applied loads are here represented by symbols as shown in Fig. 2.2.

In practice, structures often have to be analyzed for several loading conditions, or their critical combination, to which the structure might be exposed. We shall therefore try to develop analysis methods in this text that are capable of handling a number of loading conditions on a structure in an efficient manner. Matrix methods are suitable for this purpose, since by their use the duplication of calculations is minimized.

Even though in this presentation most of the loads are given, we should remember that proper selection of loads is in practice one of the most difficult and important tasks of the designer. Careful analysis of a structure with incorrectly or sloppily chosen loads is a futile undertaking.

2. *Reactions.* Reactions are the forces exerted on the structure by the supports. They are in general unknown quantities and must, for the structures of Parts II and III of this book, be determined by statics before the internal force distribution can be found.

At this stage, it is necessary to define some commonly used symbols for structural supports. Figures 2.3(a) and (e) represent, respectively, fixed supports for the plane and three-dimensional case. All displacements and rotations are prevented by suitable provisions, called *constraints.* A reactive force is needed for each restrained component of displacement, and a reactive moment is needed for each rotational constraint. The pin supports of Figs. 2.3(b) and (f) and the roller supports of Figs. 2.3(c) and (g) offer only translational, but no rotational, constraints, and thus have force, but no moment reactions. It is of course possible to provide other support conditions, such as the guided support shown in Fig. 2.3(d), which prevents rotation and one component of translation by appropriate reactive forces, but allows translation in one direction.

It will be observed that in all the supports discussed so far either the component of motion, or the corresponding component of force, is zero. Sometimes a structure is supported elastically, such as the beam end supported by the rotational spring of given stiffness k shown in Fig. 2.3(h), in which case the reactive moment M is related to the end rotation θ of the beam by the relation $M = k\theta$.

The analyst must always assume realistic support conditions that are in accord with the physical behavior of the actual structure, and, conversely, the designer must ensure that the structure as built can perform in accordance with the specified constraints. For instance, the designer who specifies a column foot fixed against rotation should be aware of the difficulty and expense of constructing a footing that will not rock or rotate under applied moment.

3. *Internal forces.* Internal forces are forces on interior sections of the structure, also called actions, necessary to keep any portion of the structure in equilibrium. They must be determined by the use of appropriate free bodies selected so that the plane or section on which the forces are desired forms an external plane of the free body.

The usual purpose of structural analysis is to determine just these forces. In design, the members must then be dimensioned sufficiently strong so as to be able to resist them with a specified factor of safety. In checking the

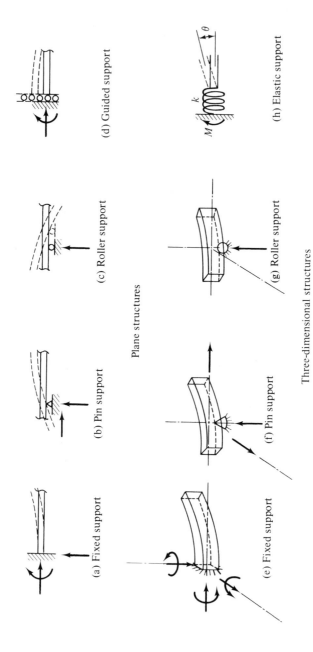

Plane structures

(a) Fixed support

(b) Pin support

(c) Roller support

(d) Guided support

Three-dimensional structures

(e) Fixed support

(f) Pin support

(g) Roller support

(h) Elastic support

Fig. 2.3 Support conditions.

adequacy of existing structures, the internal forces must be compared with the actual strength of members to ascertain the safety of the structure.

In general, there are six internal forces on a section of a structural member, as shown in Fig. 2.4: three forces and three moments. After the reactions on a statically determinate structure have been found, these forces can be determined by statics.

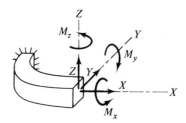

Fig. 2.4   General flexural member.

There are a number of special cases of internal-force systems of importance in structural analysis:

1. Two-force member: Ideal trusses are composed of such members, which resist only axial forces, as shown in Fig. 2.5. Thus, only one unknown force acts at any section.

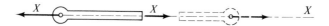

Fig. 2.5   Two-force member.

2. Plane beam: Such members have resistance to transverse forces, leading to an internal shear force V and an internal bending moment M, as shown in Fig. 2.6. Two equilibrium equations are necessary to determine these two forces.

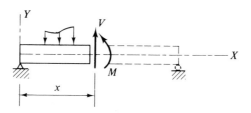

Fig. 2.6   Plane beam.

3. Planar member with axial force: Plane frame and arch-type structures are of this type. In addition to shear and moment, an axial force N is necessary to resist the applied loads, as shown in Fig. 2.7. Three equilibrium equations will solve for these three forces.

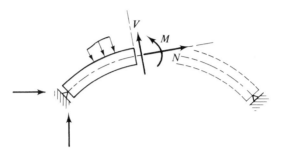

Fig. 2.7 Plane arch.

4. Member curved in plane with load normal to the plane of the structure: An example of such a structure is the bow girder shown in
   Fig. 2.8. Any internal section is subject to three forces: shear, bending

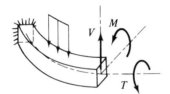

Fig. 2.8 Plane member under normal
load.

moment, and twisting moment T, and, again, three equilibrium equations are needed to determine these forces.

Other combinations of internal forces are possible. These will always
constitute special cases of the general situation shown in Fig. 2.4.

The forces acting on internal sections of the structure are the resultants
of the stresses, and are, for this reason, also called *stress resultants*. The
concept of stress resultant is further explained in Fig. 2.9. Here a section of a

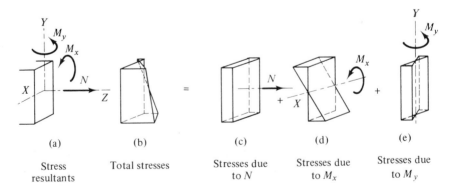

(a)	(b)	(c)	(d)	(e)
Stress resultants	Total stresses	Stresses due to N	Stresses due to M_x	Stresses due to M_y

Fig. 2.9 Stress resultants.

member is shown in Fig. 2.9(b) subjected to a linearly varying set of normal stresses. Following the methods of stress analysis covered in an earlier course, these stresses can be decomposed into the bending stress distributions shown in Figs. 2.9(d) and (e), the resultants of which are the bending moments M_x and M_y, and the uniform stress distribution shown in Fig. 2.9(c), the resultant of which is the axial force N along the longitudinal centroidal axis. The stress resultants on this section are then the forces shown in Fig. 2.9(a). Since the ultimate purpose of structural analysis will generally be to check on the magnitude of stresses, the relationship between stresses and their resultants should be clearly understood.

2.3. Special Force Systems

The purpose of structural analysis is to determine all forces acting on any free body of the structure. If the number of available equilibrium conditions is equal to the number of unknown forces, these equations are sufficient to solve the problem, and the structure is *statically determinate*. Parts II and III are concerned with such structures. If, on the other hand, the number of unknown forces is larger than the number of independent equilibrium equations, the structure is *statically indeterminate*. In this case, additional equations are needed in order to match the number of unknowns. Such structures are covered in Parts IV and V. Finally, if more equations of statics are available than unknown forces, the solution cannot be unique, and it is concluded that the structure must be unstable and cannot stand. Such structures are also called *mechanisms*. It is obvious that such unstable forms must be avoided by the designer.

To distinguish between the different cases, it is thus necessary to compare the number of unknown forces with the number of independent equations of statics of a suitably chosen free body. This latter number depends on the nature of the force system, as treated in courses on statics. Figure 2.10 reviews this material by tabulating the force system, one typical example, and the corresponding number and types of equilibrium conditions.

2.4. Static Determinacy and Indeterminacy

It is not intended here to classify all structures as to their degree of indeterminacy, since, in view of the manyfold structural forms and steadily developing new types, this would prove futile. Rather, a small number of commonly occurring structures is discussed with the aim of elucidating the foregoing and to allow the reader to develop his own facility to determine the appropriate structural type.

Figure 2.11 shows a number of different structures under load, with the number of independent equilibrium equations, *e*, the number of unknown

Type of force system	Example	No. of equations	Type of equation
Co-linear	Two-force member	1	$\Sigma F_X = 0$
Co-planar, concurrent	Plane truss joint	2	$\Sigma F_X = 0, \Sigma F_Y = 0$
Co-planar, non-concurrent, parallel	Beam under transverse load	2	$\Sigma F_Y = 0, \Sigma M_Z = 0$
Co-planar, non-concurrent, non-parallel	Arch	3	$\Sigma F_X = 0, \Sigma F_Y = 0, \Sigma M_Z = 0$

Type of force system	Example	No. of equations	Type of Equation
Non-coplanar, concurrent, non-parallel	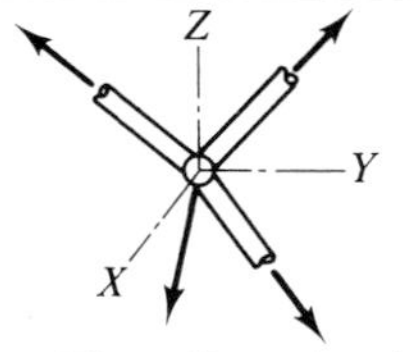 Three-dimensional truss joint	3	$\Sigma F_X = 0, \Sigma F_Y = 0, \Sigma F_Z = 0$
Non-coplanar, non-concurrent, parallel	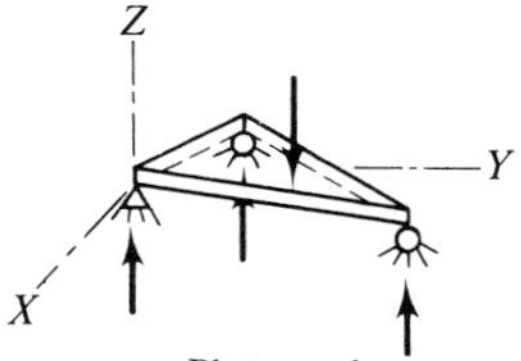 Plate under transverse load	3	$\Sigma M_X = 0, \Sigma M_Y = 0, \Sigma F_Z = 0$
Non-coplanar, non-concurrent, non-parallel (general case)	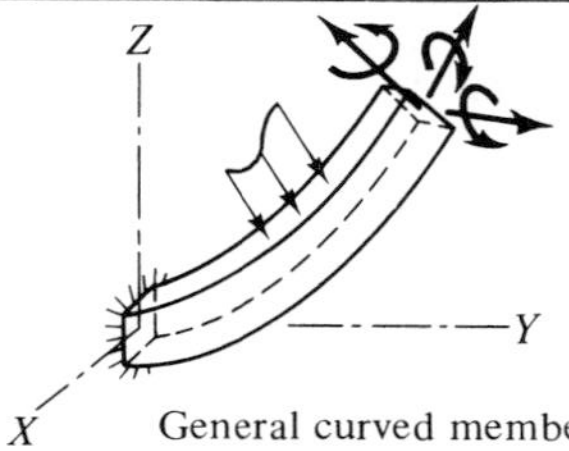 General curved member	6	$\Sigma F_X = 0, \Sigma F_Y = 0, \Sigma F_Z = 0$ $\Sigma M_X = 0, \Sigma M_Y = 0, \Sigma M_Z = 0$ or $\Sigma \mathbf{F} = 0, \Sigma \mathbf{M} = 0$

Fig. 2.10 Types of force systems.

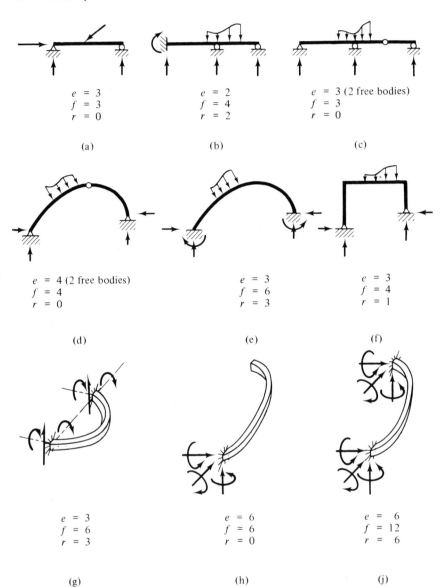

Fig. 2.11 Degrees of indeterminacy.

reactive forces, f, and the degree of static indeterminacy, $r = f - e$, specified. Those structures with $r = 0$ are statically determinate and can be analyzed by statics alone. Figures 2.11(a), (b), and (c) are, respectively, a simply supported, continuous, and articulated beam. The presence of a hinge in one span of this latter beam allows it to be split up into two separate free bodies so that there

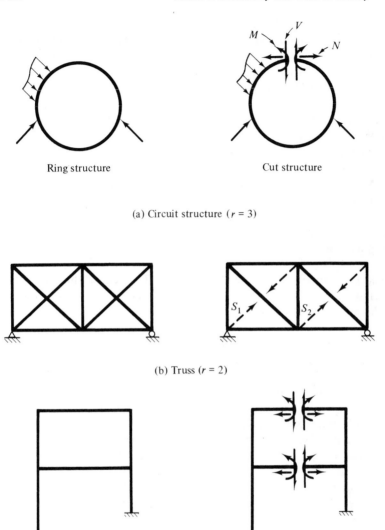

Ring structure   Cut structure

(a) Circuit structure ($r = 3$)

(b) Truss ($r = 2$)

(c) Two-story rigid frame ($r = 6$)

Fig. 2.12 Static indeterminacy.

are three independent equilibrium equations corresponding to three unknown reactions. Figures 2.11(d), (e), and (f) represent plane arch and frame-type structures, of which the former is a statically determinate "three-hinged arch," which due to its articulation can be separated into two free bodies for which four independent equilibrium conditions can be written. There is no basic theoretical difference in the treatment of the fixed arch and rigid frame of

Figs. 2.11(e) and (f), though the structural efficiency of such structures can vary greatly, depending on their shape. Figures 2.11(g), (h), and (i) represent three-dimensional structures, respectively, a statically indeterminate bow girder, a cantilevering statically determinate helicoidal girder, and a fixed-ended statically indeterminate helicoidal girder.

Many more structures could be presented here, but for the present the student is encouraged to verify the type of force system and corresponding number of statical relations, to determine the number of unknown forces, and to check on the degree of indeterminacy. Other types of structures will be discussed at appropriate stages in this text.

Another way of checking the degree of indeterminacy of a structure, which may also yield some insight into the physical nature of the problem, is to ask the following question: How many of the unknown forces can be removed before the structure becomes unstable? In the case of a statically determinate structure, all forces are necessary to ensure stability; in the other cases, a number of forces equal to the number r can be removed (or *cut*) to reduce the structure to static determinacy. The reader should apply this criterion to the structures of Fig. 2.11 to verify and understand this approach. Furthermore, Fig. 2.12 shows a number of structures whose stability is to be checked in this manner. Figure 2.12(a) shows a closed ring (or *circuit structure*) under load. In one cut, the structure is reduced to static determinacy by removing three internal forces: moment, transverse (or shear) force, and normal (or axial) force. Thus, the structure is three times statically indeterminate. Removal of any additional internal force would lead to instability. The plane truss of Fig. 2.12(b) allows removal of two two-force members, each with one axial force, before instability occurs, and is thus twice indeterminate. The two-story, one-bay rigid frame of Fig. 2.12(c) allows two cuts within the girders, each removing three internal forces, resulting in two statically determinate tree-like cantilever structures, and is thus six times indeterminate.

Since the forces thus removed are not essential for the stability of the structure, they are called *redundants*, and the degree of indeterminacy is also called *degree of redundancy*.

2.5. Reactions of Statically Determinate Structures

In the following, some examples of the calculation of statically determinate reactions are presented. Following these, the student is encouraged to work some of the problems, since complete understanding and facility in calculations are only gained by practice.

Example 2.1: Calculate all reactions of the rigid frame shown in the figure.

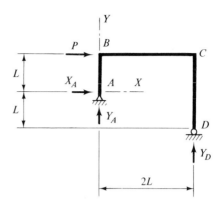

Fig. E2.1

Solution: There are three unknown reactions, X_A, Y_A, and Y_D, corresponding to three equilibrium equations. The structure is therefore statically determinate. The equilibrium equations are solved in such sequence that each involves only one unknown.

$$\Sigma F_X = 0: \qquad X_A + P = 0; \qquad \underline{X_A = -P}$$

The minus sign indicates that X_A is directed in the negative X direction.

$$\Sigma \overset{+}{M_A} = 0: \qquad P \cdot L - Y_D \cdot 2L = 0; \qquad \underline{Y_D = \frac{P}{2}}$$

$$\Sigma F_Y = 0: \qquad Y_A + Y_D = 0; \qquad \underline{Y_A = -\frac{P}{2}}$$

Example 2.2: In the three-hinged arch shown, there are four unknown reactions, corresponding to four independent equilibrium equations.
a. Calculate all reactions due to P applied as shown.

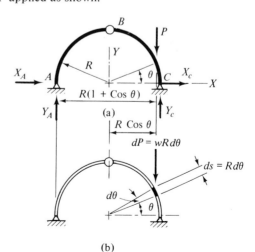

Fig. E2.2   (b)

Solution: Entire free body:

$$\Sigma \overset{+\curvearrowleft}{M_A} = 0: \qquad P \cdot R(1 + \cos\theta) - Y_C \cdot 2R = 0; \qquad \underline{Y_C = \frac{P}{2}(1 + \cos\theta)}$$

Note that when $\theta = \pi/2$ the load is at the crown, and $Y_C = P/2$ as should be expected.

Free body BC:

$$\Sigma \overset{+\curvearrowleft}{M_B} = 0: \qquad -X_C \cdot R - Y_C \cdot R + P \cdot R\cos\theta = 0;$$

$$X_C = P \cdot \cos\theta - \frac{P}{2}(1 + \cos\theta)$$

$$\underline{= \frac{P}{2}(\cos\theta - 1), \qquad \theta \le \frac{\pi}{2}}$$

Entire free body:

$$\Sigma F_X = 0: \qquad X_A + X_C = 0; \qquad \underline{X_A = -\frac{P}{2}(\cos\theta - 1), \qquad \theta \le \frac{\pi}{2}}$$

$$\Sigma F_Y = 0: \qquad Y_A + Y_C - P = 0; \qquad \underline{Y_A = \frac{P}{2}(1 - \cos\theta)}$$

b. Calculate all reactions due to the dead weight of the arch, of intensity w lb/ft of arch.

Solution: The reactions can be obtained by integrating the reactions due to the concentrated load, which, for a length $ds = R\,d\theta$ along the arch axis, is $dP = w \cdot R\,d\theta$. Using the results of Part a, and summing up all contributions along the arch, we get

$$Y_C = \int_{\theta=0}^{\pi} \frac{wR\,d\theta}{2}(1 + \cos\theta) = \frac{\pi}{2} \cdot wR = Y_A$$

$$X_A = -2\int_{\theta=0}^{\pi/2} \frac{wR\,d\theta}{2}(\cos\theta - 1) = \left(\frac{\pi}{2} - 1\right)wR = -X_C$$

Example 2.3: Calculate the reactive bending moment M, torsional moment T, and shear V due to the dead load of intensity w lb/ft of the semicircular bow girder.

Solution:

$$\Sigma M_X = 0: \qquad M + \int_{\theta=0}^{\pi} (wR\,d\theta)R\sin\theta = 0; \qquad \underline{M = -2wR^2}$$

$$\Sigma M_Y = 0: \qquad T + \int_{\theta=0}^{\pi} (wR\,d\theta)R(1 + \cos\theta) = 0; \qquad \underline{T = -\pi wR^2}$$

$$\Sigma F_Z = 0: \qquad V - \int_{\theta=0}^{\pi} wR\,d\theta = 0; \qquad \underline{V = \pi wR}$$

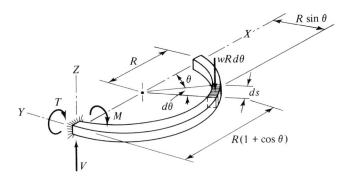

Fig. E2.3

PROBLEMS

1. Find all reactions for the beam subject to the parabolically distributed load.

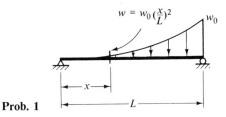

Prob. 1

2. Find the reactions due to the set of forces applied as shown.

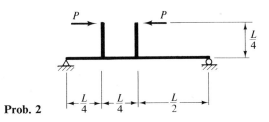

Prob. 2

3. The tank wall is braced by the longitudinal beams located as shown.
     a. Calculate the force per unit length on support beam A due to the water pressure.
     b. Calculate the location of support beam A so that the two beams are subjected to equal loads.

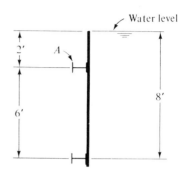

Prob. 3

4. The three-hinged gable frames are spaced at 30 ft. on center. Find the reaction components on the frame due to
   a. Roof dead load, of intensity 20 lb/ft² of roof area.
   b. Snow load, of intensity 30 lb/ft² of horizontal projection.
   c. Wind load, of value 15 lb/ft² of vertical projection.
   d. Find the critical design values for the horizontal and vertical reaction components.

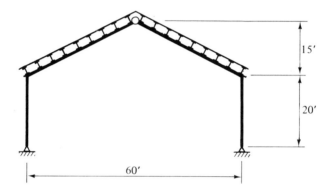

Prob. 4

Prob. 5

5. Determine the degree of indeterminacy of the structure shown. Modify the structure so as to make it statically determinate, and find reactions.

6. The parabolic three-hinged arch ABC carries a bridge of the filled-spandrel type; that is, the space between arch and roadway level is filled with a material of unit weight w lb/ft^3. If the bridge is b ft wide, compute the reactions due to the fill.

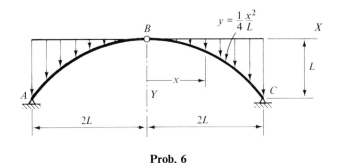

Prob. 6

7. Find all components of reaction for the semicircular three-hinged arch due to
   a. A vertical concentrated load P located at an angle θ as shown.
   b. A uniform load of intensity w lb/ft of horizontal projection.

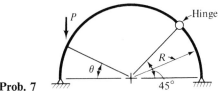

Prob. 7

8. Calculate the reactions at points A, B, and C due to the dead weight of the semicircular bow girder of weight w lb/ft.

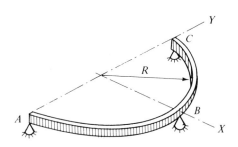

Prob. 8

9. Determine the degree of indeterminacy of the structures shown in parts (a) to (d).

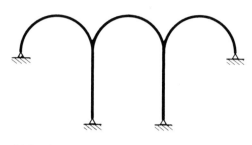

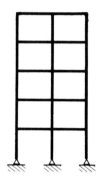

(b) Continuous arch viaduct

(a) Building frame

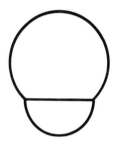

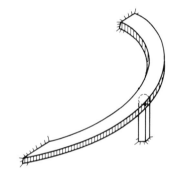

(c) Airplane fuselage

(d) Helicoidal girder with intermediate support

Prob. 9

CHAPTER 3

Internal Forces in Flexural Structures

Flexural structures are those consisting of members which can resist transverse forces through their ability to resist shears and moments, as distinct from ideal trusses whose members carry only axial forces. As defined here, they also include members called on to resist torsional moments, as occur in three-dimensional structures. The statical analysis of several statically determinate types of such structures is covered in the following sections.

3.1. Shear and Moment in Plane Beams

The basis for the analysis of beams has been laid in earlier courses and will only be reviewed at this point. It should be understood that continued practice is necessary for proficiency in carrying out efficient, dependable calculations. The drawing of shear and moment diagrams is one of the basic tasks required of every designer.

1. *Differential beam relations.* By considering the equilibrium of the differential beam element shown in Fig. 3.1, subjected to the distributed load $w(x)$, we obtain

$$\sum F_Y = 0: \qquad \frac{dV}{dx} = w \quad \text{or} \quad \Delta V = \int w\,dx \qquad (3.1)$$

and $\qquad \sum M_O = 0: \qquad \frac{dM}{dx} = V \quad \text{or} \quad \Delta M = \int V\,dx \qquad (3.2)$

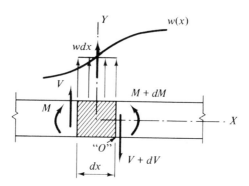

Fig. 3.1 Free body of beam element.

The first version of Eq. (3.1) states that the rate of change of shear (or the slope of the shear diagram) is equal to the load intensity. The second states that the difference of shear between any two points is equal to the total transverse load applied between these points. Equation (3.2) denotes analogous relations between moment and shear in the beam. A number of useful facts can be established from Eqs. (3.1) and (3.2), and should be reviewed in a text on statics or mechanics of deformable bodies. Load, shear, and moment are shown in their positive sense in Fig. 3.1. A downward gravity load, for instance, must be prefaced by a minus sign. Use of these statical relations is often convenient in beam analysis, as demonstrated in the subsequent example.

Example 3.1: Calculate and plot the variation of moment and shear in the simple beam subjected to the sinusoidal loading $w = -w_0 \sin(\pi x/L)$ shown in part (a), of the figure.

Solution:

$$V = \int w \, dx = w_0 \cdot \frac{L}{\pi} \cos \frac{\pi x}{L} + C_1$$

$$M = \int V \, dx = w_0 \cdot \left(\frac{L}{\pi}\right)^2 \sin \frac{\pi x}{L} + C_1 x + C_2$$

The boundary conditions to determine C_1 and C_2 are

$$M(O) = 0: \qquad 0 + 0 + C_2 = 0; \qquad C_2 = 0$$

$$M(L) = 0: \qquad w_0 \left(\frac{L}{\pi}\right)^2 \sin \pi + C_1 L + 0 = 0; \qquad C_1 = 0$$

Therefore,

$$V = w_0 L \cdot \frac{1}{\pi} \cos \frac{\pi x}{L}$$

$$M = w_0 L^2 \cdot \frac{1}{\pi^2} \sin \frac{\pi x}{L}$$

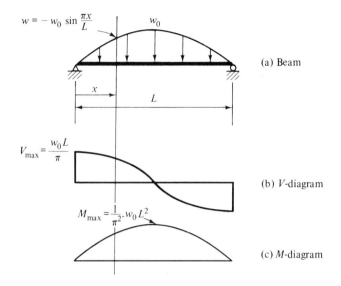

Fig. E3.1

Shear and moment diagrams are plotted in parts (b) and (c). Note that the moment diagram has the same shape as the load distribution.

For beams with discontinuities, such as concentrated loads, the integration becomes cumbersome, and the free-body technique may be preferred. In many cases, however, the above provides a good starting point.

2. *Shear and moments in beams by free bodies.* Free bodies provide the best insight into the nature of shear and moment as internal forces required for equilibrium. Correct free bodies and rigorous equilibrium equations are necessary for correct solutions. In general, the beam reactions should be found prior to determination of the internal forces.

Example 3.2: The simple beam is subject to the parabolically varying load shown. Determine reactions, write shear and moment equations, and draw the diagrams.

Solution: We first solve for the reactions at ends A and B:

$$\sum \overset{+}{M_A} = 0: \qquad \int_{x=0}^{L} \left[w_0 \left(\frac{x}{L} \right)^2 dx \right] x - R_B L = 0; \qquad R_B = \frac{1}{4} w_0 L$$

$$\sum \overset{+}{M_B} = 0: \qquad -\int_{x=0}^{L} \left[w_0 \left(\frac{x}{L} \right)^2 dx \right] (L - x) + R_A L = 0; \qquad R_A = \frac{1}{12} w_0 L$$

Turning now to the free body of part (b) of the figure, we write the equilibrium

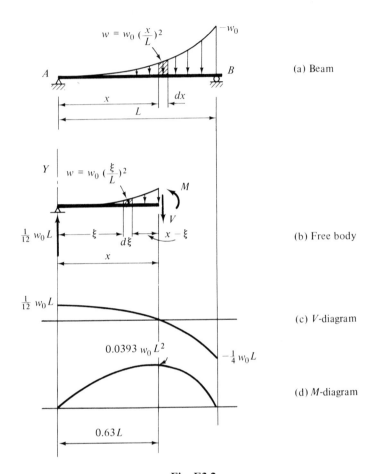

(a) Beam

(b) Free body

(c) V-diagram

(d) M-diagram

Fig. E3.2

equations:

$$\sum F_Y = 0: \quad \frac{1}{12} w_0 L - \int_{\xi=0}^{x} w_0 \left(\frac{\xi}{L}\right)^2 d\xi - V = 0;$$

$$V = \frac{w_0 L}{12}\left[1 - 4\left(\frac{x}{L}\right)^3\right]$$

$$\sum \overset{+}{M_x} = 0: \quad \left(\frac{1}{12} w_0 L\right)x - \int_{\xi=0}^{x} \left[w_0\left(\frac{\xi}{L}\right)^2 d\xi\right](x - \xi) - M = 0;$$

$$M = \frac{w_0 L^2}{12}\left[\left(\frac{x}{L}\right) - \left(\frac{x}{L}\right)^4\right]$$

The shear and moment diagrams are shown in parts (c) and (d). The maximum moment can be obtained by locating the point of zero shear, or of $dM/dx = 0$, at $x/L = 3\sqrt{\frac{1}{4}} = .631$. The moment at that point is $M_{\max} = .0393 w_0 L^2$.

Different types of beam structures and different loads require somewhat different applications of the above principles. If, for instance, only concentrated loads are involved, integration is unnecessary; for uniform or linearly varying loads, known centroidal properties may be used; for combinations of loads, superposition may be convenient; and, in all cases, the relations between the properties of load, shear, and moment diagrams may be used conveniently. Only through practice will appropriate judgment be gained for efficient computation.

3.2. Internal Forces in Plane Frames and Arches

Arches and frames are commonly occurring forms. For equilibrium, direct axial forces are necessary in addition to shears and moments. Indeed, the greater the axial force and the smaller the moments, the more efficient will be the structure, since in general the strength and stiffness of a member against axial forces are vastly greater than its flexural resistance (if buckling under compressive forces can be avoided). It is the job of the designer to devise feasible shapes of structures such that the internal moment due to a given load is minimized, and thus arch-type structures are often used for long-span bridges or roof structures, and ring-type structures in aerospace and pressure-vessel applications.

Although differential relations can be established for such structures, the emphasis here will be on the free-body approach, which will require three equilibrium equations for the determination of three unknown internal forces, as demonstrated in the following example.

Example 3.3: Draw shear and moment diagrams for the rigid frame shown.

Solution: We first find the reactions by appropriate equilibrium conditions and indicate them on the sketch of the frame. Now, considering member AB, we draw a free body of arbitrary length s_1, and, summing forces and moments about the cut section, we determine

$$\sum F_X = 0: \quad .5 \cdot s_1 - V = 0; \quad V = .5 s_1 \text{ kips}$$

$$\sum F_Y = 0: \quad 8 + N = 0; \quad N = -8 \text{ kips}$$

$$\sum \overset{+}{M_{s_1}} = 0: \quad -.5 s_1 \cdot \frac{s_1}{2} - M = 0; \quad M = -.25 s_1^2 \text{ kip-ft}$$

The equations for the internal forces enable us to sketch the appropriate diagrams; in particular, at point B, $s_1 = 20$ ft, and we find the forces shown acting on joint B and on the left end of member BC. Note that the two moments acting on opposite ends of these joints must be equal in value and opposite in sense for equilibrium.

We now proceed to member BC and, calculating similarly but noting the

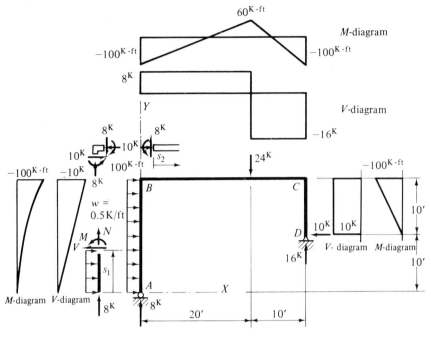

Fig. E3.3

discontinuity due to the concentrated load, obtain the shear and moment diagrams shown.

For member CD we can either proceed from joint C, after indicating the loads transferred through this joint, or begin from the reactions at point D.

Shear and moment diagrams are now drawn, thus concluding the analysis.

Example 3.4: Calculate shear, moment, and axial force in the three-hinged semicircular arch due to its dead weight w lb/ft.

Solution: The reactions have been determined earlier in Ex. 2.2 and are shown in the free body, part (b). A polar coordinate system is set up, and a convenient radial and tangential R–T set of local axes is established, as shown. Prior to summing forces along these axes, it is useful to write the R and T components of the reactions and load element, as shown in part (b).

$$\sum F_R = 0: \quad -\left(\frac{\pi}{2} - 1\right)wR\cos\theta + \frac{\pi}{2}wR\sin\theta$$

$$-\int_{\alpha=0}^{\theta} wR\,d\alpha\,\sin\theta + V = 0;$$

$$V = wR\left[\left(\frac{\pi}{2} - 1\right)\cos\theta + \left(\theta - \frac{\pi}{2}\right)\sin\theta\right]$$

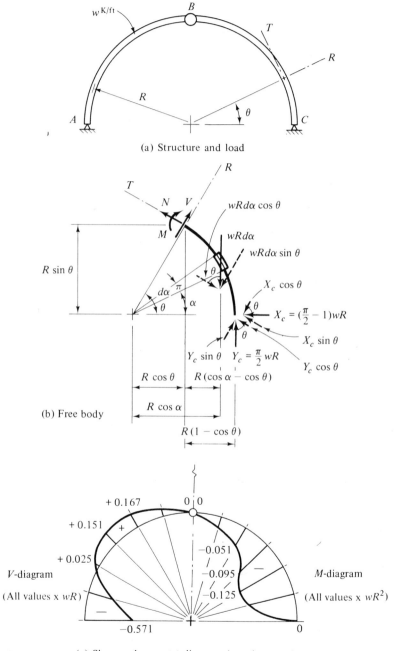

(a) Structure and load

(b) Free body

(c) Shear and moment diagrams in polar coords.

Fig. E3.4

$$\sum F_T = 0: \quad \left(\frac{\pi}{2} - 1\right)wR \sin \theta + \frac{\pi}{2} wR \cdot \cos \theta$$

$$- \int_{\alpha=0}^{\theta} wR \, d\alpha \cos \theta + N = 0;$$

$$N = wR\left[\left(\theta - \frac{\pi}{2}\right)\cos \theta - \left(\frac{\pi}{2} - 1\right)\sin \theta\right]$$

$$\sum \overset{+}{M_\theta} = 0: \quad \left(\frac{\pi}{2} - 1\right)wR \cdot R \sin \theta - \frac{\pi}{2} wR^2(1 - \cos \theta)$$

$$+ \int_{\alpha=0}^{\theta} wR \, d\alpha \cdot R(\cos \alpha - \cos \theta) + M = 0;$$

$$M = wR^2\left[\frac{\pi}{2}(1 - \sin \theta) + \left(\theta - \frac{\pi}{2}\right)\cos \theta\right]$$

Part (c) shows plots of the shear and moment in polar coordinates; the values of the functions plotted are measured as the radial distances from the arch axis to the curve.

3.3. Three-Dimensional Structures

In nonplanar structures, or planar structures under out-of-plane loading, torsional moments will occur. Since both strength and stiffness of members under torsion are often critical, the designer must be aware of this and be able to calculate the torsional response. A typical example of this structural type is a plane bow girder, and the following example will apply the basic approach to the analysis of a statically determinate version of a semicircular bow girder under transverse load.

Example 3.5: Calculate the internal forces in the horizontal bow girder of part (a) of the figure due to its dead load of intensity w lb/ft.

Solution: We refer to the plan view of the free body of part (b) and write the equilibrium conditions:

$$\sum F_Z = 0: \quad -\int_{\alpha=0}^{\theta} wR \, d\alpha + V = 0; \qquad \underline{V = wR\theta}$$

$$\sum M_R = 0: \quad -\int_{\alpha=0}^{\theta} (wR \, d\alpha)(R \sin \alpha) + M = 0; \qquad \underline{M = wR^2(1 - \cos \theta)}$$

$$\sum M_T = 0: \quad \int_{\alpha=0}^{\theta} (wR \, d\alpha)R(1 - \cos \alpha) + T = 0; \qquad \underline{T = -wR^2(\theta - \sin \theta)}$$

These internal forces are sketched in part (c).

Note that for $\theta = \pi$ the reactions already determined in Ex. 2.3 result.

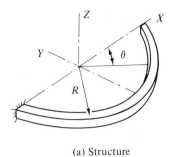

(a) Structure

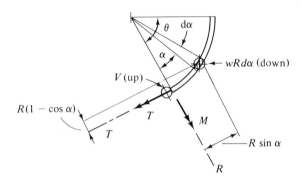

(b) Free body

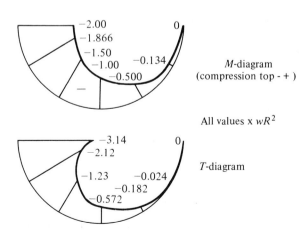

(c) Bending and torsional moment diagrams

Fig. E3.5

3.4. Curved Structures in Space

Space structures may have complicated geometry, and in such cases the use of vector analysis can serve as a useful tool for the determination of internal forces and reactions. A review of vector analysis as used in a course in vector statics may be indicated at this time. In the general case, six internal force components act at any section of the member, and the following outlines a generally valid way of finding these.

The calculations may be divided into two steps:

1. *Calculation of the member geometry.* This involves the geometry of the member axis and of the principal section axes on which the forces are required.

We consider a member whose axis is defined by the parametric set of equations

$$x = f_1(\theta); \qquad y = f_2(\theta); \qquad z = f_3(\theta) \tag{3.3}$$

as shown in Fig. 3.2. The unit vectors along the X, Y, and Z axes are denoted by **i**, **j**, and **k**. The section on which the forces are required is cut and the principal axes are indicated by N for the normal, S for the strong, and W for the weak bending axes; the corresponding unit vectors are **n**, **s**, and **w**.

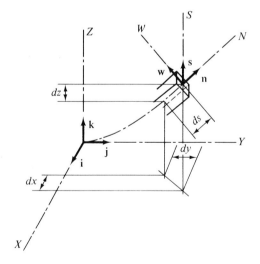

Fig. 3.2 Structure geometry.

The unit normal vector at any section is determined by

$$\mathbf{n} = \frac{(dx)\mathbf{i} + (dy)\mathbf{j} + (dz)\mathbf{k}}{ds} = \frac{(dx)\mathbf{i} + (dy)\mathbf{j} + (dz)\mathbf{k}}{(dx^2 + dy^2 + dz^2)^{1/2}} \tag{3.4}$$

The principal bending axes must be defined; that is, if the strong bending axis is parallel to the XY plane, or horizontal (as it will generally be if gravity loads are to be carried), then this axis is normal to the N and Z axes, and its unit vector is

$$\mathbf{s} = \frac{\mathbf{n} \times \mathbf{k}}{|\mathbf{n} \times \mathbf{k}|} \tag{3.5}$$

The cross product must be divided by its absolute value to obtain the unit vector. The weak bending axis is normal to the N and S axes, and thus its unit vector can be found by

$$\mathbf{w} = \mathbf{n} \times \mathbf{s} \tag{3.6}$$

Since $\mathbf{n}$ and $\mathbf{s}$ are normal unit vectors, their cross product will also be a unit vector.

2. *Equilibrium calculations.* We consider the free body of Fig. 3.3, showing a given applied load $\mathbf{P}$ acting at point A. The resultant force vector $\mathbf{F}$ and moment vector $\mathbf{M}$ acting on the cut section B are found by the vector

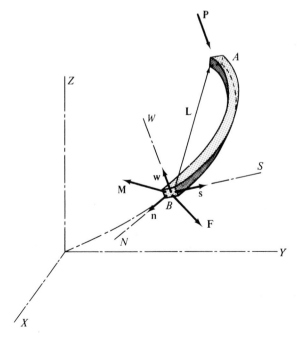

Fig. 3.3 Free body.

equilibrium equations

$$\sum F = 0: \qquad \mathbf{P} + \mathbf{F} = 0; \qquad \mathbf{F} = -\mathbf{P} \qquad\qquad (3.7)$$

$$\sum M_B = 0: \qquad \mathbf{L} \times \mathbf{P} + \mathbf{M} = 0; \qquad \mathbf{M} = -\mathbf{L} \times \mathbf{P} \qquad (3.8)$$

where $\mathbf{L}$ is the lever arm vector from B to A.

The axial and shear forces, N, V_s, and V_w, are components of the force vector $\mathbf{F}$ along the N, S, and W axes, and can be found by dot multiplication with the appropriate unit vector:

$$N = \mathbf{F} \cdot \mathbf{n}; \qquad V_s = \mathbf{F} \cdot \mathbf{s}; \qquad V_w = \mathbf{F} \cdot \mathbf{w} \qquad\qquad (3.9)$$

The torsional and bending moments, T, M_s, and M_w, are components of the moment vector $\mathbf{M}$ and are found similarly:

$$T = \mathbf{M} \cdot \mathbf{n}; \qquad M_s = \mathbf{M} \cdot \mathbf{s}; \qquad M_w = \mathbf{M} \cdot \mathbf{w} \qquad\qquad (3.10)$$

With all internal forces determined in this fashion, stresses and deformations can be calculated. Internal forces due to several applied loads on the structures can be found by summation, those due to distributed loads by integration.

An example of a cantilevered helicoidal girder under vertical point load will illustrate the use of vector analysis.

Example 3.6: Determine the internal forces N, V_s, and V_w and the internal moments T, M_s, and M_w along the helicoidal cantilevered girder shown, due to a vertical load P at its free end.

Solution: We first determine the member geometry, referring to the set of axes shown. The equation of the helix is written in terms of the angle θ as

$$x = R \cos \theta; \qquad y = R \sin \theta; \qquad z = \frac{H}{\pi} \theta$$

To find the normal unit vector $\mathbf{n}$ at any point, we first find the derivatives

$$dx = -R \sin \theta \, d\theta; \qquad dy = R \cos \theta \, d\theta; \qquad dz = \frac{H}{\pi} d\theta$$

and insert them into Eq. (3.4):

$$\mathbf{n} = \frac{-\mathbf{i} \cdot R \sin \theta + \mathbf{j} R \cos \theta + \mathbf{k}(H/\pi)}{[R^2 \sin^2 \theta + R^2 \cos^2 \theta + (H/\pi)^2]^{1/2}}$$

$$= \frac{1}{[1 + (H/\pi R)^2]^{1/2}} [-\mathbf{i} \sin \theta + \mathbf{j} \cos \theta + \mathbf{k}(H/\pi R)]$$

The denominator depends only on the helix proportions and will henceforth be designated by K.

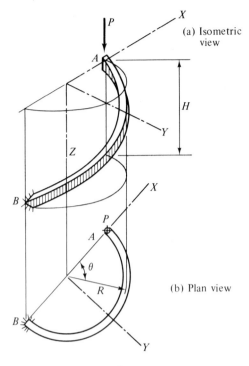

X

P

(a) Isometric view

A

H

Y

Z

B

X

P

A

θ

R

(b) Plan view

B

Y

Fig. E3.6

The strong bending axis lies in a horizontal plane, and its unit vector can thus be found by Eq. (3.5), in which

$$\mathbf{n} \times \mathbf{k} = \frac{1}{K} \begin{vmatrix} \mathbf{i} & \mathbf{j} & \mathbf{k} \\ -\sin\theta & \cos\theta & \dfrac{H}{\pi R} \\ 0 & 0 & 1 \end{vmatrix}$$

$$= \frac{1}{K}(\mathbf{i}\cos\theta + \mathbf{j}\sin\theta)$$

The absolute magnitude of this vector $|\mathbf{k} \times \mathbf{n}| = 1/K$, so that

$$\mathbf{s} = \mathbf{i}\cos\theta + \mathbf{j}\sin\theta$$

The unit vector along the weak axis is found by Eq. (3.6) as

$$\mathbf{w} = \mathbf{s} \times \mathbf{n} = \frac{1}{K} \begin{vmatrix} \mathbf{i} & \mathbf{j} & \mathbf{k} \\ \cos\theta & \sin\theta & 0 \\ -\sin\theta & \cos\theta & \dfrac{H}{\pi R} \end{vmatrix}$$

$$= \frac{1}{K}\left(\mathbf{i} \cdot \frac{H}{\pi R}\sin\theta - \mathbf{j}\frac{H}{\pi R}\cos\theta + \mathbf{k}\right)$$

This concludes the determination of the member geometry, and we turn to the equilibrium equations, Eqs. (3.7) and (3.8):

$$\Sigma F = 0: \qquad \mathbf{F} = -\mathbf{k}P$$
$$\Sigma M_B = 0: \qquad \mathbf{M} = -\mathbf{L} \times \mathbf{P}$$

in which

$$\mathbf{L} = \mathbf{i}(R - R\cos\theta) + \mathbf{j}(0 - R\sin\theta) + \mathbf{k}\left(0 - \frac{H}{\pi}\theta\right),$$

so that

$$\mathbf{L} \times \mathbf{P} = R \begin{vmatrix} \mathbf{i} & \mathbf{j} & \mathbf{k} \\ (1 - \cos\theta) & -\sin\theta & -\dfrac{\theta}{\pi} \cdot \dfrac{H}{R} \\ 0 & 0 & P \end{vmatrix}$$
$$= PR[-\mathbf{i}\sin\theta + \mathbf{j}(1 - \cos\theta)]$$

and
$$\mathbf{M} = PR[-\mathbf{i}\sin\theta + \mathbf{j}(1 - \cos\theta)]$$

The components of the force **F** and the moment **M** are obtained by appropriate dot multiplication with the unit vectors according to Eqs. (3.9) and (3.10):

$$N = \mathbf{F} \cdot \mathbf{n} = -\frac{1}{K}P\frac{H}{\pi R}$$

$$V_s = \mathbf{F} \cdot \mathbf{s} = 0$$

$$V_w = \mathbf{F} \cdot \mathbf{w} = -\frac{1}{K}P$$

$$T = \mathbf{M} \cdot \mathbf{n} = -\frac{PR}{K}(1 - \cos\theta)$$

$$M_s = \mathbf{M} \cdot \mathbf{s} = PR\sin\theta$$

$$M_w = \mathbf{M} \cdot \mathbf{w} = \frac{PH}{\pi K}(1 - \cos\theta)$$

This concludes the determination of the stress resultants.

3.5. Matrix Representation of Forces

Matrix representation is a bookkeeping scheme for keeping track of the relations between many discrete variables. For instance, if we want to list the moments at the four sections of the cantilever beam of Fig. 3.4 due to the loads $P_1, \ldots, P_4$ applied at any one of the four sections, we can represent the $(4 \times 4) = 16$ moment values in tabular form:

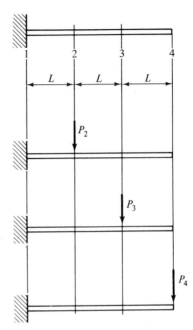

Fig. 3.4 Matrix formulation of moments.

	DUE TO LOAD			
MOMENT AT	P_1	P_2	P_3	P_4
Point 1	0	$-L \cdot P_2$	$-2L \cdot P_3$	$-3L \cdot P_4$
2	0	0	$-L \cdot P_3$	$-2L \cdot P_4$
3	0	0	0	$-L \cdot P_4$
4	0	0	0	0

A precise numbering system is essential to success here. In the table the numbering of the loads corresponds to that of the columns, and the numbering of the moments corresponds to that of the rows, so that each moment has its unambiguous place.

In matrix form, we can give the same information as

$$
\begin{Bmatrix} M_1 \\ M_2 \\ M_3 \\ M_4 \end{Bmatrix} = L \begin{bmatrix} 0 & -1 & -2 & -3 \\ 0 & 0 & -1 & -2 \\ 0 & 0 & 0 & -1 \\ 0 & 0 & 0 & 0 \end{bmatrix} \begin{Bmatrix} P_1 \\ P_2 \\ P_3 \\ P_4 \end{Bmatrix}
$$

or, in short,

$$
\{S\} = [b]\{P\}
$$

We use the symbol $\{S\}$ because in general this matrix may contain any type of action. This equation can easily be expanded to account for several, say *l*, loading conditions. We replace the $\{P\}$ column matrix by a rectangular matrix $[P]$ of *l* columns, each of which contains the combination of loads for one loading condition. Similarly, we replace the $\{S\}$ column matrix by a rectangular matrix of *l* columns, each of which contains the internal forces due to one loading condition:

$$[S] = [b][P] \tag{3.11}$$

The matrix $[b]$, which gives the actions *i* due to unit forces *j*, is called the *force transformation matrix*. It represents the equilibrium relation between the applied loads and the resulting internal actions.

Each element of the $[b]$ matrix can be denoted by the symbol b_{ij}, where the first subscript, *i*, denotes the column and the second, *j*, denotes the row within which it is located in the matrix $[b]$. In general (but not always), it is useful to think of the first subscript as designating the effect (here, the moments) and the second the cause (here, the loads).

The meaning of the symbols $[b]$ and b_{ij} is similar; in the former, attention is focused on the entire matrix, or egg crate, whereas the latter concerns the individual elements, or eggs. Mathematical operations involving matrices such as $[b]$ are based on linear matrix analysis, whereas computations in terms of elements b_{ij} use the concepts of tensor analysis. In this book we shall use matrix analysis to handle complete matrices, and a knowledge of elementary matrix algebra will be necessary to follow ensuing sections.

The emphasis on matrix methods results from two facts:

1. The use of matrix relations provides a concise, general, and easy-to-handle way of showing structural relationships.

2. Matrices are an appropriate way of furnishing information to high-speed computers. These machines also can be instructed to perform the operations of matrix analysis with great efficiency.

It is for these reasons that thinking in terms of matrices will be encouraged at an increasing rate as we get farther into the subject matter.

Example 3.7
a. Write the force transformation matrix $[b]$ relating the reactions X_1 and X_2 of the structure shown in the figure to unit values of the applied loads $P_1, P_2,$ and P_3
b. Calculate the reactions X_1 and X_2 due to three loading conditions:
    1. Vertical dead load, simulated by $P_1 = P_3 = 10$ kips.
    2. Snow load over half of the structure, simulated by $P_1 = 5$ kips.
    3. Lateral wind load, simulated by $P_2 = 4$ kips.

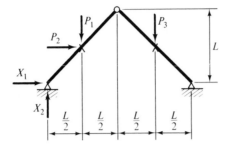

Fig. E3.7

Solution:
a. We identify the structure as a statically determinate three-hinged arch. Each column of the [b] matrix represents the two reactions due to one of the unit loads. Proceeding by basic statics, we obtain

$$\begin{Bmatrix} X_1 \\ X_2 \end{Bmatrix} = \begin{bmatrix} \frac{1}{4} & -\frac{3}{4} & \frac{1}{4} \\ \frac{3}{4} & -\frac{1}{4} & \frac{1}{4} \end{bmatrix} \begin{Bmatrix} P_1 \\ P_2 \\ P_3 \end{Bmatrix}$$

or $$\{X\} = [b]\{P\}$$

b. We expand the $\{X\}$ and $\{P\}$ matrices to three columns each to accommodate the three specified loading conditions, and perform the matrix multiplications:

$$\begin{bmatrix} X_{11} & X_{12} & X_{13} \\ X_{21} & X_{22} & X_{23} \end{bmatrix} = \begin{bmatrix} \frac{1}{4} & -\frac{3}{4} & \frac{1}{4} \\ \frac{3}{4} & -\frac{1}{4} & \frac{1}{4} \end{bmatrix} \begin{bmatrix} 10 & 5 & 0 \\ 0 & 0 & 4 \\ 10 & 0 & 0 \end{bmatrix} = \begin{bmatrix} 5.00 & 1.25 & -3.00 \\ 10.00 & 3.75 & -1.00 \end{bmatrix} \text{kips}$$

Columns 1 to 3 give the reactions due to loading conditions 1 to 3.

We can also write relations between member end forces in matrix form, as for instance the shear and moment at ends A and B of the straight beam shown in Fig. 3.5. From appropriate equilibrium relations, we write

$$\begin{Bmatrix} M_2 \\ V_2 \end{Bmatrix} = \begin{bmatrix} 1 & L \\ 0 & 1 \end{bmatrix} \begin{Bmatrix} M_1 \\ V_1 \end{Bmatrix} \tag{3.12}$$

or $$\{S_2\} = [b]\{S_1\}$$

where $\{S_1\}$ and $\{S_2\}$ are the forces at ends A and B, respectively. The equilibrium matrix [b] is in this case called the *transport matrix*.

The inverse relation, which expresses the end forces at A in terms of those at B, is

$$\{S_1\} = [b]^{-1}\{S_2\} \quad \text{or} \quad \begin{Bmatrix} M_1 \\ V_1 \end{Bmatrix} = \begin{bmatrix} 1 & -L \\ 0 & 1 \end{bmatrix} \begin{Bmatrix} M_2 \\ V_2 \end{Bmatrix}$$

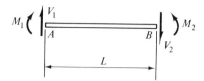

Fig. 3.5 Relations between end forces.

In further work, we shall generally designate a matrix of internal forces by the symbol $\{S\}$, and a matrix of external forces, be they loads or reactions, by the symbol $\{X\}$. The symbol $\{P\}$ is also used for loads at times. The force transformation matrix will always be called $[b]$.

3.6. Discretization of Force Systems

To handle load systems not amenable to analytic treatment, or for purposes of numerical calculation by matrix–computer methods, it is often convenient to replace continuous structural systems by discrete models.

As an example, we consider the uniformly loaded simple beam of Fig. 3.6(a). We divide this beam, along with its load, into a number of, say, four, segments, as shown, each of length $L/4$. The points denoting the segment interfaces are labeled 1 to 3. To discretize the distributed load, we consider each load segment as a simple beam that transfers its load as reactions at the beam quarter-points, as shown in Fig. 3.6(b). These reactions now constitute the statically equivalent discrete loads applied at points 1 to 3. Due to these discrete loads, we can now draw the piecewise-linear shear and moment diagrams, as in Figs. 3.6(c) and (d), which constitute an approximation to the exact solution. We note that in this case the approximate moment diagram shows exact values at the element interfaces 1 to 3, with deviations from the exact solutions in between. With increasingly finer division of the real load into statically equivalent discrete point loads, these deviations will become less and less, so that the approximate solution will converge on the exact one.

To proceed in a more general manner, we use the matrix of Eq. (3.11). The relation between unit applied forces at points 1 to 3 and the resulting moments at these points is given by the force transformation matrix $[b]$; by simple statics, we calculate

$$[b] = \frac{L}{16} \begin{bmatrix} 3 & 2 & 1 \\ 2 & 4 & 2 \\ 1 & 2 & 3 \end{bmatrix} \tag{3.13}$$

The matrix $\{P\}$ of applied loads at points 1 to 3 consists of the reactions of the simple beam load segments discussed earlier:

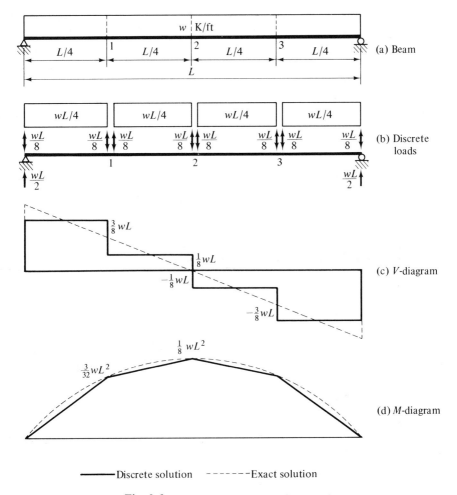

Fig. 3.6 Discrete solution for uniform load.

$$\{P\} = \frac{wL}{4} \begin{Bmatrix} 1 \\ 1 \\ 1 \end{Bmatrix}$$

The moments are now obtained by use of Eqs. (3.11) and (3.13):

$$\{S\} = [b]\{P\} = \frac{wL^2}{32} \begin{Bmatrix} 3 \\ 4 \\ 3 \end{Bmatrix}$$

that is, the same values as before.

One major problem facing the numerical analyst is the choice of the degree of precision with which the actual structure is to be simulated by the discrete model. The more pronounced the variation of the load intensity, the greater is the number of segments required to represent the load with sufficient accuracy; this will, of course, also increase the order of the matrices involved and the labor associated with their solution. Such decisions depend on the required degree of precision and the analyst's judgment.

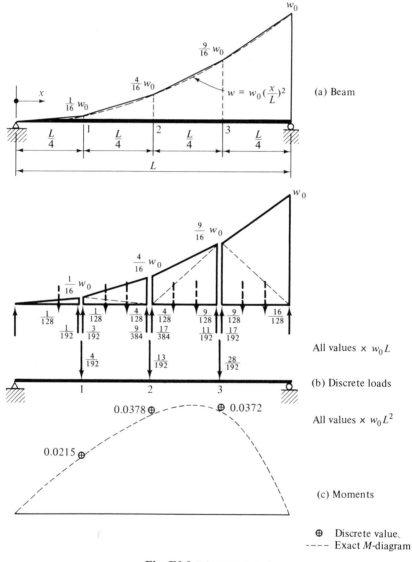

Fig. E3.8 Discrete analysis.

Example 3.8: By appropriate discretization, write the matrix $\{S\}$ of the internal moments at points 1 to 3 due to the parabolic load on the simple beam shown in part (a) of the figure.

Solution: The parabolic load is broken into four segments, each of length $L/4$, as shown in part (b). Each load segment is considered a simple beam simply supported at its ends, which coincide with the beam ends and points 1 to 3. As simplification in calculating the reactions on points 1 to 3, the load variation over each segment is assumed linear between its end points, so that the reactions can be conveniently computed by splitting the trapezoidal load segments into triangular parts of known magnitude and centroid, as in part (b). The resulting $\{P\}$ matrix of discrete loads at points 1 to 3 is

$$\{P\} = \frac{w_0 L}{192} \begin{Bmatrix} 4 \\ 13 \\ 28 \end{Bmatrix}$$

Because the same nodal points are involved as in the previous case discussed, we can use the same $[b]$ matrix, Eq. (3.13), and write

$$\{S\} = [b]\{P\} = \frac{w_0 L}{192} \frac{L}{16} \begin{bmatrix} 3 & 2 & 1 \\ 2 & 4 & 2 \\ 1 & 2 & 3 \end{bmatrix} \begin{Bmatrix} 4 \\ 13 \\ 28 \end{Bmatrix} = \frac{w_0 L^2}{3072} \begin{Bmatrix} 66 \\ 116 \\ 114 \end{Bmatrix} = \begin{Bmatrix} .0215 \\ .0378 \\ .0372 \end{Bmatrix} w_0 L^2$$

The resulting discrete moment values at points 1 to 3 of the beam are compared in part (c) with the exact moment diagram obtained from Ex. 3.2. The comparison is good, but we note that the maximum moment value of $.0393 w_0 L^2$ (see Ex. 3.2) is not obtained simply because it is away from any of the discrete sections. It could, however, be obtained with fair accuracy by French-curve interpolation of the discrete moment values. The fact that the discrete moment values are on the high side could have been predicted because of the over-estimation of the applied load due to the linear approximation between discrete points.

Example 3.9: Find moment values at sections 45° apart in the cantilever bow girder of Ex. 3.5 without performing integrations.

Solution: The uniform load is discretized by replacing it with four concentrated loads located as shown in part (a) of the figure, each of magnitude $p = (\pi/4)wR = .785wR$ (note that these resultants are not at the centroid of the load segments, and thus are not statically equivalent). The bending and torsional moments $\{S\}$ at the discrete sections A to E due to unit values of the concentrated loads are shown in the $[b]$ matrix, and the discrete moment values due to the discrete loads $\{P\}$ are calculated by the matrix multiplication

$$\{S\} = [b]\{P\}$$

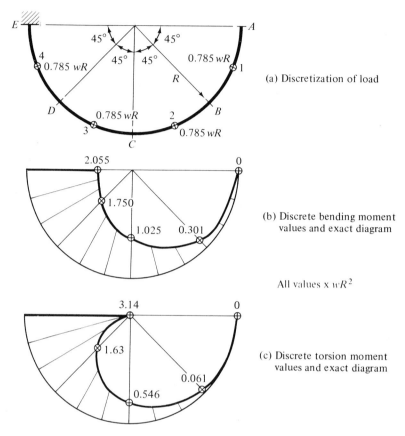

(a) Discretization of load

(b) Discrete bending moment values and exact diagram

All values x wR^2

(c) Discrete torsion moment values and exact diagram

Fig. E3.9 Discrete analysis of bow girder under dead load.

or

$$
\begin{Bmatrix} M_A \\ T_A \\ M_B \\ T_B \\ M_C \\ T_C \\ M_D \\ T_D \\ M_E \\ T_E \end{Bmatrix} = \begin{bmatrix} 0 & 0 & 0 & 0 \\ 0 & 0 & 0 & 0 \\ -.383 & 0 & 0 & 0 \\ -.077 & 0 & 0 & 0 \\ -.923 & -.383 & 0 & 0 \\ -.618 & -.077 & 0 & 0 \\ -.923 & -.923 & -.383 & 0 \\ -1.383 & -.618 & -.077 & 0 \\ -.383 & -.923 & -.923 & -.383 \\ -1.923 & -1.383 & -.618 & -.077 \end{bmatrix} R \begin{Bmatrix} 1 \\ 1 \\ 1 \\ 1 \end{Bmatrix} .785wR = \begin{Bmatrix} 0 \\ 0 \\ .301 \\ .061 \\ 1.025 \\ .546 \\ 1.750 \\ 1.630 \\ 2.055 \\ 3.142 \end{Bmatrix} wR^2
$$

Parts (b) and (c) show the moment values compared to exact curves.

PROBLEMS

In all problems, attempt to draw the deflected shape of the structure due to the given loads prior to analysis, and try to reconcile this with the internal moments obtained by analysis.

1. Compute shear and moment expressions by integration of the load function. Compare the effort with that of Ex. 3.2 and draw conclusions.

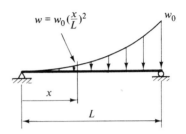

Prob. 1

2. Find reactions for the beam shown. Draw shear and moment diagrams, indicating all important points and values. Draw the approximate deflected shape of the structure. Write the moment expression for portion *BC* of the beam, using point *B* as the origin.

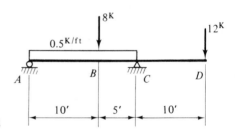

Prob. 2

3. Draw shear and moment diagrams for a vertical strip 1 ft wide of the tank wall due to the water pressure. Indicate all critical points and values. Draw the approximate deflected shape of the wall.

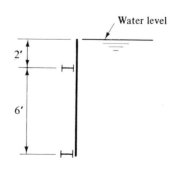

Prob. 3

4. For the overhanging beam under uniform load w, determine the ratio of the distances a/b so that maximum positive and negative moments are of equal magnitude.

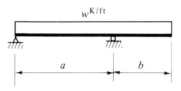

Prob. 4

5. Draw shear and moment diagrams for the articulated beam under uniform load. Draw the approximate deformed shape of the beam.

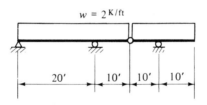

Prob. 5

6. The stairway frame shown is to be completely analyzed. The stairs are 4 ft wide and the total vertical load is 150 lb/ft² of the structure. Draw shear and moment diagrams and indicate all critical points.

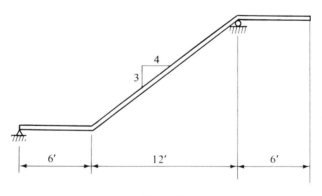

Prob. 6

7. Draw shear and moment diagrams for the stairway structure shown. The load is .6 kip/ft of projected length. Indicate the magnitudes and location of all significant points, and draw the deflected shape.

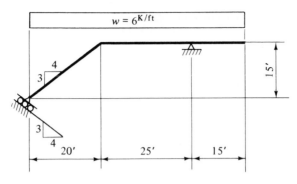

Prob. 7

8. Find reactions for the structure shown. Draw shear and moment diagrams. Indicate the deflected shape of the structure.

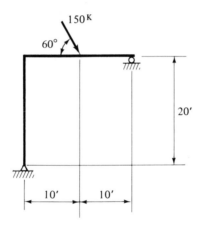

Prob. 8

9. For the rigid frame under the uniform lateral load shown,
   a. Draw shear and moment diagrams.

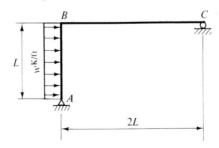

Prob. 9

b. Write the moment equation for member *AB*.

c. Draw the deflected shape of the structure.

10. Draw shear and moment diagrams for the structure under the load shown, and show the approximate deflected shape.

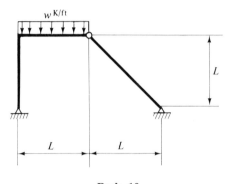

Prob. 10

11. For the rigid frame under the loads shown,

a. Draw shear and moment diagrams; indicate all critical values.

b. Write shear and moment equations for the left-hand column.

c. Carefully draw the approximate deflected shape of the structure.

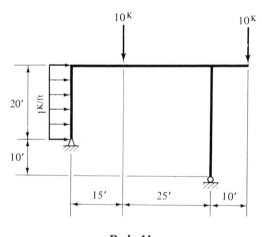

Prob. 11

12. Draw shear and moment diagrams for the symmetrical three-hinged frame shown. Indicate all critical values, and draw the approximate deformed shape.

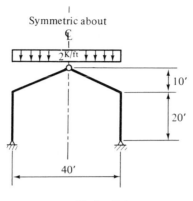

Prob. 12

13. Find reactions, and draw shear and moment diagrams for the rigid frame shown. Denote all important values, and draw the deformed shape.

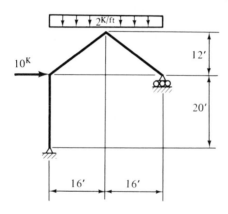

Prob. 13

14. Completely analyze the structure. Draw plots of the variation of shear, moment, and axial force in the left-hand member.

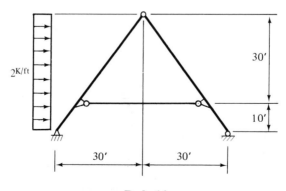

Prob. 14

15. The portal frame shown consists of upright members that are stiff in bending, to which are pinned diagonal and horizontal two-force members. Analyze the structure, draw shear and moment diagrams where applicable, and indicate the axial forces.

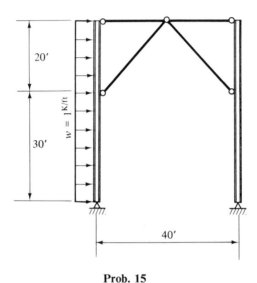

Prob. 15

16. In the three-hinged parabolic arch shown, find the variation of moments as a function of the coordinate x due to the load between arch and deck, of unit weight w kips/ft². Draw the moment diagram and comment on the appropriateness of the arch shape to resist this load.

Prob. 16

17. Write expressions for moment, shear, and axial force in the semicircular three-hinged arch due to the uniform load of intensity w as a function of the angle θ. Plot the variation of these actions in polar coordinates for half the arch and denote critical values.

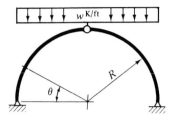

Prob. 17

18. The semicircular bow girder lies in the horizontal plane. Adopt a suitable coordinate system and write equations for the variation of the bending and torsional moments in the structure due to the concentrated vertical load *P*.

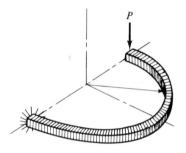

Prob. 18

19. For the semicircular bow girder lying in the horizontal plane,
    a. Compute bending and torsional moments at section θ due to a vertical load *P* at point α.
    b. By integration of the effects of the load $dP = w \cdot ds = wR\, d\alpha$, compute the internal moments due to a uniform load w; compare results with those of Ex. 3.5.

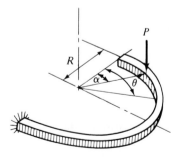

Prob. 19

20. Compute and plot shears, bending, and torsional moments in the horizontal bent cantilever structure due to its own dead weight of intensity w K/ft.

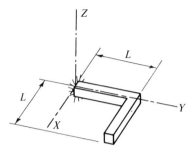

Prob. 20

21. The bent girder in space has its strong bending axis in the horizontal (XY) plane.
    a. Compute and plot the internal member forces due to a concentrated downward (parallel to the Z axis) load P applied at point C.
    b. Compute the internal moments in member AB due to the weight $w\,ds$ of the portion ds of the structure, as shown.
    c. By integration of the results of Part b, compute the internal moments in member AB due to the dead weight of the member BC.

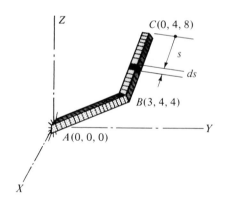

Prob. 21

22. a. In the helicoidal girder shown (which is identical to that of Ex. 3.6), compute the internal forces in the structure due to concentrated force $\mathbf{P} = P_x\mathbf{i} + P_y\mathbf{j} + P_z\mathbf{k}$. Check with Ex. 3.6 for the special case $P_x = P_y = 0$.
    b. Compute the internal moments at any section subtending a horizontal angle θ due to the dead weight of the portion of the structure subtended by the horizontal angle $d\alpha$. Unit weight of the girder is w lb/ft of length.
    c. By integration of the results of Part a, determine the variation of the internal moments due to the dead weight of the girder.

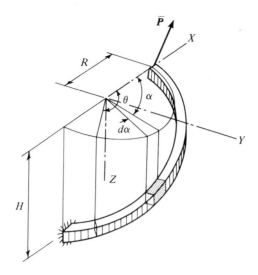

Prob. 22

23. a. For the simple beam shown, compute the force transformation matrix $[b]$ defined by

$$\begin{Bmatrix} M_1 \\ M_2 \\ \cdot \\ \cdot \\ M_5 \end{Bmatrix} = \begin{bmatrix} b_{11} & b_{12} & \cdots & b_{15} \\ b_{21} & & & b_{25} \\ \cdot & & & \cdot \\ \cdot & & & \cdot \\ b_{51} & b_{52} & \cdots & b_{55} \end{bmatrix} \begin{Bmatrix} P_1 \\ P_2 \\ \cdot \\ \cdot \\ P_5 \end{Bmatrix}$$

where P_j is an upward load at point j.

   b. Using the matrix $[b]$, compute the (5×3) matrix $[M]$ due to three loading conditions:

   1. $P_2 = -5$ kips.
   2. $P_3 = -2$ kips; $P_4 = -1$ kip.
   3. $P_2 = -5$ kips; $P_3 = -2$ kips; $P_4 = -1$ kip.

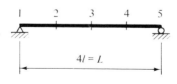

Prob. 23

24. a. Write the $[b]$ matrix of order (5×5) relating the moments at point i due to unit loads at point j; upward is positive.

   b. Write the (5×2) matrix of moments due to two loading conditions:

   1. $P_1 = P_5 = \frac{1}{2}wl$; $P_2 = P_3 = P_4 = wl$.
   2. $P_5 = 10$ kips downward.

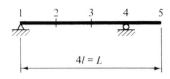

Prob. 24

25. Calculate the elements of the [b] matrix defined by the moment–load matrix equation:

$$\begin{Bmatrix} M_2 \\ M_3 \end{Bmatrix} = \begin{bmatrix} b_{11} & \cdots & b_{16} \\ b_{21} & \cdots & b_{26} \end{bmatrix} \begin{Bmatrix} P_1 \\ P_2 \\ \cdot \\ \cdot \\ \cdot \\ P_6 \end{Bmatrix}$$

where P_i is a downward vertical load applied at point i.

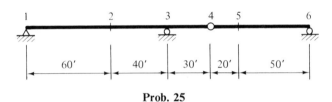

Prob. 25

26. For the rigid frame shown, write the force transformation matrix [b] relating the loads $\{P\} = \begin{Bmatrix} P_1 \\ P_2 \\ M_3 \end{Bmatrix}$ and the actions $\{S\} = \begin{Bmatrix} R_4 \\ R_5 \\ M_6 \end{Bmatrix}$, all defined by the numbered nodes shown:

$$\{S\} = [b]\{P\}$$

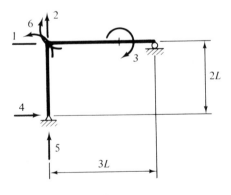

Prob. 26

27. Discretize the sinusoidally varying load by means of concentrated loads applied at the numbered points, and, by use of the force transformation matrix, Eq. (3.13), compute the moments at points 1, 2, and 3. Superimpose upon a plot of the exact moments (see Ex. 3.1), and comment on accuracy.

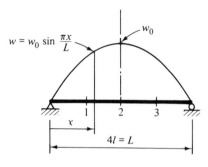

$$w = w_0 \sin \frac{\pi x}{L}$$

$$4l = L$$

Prob. 27

28. a. Write the force transformation matrix $[b]$ between the vertical forces and the resulting moments at the numbered points of the simple beam.

   b. Discretize the parabolically varying load of Ex. 3.8 by means of concentrated loads at points 1 to 5, and, using the force transformation matrix of Part a, compute the moments at these points.

   c. Compare the amount of work and accuracy of the results with Ex. 3.8, and draw engineering conclusions regarding the convergence of solution with increasing number of segments.

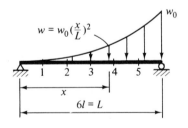

$$w = w_0 \left(\frac{x}{L}\right)^2$$

$$6l = L$$

Prob. 28

29. a. Write the $[b]$ matrix relating the bending and twisting moments at points 1 to 5 of the bent horizontal girder due to vertical loads applied at these points.

   b. Discretize the dead load, of intensity w kips/ft, by means of concentrated loads at points 1 to 5, and compute the bending and torsional dead-load moments at points 1 to 5.

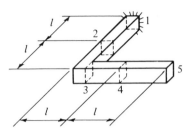

Prob. 29

30. a. For the three-hinged roof arch shown, calculate and write the [*b*] matrix relating the matrix [*M*] of internal moments at points 1 to 6 to the matrix [*P*] of vertical and horizontal concentrated loads at these points.

   b. By discretizing the specified loads and using the [*b*] matrix of Part a, calculate the moments at these points due to
   1. The roof dead load, of intensity .4 kip/ft of length of arch rib.
   2. The snow load, of intensity .6 kip/ft of horizontal projection.
   3. The wind load from the left, represented by a pressure of .3 kip/ft of arch rib on the window arch and a suction of .15 kip/ft of arch rib on the leeward side.
   4. A similar wind load from the right.
   5. The design moments due to the critical combination of loads 1 to 4.

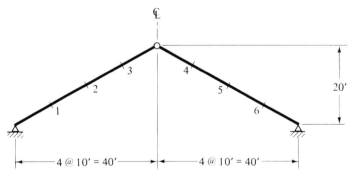

Prob. 30

CHAPTER 4

Forces in Trusses

Ideal trusses are assemblies of two-force members connected by joints capable of resisting forces but not moments. To be stable, the members must be so arranged that they subdivide the truss into triangular panels, and the loads should be applied at the joints, or *panel points*. Trusses are efficient structural forms for transmitting loads and have in the past been widely used for bridges. Although for this purpose they have for all but very long spans been superseded by other structural types, they continue to be used for roof structures, transmission towers and microwave dishes, in aerospace applications, and so forth. In many of these uses the structure is three-dimensional, and the analyst should be capable of calculating stresses in both plane and space trusses.

It should be understood that the idealization of trusses as assemblies of two-force members connected by perfect pins may not conform exactly to reality. Here, as in all engineering, the designer must strike a happy medium between rigor and efficiency. Actually, truss joints are usually welded, some loads may act transverse to the members, and buckling of compression members may be a problem. Such considerations are, in our context, called *secondary*, but in actual practice they should not be overlooked.

It is not intended here to give a survey of the many different truss types that have been used for different purposes. Rather, an attempt will be made to give generally applicable methods for both two- and three-dimensional trusses.

Accordingly, the analysis of plane trusses will be reviewed in the next section. In Sec. 4.2 a generally valid, formal matrix analysis will be presented, and in Sec. 4.3 this will be implemented by a computer routine.

4.1. Plane Trusses

The calculation of member forces in plane statically determinate trusses involves straightforward application of equilibrium equations to appropriate free bodies. These free bodies must always be so cut from the structure that the unknown forces to be determined are external to the free body. The equations of statics should, if possible, be so selected that only one unknown at a time appears in any one equation, so as to avoid the necessity of solving simultaneous equations.

For plane trusses, two methods are widely used for longhand calculation: the *method of joints* and the *method of sections*. For computer use, the method shown in Sec. 4.2 for the analysis of space trusses leads to a convenient formulation of the problem.

1. *Method of joints.* In the method of joints, each joint is used as a free body; it forms a coplanar, concurrent force system for which two independent equilibrium equations can be written. Accordingly, we select a joint with two unknown member forces (such a joint can always be found after the truss reactions have been determined), solve for these unknown forces, and proceed to the next joint until the truss is traversed. In this process, all member forces are determined, and, accordingly, this method is useful if all member forces are required. At the end of the calculation a check is obtained with a previously found reaction. One word of caution: since the member forces determined in this process are used sequentially, one error will void all subsequent work.

Example 4.1: Find all member forces in the truss due to the loading shown shown in the figure.

Solution: We first determine the reactions and indicate them in the sketch as shown. Proceeding to joint A, we first calculate equilibrium in the Y direction, since this involves only the unknown force AE; then the force AB is determined by $\sum F_x = 0$. Next we proceed to joint E, which has the two unknown forces ED and EB acting on it. Note that the axis perpendicular to ED, Y', will involve only one unknown, EB; so we sum forces along the orthogonal axes Y' and X' sequentially to solve for the two unknowns. We next proceed in order through joints D, B, and C as shown, and obtain a check with the previously determined reaction at C, thus concluding the problem. The member forces are indicated on the truss members with a $+$ sign indicating tension and a $-$ sign, compression. Since all unknown member forces were

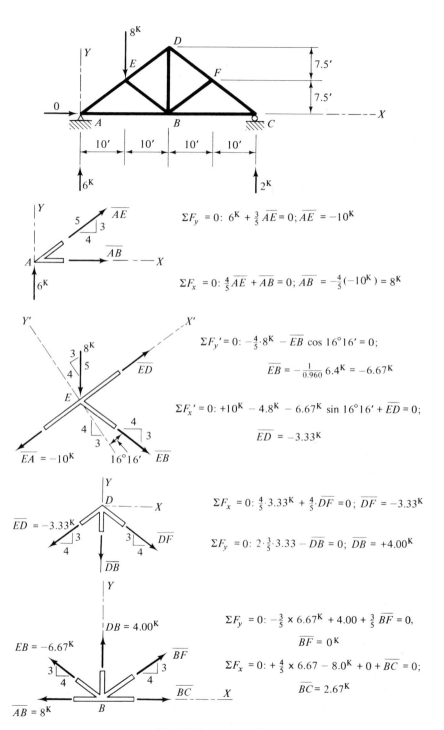

$\Sigma F_y = 0$: $6^K + \frac{3}{5}\,\overline{AE} = 0$; $\overline{AE} = -10^K$

$\Sigma F_x = 0$: $\frac{4}{5}\,\overline{AE} + \overline{AB} = 0$; $\overline{AB} = -\frac{4}{5}(-10^K) = 8^K$

$\Sigma F_y' = 0$: $-\frac{4}{5}\cdot 8^K - \overline{EB}\cos 16°16' = 0$;

$\overline{EB} = -\frac{1}{0.960}\,6.4^K = -6.67^K$

$\Sigma F_x' = 0$: $+10^K - 4.8^K - 6.67^K \sin 16°16' + \overline{ED} = 0$;

$\overline{ED} = -3.33^K$

$\Sigma F_x = 0$: $\frac{4}{5}\cdot 3.33^K + \frac{4}{5}\cdot\overline{DF} = 0$; $\overline{DF} = -3.33^K$

$\Sigma F_y = 0$: $2\cdot\frac{3}{5}\cdot 3.33 - \overline{DB} = 0$; $\overline{DB} = +4.00^K$

$\Sigma F_y = 0$: $-\frac{3}{5} \times 6.67^K + 4.00 + \frac{3}{5}\,\overline{BF} = 0$,

$\overline{BF} = 0^K$

$\Sigma F_x = 0$: $+\frac{4}{5} \times 6.67 - 8.0^K + 0 + \overline{BC} = 0$;

$\overline{BC} = 2.67^K$

Fig. E4.1 Method of joints.

77

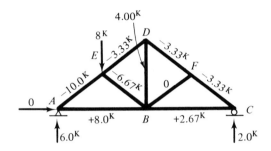

$$\Sigma F_x = 0: -2.67^K - \tfrac{4}{5}\overline{CF} = 0, \ \overline{CF} = -3.33^K$$

$$\Sigma F_y = 0: -\tfrac{3}{5} \cdot 3.33 + R_c = 0; \ R_c = +2.00^K \quad \text{Check.}$$

Fig. E4.1 (*Continued*)

initially assumed tensile, the resulting forces will automatically assume their proper signs in the solution.

With an understanding of the method of joints, it is also possible to establish a criterion for static determinacy of planar trusses. For a truss with j joints, each with two independent equilibrium equations, $2j$ conditions of statics are available, which permit a like number of unknown forces to be determined. These forces may be either member forces or reactive components. A criterion for static determinacy of the truss is thus

$$m + r = 2j \qquad (4.1)$$

where m is the number of members and r is the number of reactive components. Equation (4.1) is not an absolute criterion of static determinacy, since the members may be arranged in an unsuitable configuration, as shown, for instance, in the unstable truss of Fig. 4.1. A more comprehensive check for stability is furnished in Sec. 4.2.

2. *Method of sections.* If only one, or a few, member forces in a truss are required, the method of sections is useful. The procedure is to cut a free body containing the desired member forces externally and to write one or more appropriate equilibrium equations. Often it is possible to select an equilibrium equation containing the desired quantity as the only unknown, thus minimizing the work.

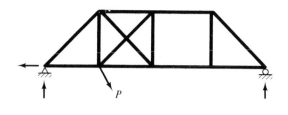

$$j = 8$$
$$m = 13 \quad m + r = 2j, \text{ but unstable}$$
$$r = 3$$

Fig. 4.1 Unstable truss.

Example 4.2: In the truss shown in part (a) of the figure find the member forces $\overline{U_1 U_2}$, $\overline{U_1 L_2}$, and $\overline{U_2 L_3}$.

Solution: The reactions are found first, and shown in part (a). Part (b) shows an appropriate free body for calculation of $U_1 U_2$. Of the three unknown member forces acting on this free body, $U_1 L_2$ and $L_1 L_2$ pass through the lower chord panel point L_2, so a moment equilibrium equation about this point as moment center will contain only the desired unknown $U_1 U_2$. The calculation is shown in part (b).

A suitable free body for the determination of $U_1 L_2$ is shown in part (c). The indicated moment center O lies on the intersection of the lines of action of forces $L_1 L_2$ and $U_1 U_2$; thus the moment equilibrium equation about O shown in part (c) will only contain the desired unknown $U_1 L_2$.

Part (d) shows a force equilibrium condition in the Y direction (or shear condition) to find $\overline{U_2 L_3}$.

As before, it is convenient to assume all unknown forces tensile, so that a resulting positive sign will always indicate tension and a negative sign compression in the member.

4.2. Three-Dimensional Trusses

Each joint of an ideal space truss is a non-coplanar, concurrent force system, so that for each we can write three independent force equilibrium conditions. Therefore, for a space truss with j joints we can write $3j$ independent equilibrium equations. For a statically determinate truss this number must equal the number of unknown forces, which may be either member forces or reactive components. Therefore, a condition for static determinacy of a truss with m members and r reactive components is

$$m + r = 3j \tag{4.2}$$

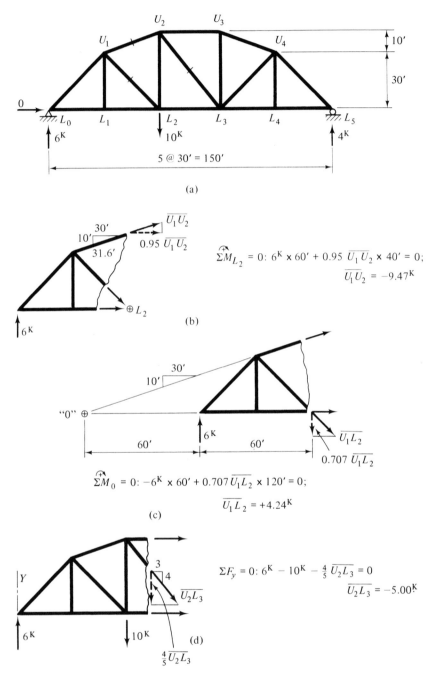

Fig. E4.2 Method of sections.

Although this condition is necessary, it may not be sufficient, because even if enough members have been provided, they may still be arranged in an unstable configuration. A more comprehensive check will be indicated later.

The analysis of such trusses will be carried out by the method of joints in these major steps: first, the components of each member force along a previously selected set of coordinate axes are computed, and then three force equilibrium equations are written for each joint, for a total of $3j$ equations. Solution of this set yields $3j$ unknown forces.

To begin, we consider a member connecting joint 1 of coordinates x_1, y_1, and z_1 and joint 2 of cordinates x_2, y_2, and z_2, as shown in Fig. 4.2. The member force, assumed tensile, is called S_{12}, and its components along the X, Y, and Z axes are, in terms of the direction cosines l_{12}, m_{12}, and n_{12},

$$S_{12x} = S_{12}l = S_{12}\frac{x_2 - x_1}{L}$$

$$S_{12Y} = S_{12}m = S_{12}\frac{y_2 - y_1}{L} \qquad (4.3)$$

$$S_{12z} = S_{12}n = S_{12}\frac{z_2 - z_1}{L}$$

where L = member length = $\sqrt{(x_2 - x_1)^2 + (y_2 - y_1)^2 + (z_2 - z_1)^2}$.

Next, we consider two joints 1 and 2 of the truss of which a portion is shown in Fig. 4.3. Force S_{12}, acting on joint 1, is equal in magnitude and sense (assumed tensile) to force S_{21} acting on joint 2, so that

$$S_{12} = S_{21} \quad \text{or, generally,} \quad S_{ij} = S_{ji} \qquad (4.4)$$

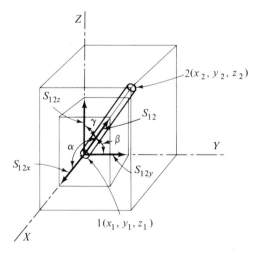

Fig. 4.2 Member force components.

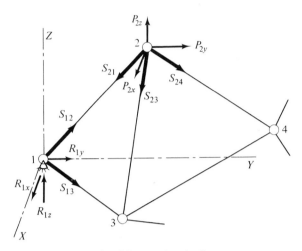

Fig. 4.3 Joint free bodies.

The direction of S_{12}, however, is opposite to that of S_{21}, so that the direction cosines of one force are equal in magnitude but opposite in sign to those of the other:

$$l_{12} = -l_{21}; \qquad m_{12} = -m_{21}; \qquad n_{12} = -n_{21}$$

or, generally,

$$l_{ij} = -l_{ji}; \qquad m_{ij} = -m_{ji}; \qquad n_{ij} = -n_{ji} \qquad (4.5)$$

We now sum the components of all forces acting on each joint, one after the other, placing each force into a separate column. For the first two joints of the truss of Fig. 4.3, these equations take the form of Eq. (4.6).

$$
\begin{array}{lll}
& \overbrace{\hspace{5cm}}^{m \text{ columns}} & \overbrace{\hspace{2cm}}^{r \text{ columns}} \\
\text{joint 1}\begin{cases} \sum F_x = 0: & S_{12}l_{12} + S_{13}l_{13} \quad\cdots + R_{1X} \quad\cdots = 0 \\ \sum F_y = 0: & S_{12}m_{12} + S_{13}m_{13} \qquad\qquad + R_{1Y} \quad = 0 \\ \sum F_z = 0: & S_{12}n_{12} + S_{13}n_{13} \qquad\qquad + R_{1Z} = 0 \end{cases} \\
\text{joint 2}\begin{cases} \sum F_x = 0: & -S_{12}l_{12} \qquad\quad + S_{23}l_{23} + S_{24}l_{24} \qquad = -P_{2X} \\ \sum F_y = 0: & -S_{12}m_{12} \qquad + S_{23}m_{23} + S_{24}m_{24} \qquad = -P_{2Y} \\ \sum F_z = 0: & -S_{12}n_{12} \qquad + S_{23}n_{23} + S_{24}n_{24} \cdots \qquad = -P_{2Z} \end{cases}
\end{array}
$$

j joints

$\underline{\hspace{2cm}}$
$3j$ rows $\underbrace{\hspace{6cm}}_{(m + r) = 3j \text{ columns}}$ know
                                                                                                      load

(4.

The resulting $3j$ equations can now be solved for the $3j$ unknown forces, since the direction cosines are known properties of the truss and the applied loads are specified. A longhand solution of such a set of equations becomes prohibitive for all but the simplest cases. Rather, use of a high-speed computer is indicated here; therefore, to express the calculations in a form acceptable to computers, we shall recast Eq. (4.6) in matrix form:

$$
\begin{bmatrix}
l_{12} & l_{13} & & & \cdots & 1 & \\
m_{12} & m_{13} & & & & 1 & \\
n_{12} & n_{13} & & & & & 1 & \cdots \\
-l_{12} & & l_{23} & l_{24} & & & \\
-m_{12} & & m_{23} & m_{24} & & & \\
-n_{12} & & n_{23} & n_{24} & & & \\
\cdot & & & & & & \\
\cdot & & & & & & \\
\cdot & & & & & & \\
\cdot & & & & & & \\
\cdot & & & & & & \\
\cdot & & & & & & \\
\end{bmatrix}
\begin{Bmatrix}
S_{12} \\ S_{13} \\ S_{23} \\ S_{24} \\ \cdot \\ \cdot \\ \cdot \\ R_x \\ R_y \\ R_z \\ \cdot \\ \cdot
\end{Bmatrix}
=
\begin{Bmatrix}
0 \\ 0 \\ 0 \\ -P_x \\ -P_y \\ -P_z \\ \cdot \\ \cdot \\ \cdot \\ \cdot \\ \cdot
\end{Bmatrix}
$$

$$(4.7)$$

or
$$[DC\emptyset S]\{S\} = -\{P\}$$

Solving for the vector of unknown forces, we write

$$\{S\} = -[DC\emptyset S]^{-1}\{P\}$$

If, as occurs often in practice, a number, say l, of loading conditions must be considered, the single column of unknown forces is replaced by l columns, and, correspondingly, one column of applied loads P is written for each separate loading condition:

$$[S] \quad = \quad -[DC\emptyset S]^{-1} \quad [P] \qquad (4.8)$$
$$(3j \times l) \qquad (3j \times 3j) \quad (3j \times l)$$

Note that the $[DC\emptyset S]$ matrix must be inverted only once, so that additional loading conditions require only additional matrix multiplications, which are done very rapidly by the computer. The matrix $-[DC\emptyset S]^{-1}$ relates the truss member forces to the applied loads, and is thus a force transformation matrix $[b]$ in the sense of Eq. (3.11).

The possibility that the truss may be unstable was mentioned earlier. In such a case, no set of forces exists that can satisfy the equilibrium equations, and, accordingly, no solution can be expected. This is shown by the fact that the [DCØS] matrix is singular and cannot be inverted, as indicated by a zero determinant of this matrix. Thus, the computation of this determinant can serve as a check on the stability of the structure.

The method outlined in this section is not suitable for longhand calculation, and serves here mainly as an example of the way in which structures problems are cast in matrix form to serve as a basis for computer programs; accordingly, this approach forms the necessary background for the computer program presented in the next section. Nevertheless, we shall add a small example problem to illustrate the procedure.

Example 4.3: Calculate the member forces in the truss shown in the figure, due to the loading indicated, by use of Eq. (4.8).

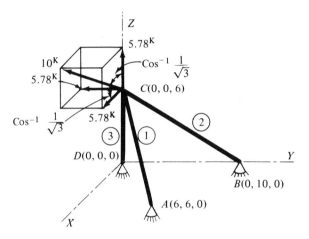

Fig. E4.3 Space truss.

Solution: Since all three member forces as well as the loads on this truss meet at a common joint C, it is sufficient to write the three force equilibrium conditions of this joint, which, in matrix form, are

$$[\text{DCØS}]\{S\} = -\{P\}$$

A convenient coordinate system is introduced, and the direction cosines are determined from the joint coordinates, whereby the signs must be found for member directions away from joint C. The load components parallel to the coordinate axes are calculated, and all values are substituted in the equilibrium equations:

$$\begin{bmatrix} .576 & 0 & 0 \\ .576 & .857 & 0 \\ -.576 & -.515 & -1.0 \end{bmatrix} \begin{Bmatrix} S_1 \\ S_2 \\ S_3 \end{Bmatrix} = - \begin{Bmatrix} 5.78 \\ -5.78 \\ 5.78 \end{Bmatrix} \text{kips}$$

The inverse of the [DCØS] matrix is calculated:

$$[DCØS]^{-1} = \begin{bmatrix} 1.732 & 0 & 0 \\ -1.167 & 1.167 & 0 \\ -.398 & -.600 & -1.00 \end{bmatrix}$$

so that the solution for the desired member forces is

$$\begin{Bmatrix} S_1 \\ S_2 \\ S_3 \end{Bmatrix} = - \begin{bmatrix} 1.732 & 0 & 0 \\ -1.167 & 1.167 & 0 \\ -.398 & -.600 & -1.00 \end{bmatrix} \begin{Bmatrix} 5.78 \\ -5.78 \\ 5.78 \end{Bmatrix} = \begin{Bmatrix} -10.00 \\ 13.50 \\ 4.63 \end{Bmatrix} \text{ kips}$$

4.3. Computer Program for Analysis of Statically Determinate Trusses

In Appendix A, a program TRUSS is presented that implements the calculations of the preceding section. It is assumed that the reader is familiar with the FORTRAN language and its input and output, statements, and commands, so that the emphasis here can be directed toward the relation of the program to the calculations outlined in Sec. 4.2. To this end, the reader should go through the program step by step.

To help with this, Fig. 4.4 shows a flows a flow chart that provides a systematic overview of the entire sequence of operations. The actual commands for each step are contained in the program listing in Appendix A. The comments in this listing are intended as a help for matching this with the flow chart.

The diamond-shaped branching steps in the flow chart, which control the sequence of computations, are carried out by means of DØ-loops. The bulk of the program is taken up with the formation of the matrices. Of particular importance is the manner in which the direction cosines are inserted in the [DCØS] matrix of the truss; this is performed by an appropriate double-subscript numbering system, as indicated in the portion of the program labeled "Insertion of Direction Cosines . . . ." The best manner of understanding such a program is to verify it with actual numbers in terms of a suitably small example problem.

The inversion of the [DCØS] matrix is performed by calling a subroutine MATINV, which is listed separately in Appendix A. The use of subroutines is an important device for reducing the labor of writing a program, as well as its bulk.

Prior to furnishing input for the program, the truss to be analyzed should be sketched and a convenient set of reference axes selected. Each joint must be numbered and its coordinates calculated. Each member is numbered, following which the reactions are numbered consecutively and their given direction cosines computed. Finally, the applied loads, along with their direction cosines, are indicated for each specified loading condition.

The headings in the program labeled "Input" suggest that at these points the computer must be furnished data cards to supply the necessary information, which is to be provided in the following order:

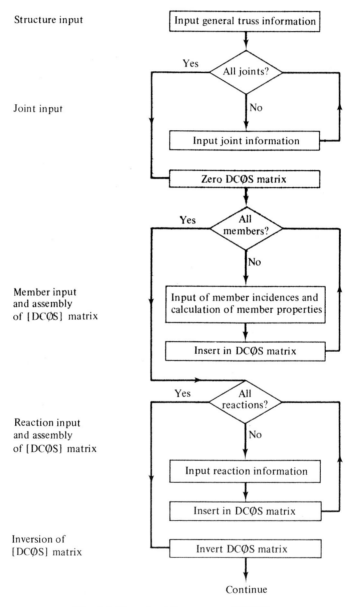

Fig. 4.4 Flow chart for program TRUSS.

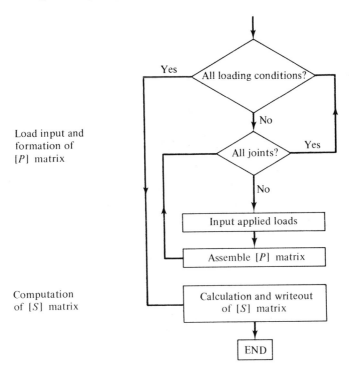

Fig. 4.4 (*Continued*)

1. General truss information: Number of joints, members, reactive components, and loading conditions.
2. Joint information: Joint number and joint coordinates.
3. Member information: Member number and end joints, or "incidences."
4. Reaction information: Reaction number, joint, and direction cosines.
5. Load information: Load, joint, and direction cosines.

All other quantities required in the analysis are computed from the above information.

An essential supplement to the program is a set of exact instructions, or "cookbook," specifying the precise format and sequence of the data cards required. Problem 4.13 asks for this information.

This routine was written to be as systematic as possible; thus, simplicity rather than elegance is emphasized, with the result that some of the steps are not as convenient as they might be. This will become evident to the user of the program. Several improvements are suggested in Prob. 4.16.

Example 4.4: Analyze the truss shown in part (a) of the figure by use of program TRUSS of Appendix A for the following loading conditions:

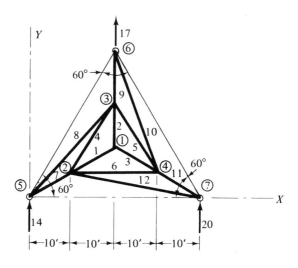

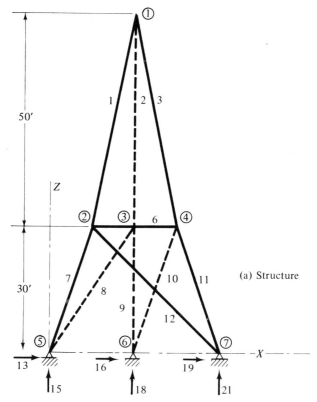

(a) Structure

Fig. E4.4

NUMBER OF JOINTS,MEMBERS,REACTIONS,LOADING CONDITIONS

```
7  12   9   2
```

JOINT NUMBER AND X,Y,AND Z COORDINATES

```
1      20.00      11.55      80.00
2      10.00       5.77      30.00
3      20.00      23.08      30.00
4      30.00       5.77      30.00
5       0.00       0.00       0.00
6      20.00      34.64       0.00
7      40.00       0.00       0.00
```

MEMBER NUMBER,NEAR AND FAR ENDS,LENGTH,DIRECTION COSINES

```
 1   1   2   51.32   -.194868   -.112634   -.974341
 2   1   3   51.31   0.000000    .224703   -.974427
 3   1   4   51.32    .194868   -.112634   -.974341
 4   2   3   19.99    .500228    .865894   0.000000
 5   3   4   19.99    .500228   -.865894   0.000000
 6   4   2   20.00  -1.000000   0.000000   0.000000
 7   2   5   32.14   -.311092   -.179500   -.933275
 8   3   5   42.81   -.467182   -.539128   -.700773
 9   3   6   32.15   0.000000    .359563   -.933121
10   4   6   42.82   -.233541    .674232   -.700622
11   4   7   32.14    .311092   -.179500   -.933275
12   2   7   42.82    .700657   -.134760   -.700657
```

REACTION NUMBER,JOINT,DIRECTION COSINES

```
13   5   1.000000  -0.000000  -0.000000
14   5  -0.000000   1.000000  -0.000000
15   5  -0.000000  -0.000000   1.000000
16   6   1.000000  -0.000000  -0.000000
17   6  -0.000000   1.000000  -0.000000
18   6  -0.000000  -0.000000   1.000000
19   7   1.000000  -0.000000  -0.000000
20   7  -0.000000   1.000000  -0.000000
21   7  -0.000000  -0.000000   1.000000
```

LOADS AND DIRECTIONS FOR LOADING CONDITION   1

```
1     -5.00  -0.000000  -0.000000   1.000000
2    -20.00  -0.000000  -0.000000   1.000000
3    -20.00  -0.000000  -0.000000   1.000000
4    -20.00  -0.000000  -0.000000   1.000000
5     -0.00  -0.000000  -0.000000  -0.000000
6     -0.00  -0.000000  -0.000000  -0.000000
7     -0.00  -0.000000  -0.000000  -0.000000
```

LOADS AND DIRECTIONS FOR LOADING CONDITION   2

```
1      2.00  -0.000000   1.000000  -0.000000
2      5.00  -0.000000   1.000000  -0.000000
3      5.00  -0.000000   1.000000  -0.000000
4      5.00  -0.000000   1.000000  -0.000000
5     -0.00  -0.000000  -0.000000  -0.000000
6     -0.00  -0.000000  -0.000000  -0.000000
7     -0.00  -0.000000  -0.000000  -0.000000
```

Fig. E4.4 (*Continued*)

MEMBER FORCES AND REACTIONS

ROW/COLUMN	1	2
1	-1.70908E+00	2.96457E+00
2	-1.71337E+00	-5.92862E+00
3	-1.70908E+00	2.96457E+00
4	-4.58999E+00	-5.51841E+00
5	-4.59929E+00	2.36874E+00
6	-4.59265E+00	3.14332E+00
7	-2.32142E+01	3.09319E+00
8	-9.95953E-03	8.44504E+00
9	-2.32152E+01	-1.25333E+01
10	9.95593E-03	-8.44199E+00
11	-2.32217E+01	9.43253E+00
12	-2.87563E-06	2.43835E-03
13	7.22639E+00	-4.90763E+00
14	4.17231E+00	-5.10818E+00
15	2.16722E+01	-8.80485E+00
16	-2.32511E-03	1.97155E+00
17	-8.34060E+00	-1.01984E+01
18	2.16556E+01	1.76097E+01
19	-7.22407E+00	2.93609E+00
20	4.16829E+00	-1.69347E+00
21	2.16722E+01	-8.80485E+00

Fig. E4.4 (*Continued*)

a. Dead load, of amount $P_{1z} = -5$ kips, and $P_{2z} = P_{3z} = P_{4z} = -20$ kips.
b. Wind load in the Y direction, of amount $P_{1y} = 2$ kips, and $P_{2y} = P_{3y} = P_{4y} = 5$ kips.

Solution: This truss has 12 members and 9 reactive components, for a total of 21 unknowns. There are 7 joints, for a total of $3 \times 7 = 21$ equilibrium equations. If the arrangement of the members is adequate, this truss will be statically determinate.

The truss is drawn as in part (a), and the joints are numbered as shown by the circled numbers; similarly, the members are numbered. Next, the joint coordinates are calculated within the system of X, Y, and Z reference axes shown, and all truss information is listed on data cards as called for by the program.

The reactive components are numbered consecutively with the members and entered on data cards as required; note that the successive listing of the X, Y, and Z components leads to a unit matrix of direction cosines for each supported joint.

Next, the load information is supplied in accordance with the specified loading conditions. Even though only a few joints are loaded, a load card is required for each joint. This is an undesirable feature of the program, which is to be remedied in Prob. 4.16.

The listing of input data shown in part (b) corresponds exactly to the sequence of data cards, with the exception of the member lengths and direction cosines, which were automatically computed. Such a listing can help identify and check the problem. The program also provides an output of the [DCØS] matrix and its inverse. This portion of the output, however, is deleted here to save space.

Further computer calculations result in the listing of the member and reactive forces shown in part (c). Each column represents one loading condition; the first 12 lines are the member forces 1 through 12; the following 9 lines furnish the reactive components 13 through 21 in the sequence numbered, that is, starting with the X component of reaction 5 and terminating with the Z component of reaction 7. This concludes the solution of the problem.

PROBLEMS

1. Analyze the truss. Indicate all member forces on a sketch of the truss.

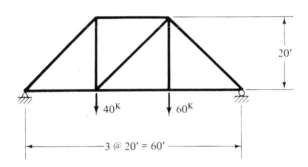

Prob. 1

2. Analyze the truss. Indicate all member forces on a sketch of the truss.

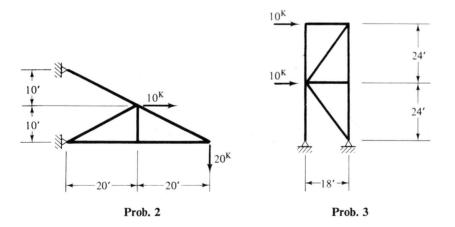

Prob. 2   **Prob. 3**

3. Analyze the truss. Indicate all member forces on a sketch of the truss.

4. Analyze the trussed three-hinged arch shown. Indicate member forces on a sketch of half the structure.

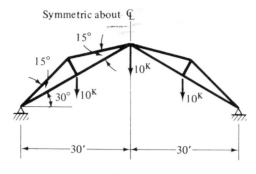

Prob. 4

5. Find forces in members U_2U_3, U_3L_3, U_3L_4, and L_3L_4 due to a uniform load of 2 kips/ft of truss.

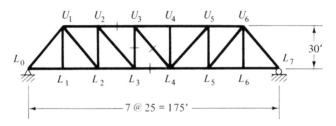

Prob. 5

6. Find forces in members U_2U_3, U_2L_2, and U_2L_3 due to the dead weight of the truss, of intensity 1 kip/ft.

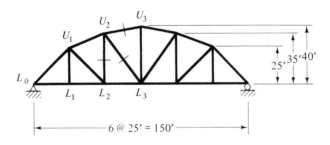

Prob. 6

7. Find forces in members *DF*, *DG*, and *FG* due to a horizontal wind force of .5 kip/ft of vertical projection of the truss.

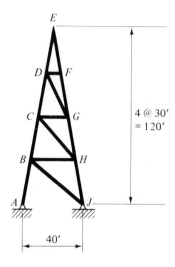

Prob. 7

8. A typical interior roof truss of a building is shown. The trusses are spaced at 20-ft centers and the loads are carried to them by means of purlins to the panel points of the top chord. Vertical loads only are to be considered; the roof dead load is 20 lb/ft² of roof area, and the snow load (over the *entire* roof) is 30 lb/ft² of horizontal projection. Find the member forces in the truss (indicate all values on a sketch of the truss).

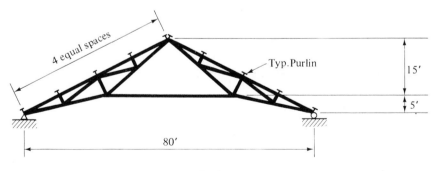

Prob. 8

9. Calculate member forces $U_2 L_2$, $U_2 L_3$, and $U_2 U_3$ of the grandstand roof shown.

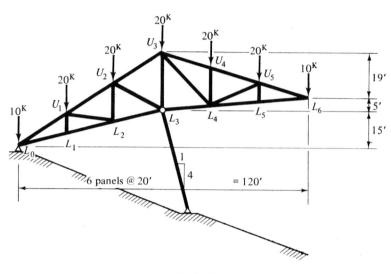

Prob. 9

10. Compute the force transformation matrix $[b]$, of order (5×3), relating the member forces to the external loads 1 and 2 and to the set of self-equilibrating forces 3.

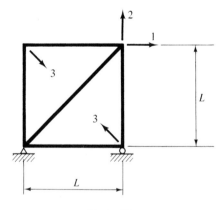

Prob. 10

11. Compute the force transformation matrix $[b]$, of order (3×3), relating the member forces to the loads applied along nodes 1, 2, and 3.

12. The space truss shown is pinned at all joints, and supported as shown.
   a. Discuss the degree of determinacy of the truss.
   b. Set up a matrix equation for the member forces and reactions.

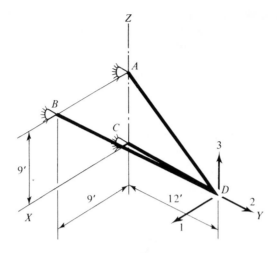

Prob. 11

c. Verify by computing the value of the determinant that the matrix of the direction cosines of the forces is singular, and therefore cannot be inverted; discuss the physical reason for this.

d. To make the structure stable, change the direction of the roller support at point A in an appropriate fashion, change the matrix equation already established in Part b accordingly, and solve.

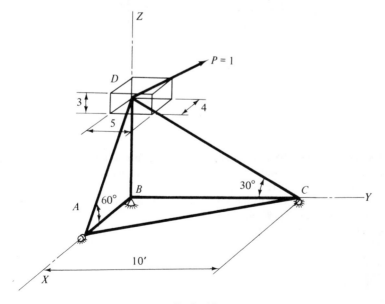

Prob. 12

13. Write a complete set of instructions for use of program TRUSS. State all necessary steps in outline form, for easy reference. Give, in proper sequence, the required data input cards and their format.

14. Use program TRUSS to analyze the truss shown for the following loading conditions:
a. 10 kips along node 1.
b. 5 kips along node 2.
c. 5 kips each along nodes 2 and 3.

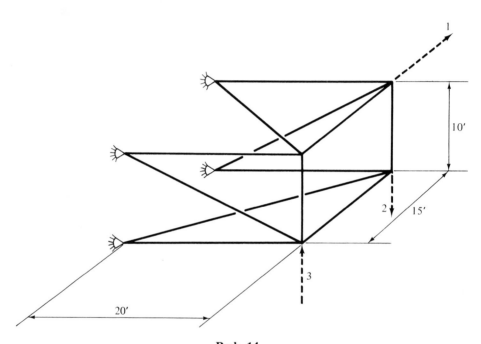

Prob. 14

15. Use program TRUSS to calculate the member forces in the truss shown due to
a. The truss dead load, represented by vertical loads of 5 kips each at the upper-level joints ($z = 16$ ft).
b. Wind loads parallel to the x axis, represented by equal loads of 2 kips each, acting on the upper-level joints.
c. The combination of loads a and b.

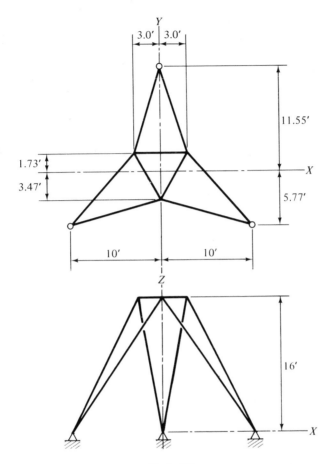

Prob. 15

16. Note that the input of applied loads in program TRUSS of Sec. 4.3 is cumbersome. Reprogram this portion of the program so that input cards are required only for those joints which are actually loaded. This should decrease the number of data cards considerably.

17. Discuss the applicability of the computer program of Sec. 4.3 to the analysis of plane trusses. Then use it to calculate all member forces of the truss of Prob. 9.

CHAPTER 5

The Theorem of Virtual Work

The theorem of virtual work is one of the most important and useful relationships of structural mechanics and will be used throughout this book. One of its attractive features is its ability to solve both statical and geometrical problems in a unified manner. Its discussion, therefore, does not strictly speaking belong in a chapter that is otherwise devoted only to statical relations between forces. Nevertheless, it seems useful to introduce both its statical and geometrical features together for the sake of a coherent presentation. Whereas the former will be pursued here, the latter will be taken up again in chapter 8.

The theorem of virtual work is based on abstract relationships and can be derived in a completely formal manner. This is shown in Sec. 15.5. It is perhaps the most basic of the energy methods, and, in slightly different form, goes by a number of different names. Some of these various versions will be touched on in Secs. 15.2 to 15.4, but it is the theorem of virtual work in its basic form that will be used throughout this book.

Energy methods do not, generally, give as much insight into the physical action of the structure as, for instance, is obtained by the consideration of free bodies or deformed shapes. On the other hand, they provide a systematic, formal method of solution, and are therefore particularly useful for complicated problems in which the physical action cannot be grasped by the analyst's mind. The theorem of virtual work, for instance, will be relied on in the chapter dealing with the finite element method, which is a completely general approach to any structural problem, including those intractable by more conventional means.

5.1. General Formulation

In a structure to be analyzed, we shall consider two distinct and separate systems:

1. *Any set of statically compatible forces* (that is, in equilibrium). For instance, for the continuous beam shown in Fig. 5.1(a), two moment diagrams are drawn. Either is statically compatible because the internal forces represented by it are in equilibrium with the applied load.

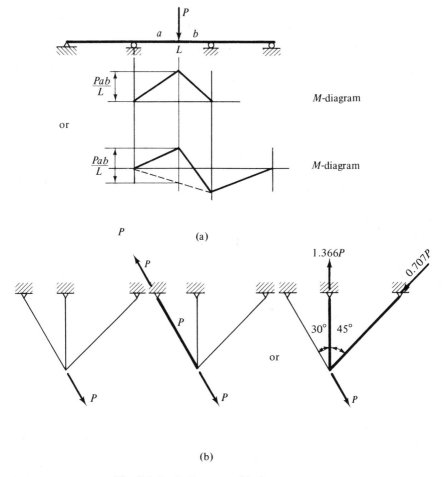

Fig. 5.1 Statically compatible force systems.

Likewise, in the statically indeterminate three-member truss of Fig. 5.1(b), either of the indicated member force distributions is statically compatible, as can be checked by appropriate equilibrium calculations.

It is important to observe that the statically compatible force systems do not have to satisfy any conditions of geometry. Thus, for the statically indeterminate structures shown there is an infinite number of such force systems.

In what follows, the external forces will be called P, and the stresses due to the statically compatible internal forces will be called σ.

2. *Any set of geometrically compatible deformations.* For the beam of Fig. 5.2(a), for instance, any of the three deformed shapes shown is a geometrically compatible deformation. Similarly, the deformed shape of the three-member truss of Fig. 5.2(b) is geometrically compatible.

Note that these compatible deformations are not necessarily related to the statically compatible forces, nor do they have to satisfy conditions of external or internal constraint. There does not have to be any cause-and-effect relationship between the force and displacement system. We shall call the displacements Δ and the corresponding geometrically compatible internal strains ϵ.

We now let the external forces of the first system, P, act through the corresponding displacements, Δ, of the second system, and call the resulting product, $\sum P \cdot \Delta$, the *external virtual work* (EVW). Likewise, the internal forces of the first system, equal to the stresses σ acting over the fiber cross section, dA, ride through the corresponding fiber elongations of the second system, furnished by the strains ϵ times the element of length dx. The resulting product for all fibers of the structure, $\int_L \int_A \sigma \, dA \cdot \epsilon \, dx = \int_{\text{vol}} \sigma\epsilon \, dV$, will be called the *internal virtual work* (IVW).

The theorem of virtual work states that for any body in equilibrium, the internal virtual work is equal to the external virtual work:

$$\sum P \cdot \Delta \quad = \quad \int_{\text{vol}} (\sigma \, dA) \cdot (\epsilon \, dx) \qquad (5.1)$$

statically compatible

geometrically compatible

The lines connecting the force quantities on both sides indicate that they belong to the statically compatible force system; those connecting the displacement quantities denote their geometrical dependence. However, the force and the displacement systems can be entirely independent of each other.

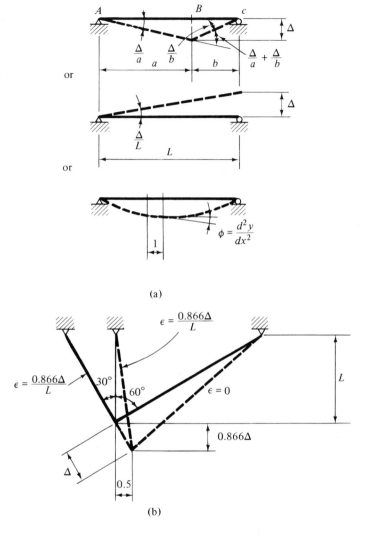

(a)

(b)

Fig. 5.2 Geometrically compatible displacement systems.

Thus, for purposes of solving particular problems we are free to select them entirely for our convenience.

To avoid confusion between real and imaginary, or *virtual*, systems, we shall henceforth, whenever it seems useful, attach a superscript star * to all quantities of the real system, and a superscript double star ** to all quantities of the virtual system.

5.2. Theorem of Virtual Displacements

To obtain real relations between forces, we have to select a real force system P, σ. The displacement system Δ, ϵ, however, can be chosen entirely for convenience in the calculations; that is, it can be imaginary, or *virtual*. In this form, which is called the *theorem of virtual displacements*, the theorem can be used to determine equilibrium relations, as will be shown in the following examples.

Example 5.1: a. Find the reaction R_B of the simple beam shown in part (a) of the figure.

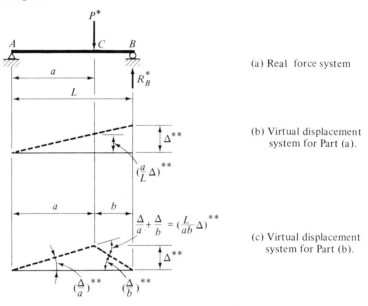

(a) Real force system

(b) Virtual displacement system for Part (a).

(c) Virtual displacement system for Part (b).

Fig. E5.1

Solution: The virtual displacement shown in part (b) is selected, since it permits only the desired unknown R_B to do virtual work, and since due to its rigid body motion all internal strains ϵ, and therefore the internal virtual work, are zero. According to Eq. (5.1), we set

$$EVW = IVW$$

$$-P^* \cdot \left(\frac{a}{L}\Delta\right)^{**} + R_B^* \cdot \Delta^{**} = 0$$

$$R_B = \frac{a}{L}P$$

b. For the same beam, find the bending moment at C.

Solution: A convenient virtual displaced shape that allows only the desired moment to do virtual work is shown in part (c). Note that if the real moment M_C^* is assumed positive so as to cause compression in the top fiber, then its sense is opposite to that of the virtual rotation (or kink) through which it turns, so that the resulting internal virtual work is negative. Again, applying Eq. (5.1),

$$EVW = IVW$$

$$-P^* \cdot \Delta^{**} = -M_C^* \left(\frac{L}{ab}\Delta\right)^{**}$$

Therefore,

$$M_C = \frac{Pab}{L}$$

We note that in all cases the quantity Δ^{**} defining the magnitude of the virtual displacements is common to all terms on both sides of the equation, and can thus be factored out. Thus, the actual value of the virtual displacement is unimportant and could, for convenience, be set equal to unity to avoid further cancellation.

In using the theorem of virtual displacements, geometrical visualization of the virtual deformed shape takes the place of statical calculations of free bodies. Its appeal will thus depend on individual preference for one or the other. In Ex. 5.1, a virtual displacement could be found that allowed only the desired force to do virtual work, thus eliminating all other unknowns. This is always possible in statically determinate structures, although the geometry may not always be as simple as in this case.

5.3. Theorem of Virtual Forces

To determine real geometrical relations, we have to select a real displacement system Δ, ϵ. The force system P, σ, however, can be chosen for convenience in the calculations; that is, it can be imaginary, or virtual. In this form, which is called the *theorem of virtual forces*, the theorem can be used to determine displacement relations, as will be shown in the following examples; as in Sec. 5.2, the real system will be identified by a single- and the virtual system by a double-star superscript.

Example 5.2: Idealized Beam
Find the rotation of the spring at *B* corresponding to deflection Δ.

Solution: A convenient virtual force system is a concentrated load *P* at the point of the deflection Δ, shown in part (b) of the figure. The virtual moment at the hinge point is $PL/4$, as shown in the virtual-moment diagram. Again applying Eq. (5.1),

$$EVW = IVW$$

$$P^{**} \cdot \Delta^* = \frac{PL^{**}}{4} \cdot \theta^*$$

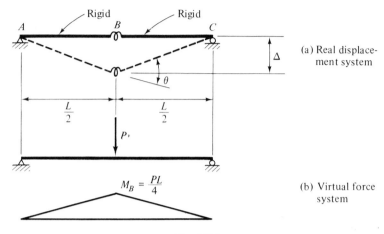

(a) Real displacement system

(b) Virtual force system

Fig. E5.2

Therefore,

$$\theta = \frac{4\Delta}{L}$$

Example 5.3: Idealized Truss
Find the horizontal and vertical components of deflection at C due to specified bar elongations of members.

Solution:

$$\text{EVW} = \text{IVW}$$

1. $P^{**} \cdot \Delta_V^* = .73P^{**} \times .10^* \text{ in.} + .52P^{**} \times .20^* \text{ in.}; \qquad \Delta_V = .177 \text{ in.}$

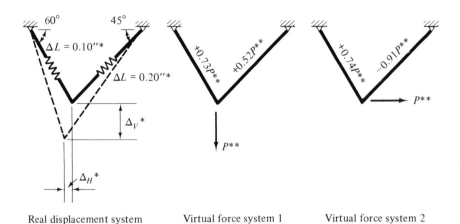

Real displacement system   Virtual force system 1   Virtual force system 2

Fig. E5.3

2. $P^{**} \cdot \Delta_H^* = .74 P^{**} \times 10^* \text{ in.} - .91 P^{**} \times .20^* \text{ in.};$ $\Delta_H = -.108 \text{ in.}$

The minus sign indicates a displacement opposite in sense to the virtual force. Note that in each case, a virtual concentrated load at the point and in the direction of the desired deflection component is found convenient.

The use of the theorem of virtual forces will be pursued in the chapter dealing with deformations of structures.

5.4. Matrix Formulation of the Theorem of Virtual Displacements

When a number of forces due to several applied loads is required, the long-hand formulation of the theorem of virtual displacements of Sec. 5.2. is unhandy. In such cases, a matrix formulation of the relations becomes useful. In the following we consider the case of a simple beam, shown in Fig. 5.3(a), for which the internal moments at three sections 1, 2, and 3 due to loads P_1, P_2, and P_3 are to be determined.

We consider first only the moment M_1 due to the three loads. The virtual displacement for determination of M_1 is generated by introducing a unit kink at section 1, as shown in Fig. 5.3(b); the virtual displacements associated with the applied loads are designated by d_{ij}^{**}, where i indicates the location of the load and j the location of the unit kink corresponding to the desired force. These elements constitute the column matrix

$$\{d_1^{**}\} = \begin{Bmatrix} d_{11} \\ d_{21} \\ d_{31} \end{Bmatrix} = \frac{L}{16} \begin{Bmatrix} 3 \\ 2 \\ 1 \end{Bmatrix}$$

Similarly, the loads can be contained in a load matrix:

$$\{P^*\} = \begin{Bmatrix} P_1 \\ P_2 \\ P_3 \end{Bmatrix}$$

We write now the virtual work equation:

$$\text{IVW} = \text{EVW}$$

$$1^{**} \cdot M_1^* = d_{11}^{**} P_1^* + d_{21}^{**} P_2^* + d_{31}^{**} P_3^*$$

$$= \tfrac{3}{16} L \cdot P_1 + \tfrac{2}{16} L \cdot P_2 + \tfrac{1}{16} L \cdot P_3$$

In terms of the previously defined matrices, this can be written as

$$M_1^* = \{d_1^{**}\}^T \{P^*\}$$

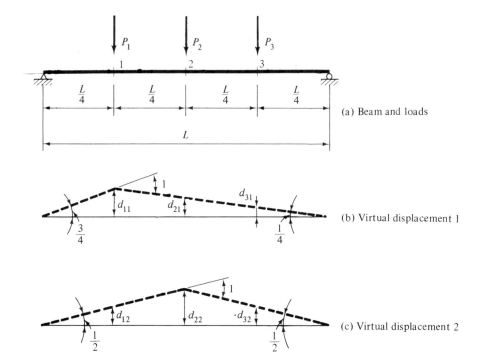

(a) Beam and loads

(b) Virtual displacement 1

(c) Virtual displacement 2

(d) Virtual displacement 3

Fig. 5.3 Virtual displacement relations.

Proceeding similarly for moments M_2 and M_3, we draw the appropriate virtually displaced shapes in Figs. 5.3(c) and (d), and write

$$M_2^* = \{d_2^{**}\}^T\{P^*\}, \qquad \text{where } \{d_2^{**}\} = \frac{L}{16}\begin{Bmatrix} 2 \\ 4 \\ 2 \end{Bmatrix}$$

and

$$M_3^* = \{d_3^{**}\}^T\{P^*\}, \qquad \text{where } \{d_3^{**}\} = \frac{L}{16}\begin{Bmatrix} 1 \\ 2 \\ 3 \end{Bmatrix}$$

We can now combine the relations for all moments in compact matrix

form by defining the matrix $\{S\} = \begin{Bmatrix} M_1 \\ M_2 \\ M_3 \end{Bmatrix}$, and writing

$$\begin{Bmatrix} M_1 \\ M_2 \\ M_3 \end{Bmatrix} = \begin{bmatrix} d_{11} & d_{21} & d_{31} \\ d_{12} & d_{22} & d_{32} \\ d_{13} & d_{23} & d_{33} \end{bmatrix} \begin{Bmatrix} P_1 \\ P_2 \\ P_3 \end{Bmatrix}$$

or
$$\{S^*\} = [d^{**}]^T \{P^*\} \tag{5.2}$$

where, in our case,

$$[d^{**}] = \begin{bmatrix} d_{11} & d_{12} & d_{13} \\ d_{21} & d_{22} & d_{23} \\ d_{31} & d_{32} & d_{33} \end{bmatrix} = \frac{L}{16} \begin{bmatrix} 3 & 2 & 1 \\ 2 & 4 & 2 \\ 1 & 2 & 3 \end{bmatrix}$$

If several, say l, loading conditions are to be accommodated, we write a $[P]$ matrix containing l columns, one for each loading condition, and, similarly, an $[S]$ matrix of l columns, so that Eq. (5.2) expands into

$$[S^*] = [d^{**}]^T [P^*] \tag{5.3}$$

Each column of the $[S]$ matrix contains the moments due to one loading condition.

We see by comparing Eqs. (3.11) and (5.3) that the force transformation matrix $[b]$ and the matrix of virtual displacements $[d]$ are related by

$$[b] = [d]^T \tag{5.4}$$

Example 5.4: By matrix formulation of the theorem of virtual displacements, calculate the moments at A and B and the shear in member AB of the structure shown in part (a) of the figure, due to
a. The vertical loads $P_2 = P_3 = 5$ kips.
b. The lateral load $P_1 = 3$ kips.
c. All loads jointly.

Solution: We draw the appropriate virtual displacements generated by introducing a unit deformation corresponding to the unknown force, as shown in parts (b) to (d), and write the $[d]$ matrix:

$$[d^{**}] = \begin{bmatrix} L & 0 & \dfrac{1}{\sqrt{2}} \\[2mm] \dfrac{3}{2}L & \dfrac{1}{2}L & \dfrac{1}{\sqrt{2}} \\[2mm] 2L & L & \dfrac{1}{\sqrt{2}} \end{bmatrix}$$

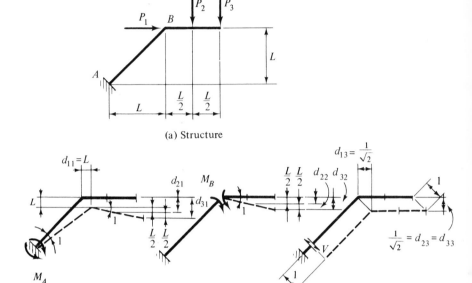

(a) Structure

(b) Virtual displacement 1   (c) Virtual displacement 2   (d) Virtual displacement 3

Fig. E5.4

The virtual work equation (5.3) then becomes

$$
\begin{bmatrix} M_{A1} & M_{A2} & M_{A3} \\ M_{B1} & M_{B2} & M_{B3} \\ V_{AB1} & V_{AB2} & V_{AB3} \end{bmatrix} = \begin{bmatrix} L & \dfrac{3}{2}L & 2L \\ 0 & \dfrac{1}{2}L & L \\ \dfrac{1}{\sqrt{2}} & \dfrac{1}{\sqrt{2}} & \dfrac{1}{\sqrt{2}} \end{bmatrix} \begin{bmatrix} 0 & 3 & 3 \\ 5 & 0 & 5 \\ 5 & 0 & 5 \end{bmatrix}
$$

$$
= \begin{bmatrix} 17.5L & 3.0L & 20.5L \\ 7.5L & 0 & 7.5L \\ 7.07 & 2.12 & 9.19 \end{bmatrix}
$$

PROBLEMS

In all problems show clearly the appropriate real and virtual systems in separate sketches.

1. For the simple beam shown, calculate by the theorem of virtual displacements
   a. The bending moment at *C*.
   b. The shear force at *C*.

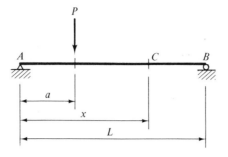

Prob. 1

2. Calculate by use of the theorem of virtual displacements
   a. The reaction at *A*.
   b. The shear force at *C*.
   c. The moment at *C*.

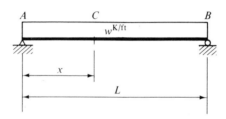

Prob. 2

3. Calculate by the theorem of virtual displacements
   a. The reaction at *B*.
   b. The force transmitted by the pin at *C*.
   c. The bending moment at *B*.

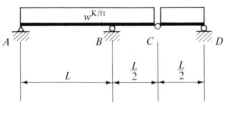

Prob. 3

4. By use of the theorem of virtual forces, compute the displacement *u* at point A
   of the spring assembly with specified stiffnesses due to the applied force *P*.

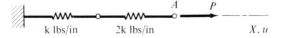

Prob. 4

5. a. By the theorem of virtual displacements, compute the forces in members *AC* and *BC* due to the applied force *P*.

   b. By the theorem of virtual forces, compute the displacements *u* and *v* of point *C* due to the applied force *P*.

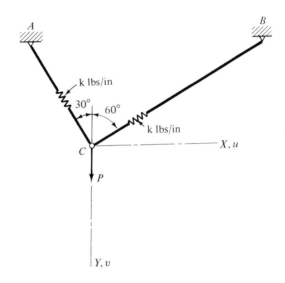

Prob. 5

6. By use of the theorem of virtual displacements in its matrix form, compute the matrix of bending moments [b], of order (4 × 4), in which b_{ij} is the bending moment at *i* due to a vertical unit downward load at *j*.

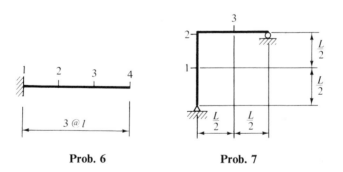

Prob. 6   **Prob. 7**

7. By use of the theorem of virtual displacements in matrix form, compute the bending moment matrix [b], of order (3 × 3), in which b_{ij} is the moment at point *i* due to a unit transverse load applied at point *j*. Unit loads at Points 1 and 2 are to the right, at point 3 downward. Moments are positive when compression side is on outside of frame.

CHAPTER 6

Analysis for Moving Loads

In many structures a number of different loading conditions are possible. For instance, a traffic load moving across a bridge will cause different forces, depending on its location; similarly, a warehouse structure designed to carry heavy live loading may have this load distributed over different bays or different stories at any one time, and it is not always obvious which of these distributions will cause the maximum stresses in a member. In such cases, the analyst must determine that load position or combination which will be critical in its effect on the forces; influence lines are a useful tool to accomplish this. Several methods of determining influence lines are covered in Secs. 6.1 and 6.2. In Secs. 6.3 and 6.4 these influence lines are used to determine critical member forces at given sections of the structure. Finally, Sec. 6.5 presents the concept of envelopes of forces by which the designer can determine the maximum design forces at all sections of the structure due to critical positioning of the live loads.

6.1. Influence Lines

An *influence function* (also called *Green's function* in mathematics) denotes the effect at one specified point as a function of the position of the cause, of unit value. For instance, the value of a simple beam reaction due to a unit load as a function of the position of this load would be an influence function.

An *influence line* denotes the plot of the influence function. Thus, the influence line of a specified structural effect (such as internal force, reaction, or deflection) is the value of this effect plotted as a function of the position of the unit load that causes it. In statically determinate structures, influence lines for forces can be found by simple statics. In the following, several ways of performing these statical calculations will be shown.

The units of influence ordinates are the units of the function for which the influence line is plotted, divided by the units of the applied load. Thus, the units of an influence line for beam shear due to a unit force will be kips/kip, a dimensionless ratio; those of a moment influence line due to a moving unit force will be kip-ft/kip, or, canceling, feet. However, to grasp the physical meaning more clearly, it is advisable to attach the uncanceled units to the influence values.

1. *Influence lines by free bodies.* In this basic approach the influence line for the desired force is obtained by considering the equilibrium of appropriate free bodies to which the unit load is applied at different discrete points. More conveniently, the unit load is applied at a point defined by the variable x, and the desired force is computed as a function of x and plotted along the structure.

Example 6.1: Determine the influence lines for
a. The reaction at A.
b. The moment at B.
c. The shear at B, of the simple beam shown in part (a) of the figure.

Solution: a. We consider the entire beam as free body and sum moments about C:

$$\overset{+}{\sum}M_C = 0: \qquad R_A \cdot L - 1 \cdot x = 0; \qquad R_A = \frac{x}{L}$$

The plot of this function is shown in part (b).
b. We consider two free bodies AB and BC, shown in part (c), such that the desired moment is an external force. The reactions are determined first. It is convenient to take the equilibrium of the unloaded free body; thus, for a unit load between A and B,

$$\sum M_B \text{ (of free body } BC) = 0: \qquad M_B = \frac{3}{4}L\left(1 - \frac{x}{L}\right), \qquad \left(\frac{3}{4}L \leq x \leq L\right)$$

and for the load between B and C,

$$\sum M_B \text{ (of free body } AB) = 0: \qquad M_B = \frac{x}{4}, \qquad \left(0 \leq x \leq \frac{3}{4}L\right)$$

The influence line is shown in part (d). Note that the mathematical discontinuity at B requires two separate calculations.

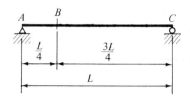

(a) Beam

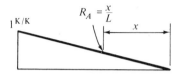

(b) Influence line for R_A

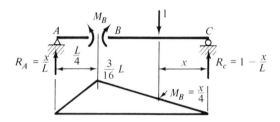

(c) Free bodies for M_B

(d) Influence line for M_B

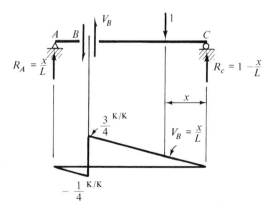

(e) Free bodies for V_B

(f) Influence line for V_B

Fig. E6.1

c. We consider again two free bodies AB and BC, as shown in part (e), and sum vertical forces of the unloaded body:
For $\frac{3}{4}L \leq x \leq L$,

$$\sum F_Y \text{ (of } BC) = 0: \qquad V_B = \left(\frac{x}{L} - 1\right)$$

For $0 \le x \le \frac{3}{4}L$,

$$\Sigma F_Y \text{ (of } AB\text{)} = 0: \qquad V_B = \frac{x}{L}$$

The influence line is plotted in part (f); note that it has a unit jump in shear as the unit load crosses the critical section.

2. *Influence lines by the theorem of virtual displacements.* We recall that the theorem of virtual displacements is just an alternative way of stating equilibrium conditions. It is particularly convenient for establishing influence lines. An appropriate virtual displacement y^{**} for this purpose is one with a unit value of distortion corresponding to the desired action, F^*, and zero displacement corresponding to all other forces except the applied unit load.

If the desired force F^* is a reaction, its contribution will be to the external virtual work, so that the virtual work Eq. (5.1) becomes

$$EVW = IVW,$$

or

$$1^* \cdot y^{**} + F^* \cdot 1^{**} = 0$$

from which

$$F^* = -y^{**} \tag{6.1}$$

where F^* is the real desired force due to an applied unit load, and y^{**} is the displacement resulting from an induced unit distortion at the point and in the sense of the desired force.

Should the desired force F^* be internal, such as an internal shear or moment, its contribution will be to the internal virtual work, and Eq. (5.1) becomes

$$EVW = IVW$$

$$1^* \cdot y^{**} = F^* \cdot 1^{**}$$

from which the desired force quantity

$$F^* = y^{**} \tag{6.2}$$

In either case, it is concluded that the influence line for any force is generated by the virtual displacements resulting from a unit deformation corresponding to the desired force. In this form, the method goes by the name of *Mueller–Breslau's principle*. We note that in this method the emphasis is on geometrical, rather than statical, calculations.

Example 6.2: Determine the influence lines for
a. The reaction at C.
b. The moment at D of the articulated beam shown in part (a) of the figure.

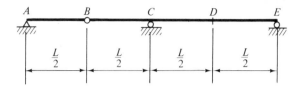

(a) Beam

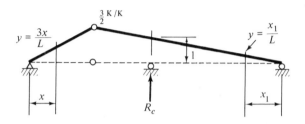

(b) Influence line for R_c

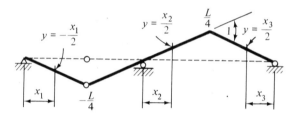

(c) Influence line for M_D

Fig. E6.2

Solution: a. We induce a unit vertical translation of point *C* of the beam, corresponding to the desired vertical force R_C; the resulting beam deformation, equal to the influence line, is as shown in part (b).

b. We induce a unit relative rotation (or "kink") at point *D*, corresponding to the desired moment M_D; the resulting beam deformation, equal to the influence line, is as shown in part (c).

3. *Influence lines by use of force transformation matrix.* We recall that the element b_{ij} in the force transformation matrix [*b*] represents the force at *i* due to a unit load at *j*; therefore, the elements of the *i*th row $(b_{i1}, b_{i2}, \ldots, b_{ij}, \ldots)$ represent the influence ordinates for the force at *i* due to unit loads at points $1, 2, \ldots, j, \ldots$.

Example 6.3: For the articulated beam, shown in part (a) of the figure, sketch the influence lines for M_3 and M_4 from the force transformation matrix [*b*], defined by $\{M\} = [b]\{P\}$.

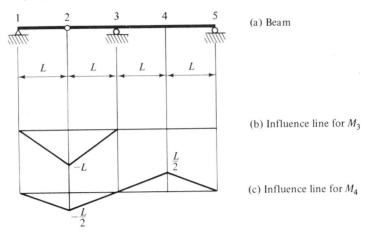

(a) Beam

(b) Influence line for M_3

(c) Influence line for M_4

Fig. E6.3

Solution: The force transformation matrix for the beam is

$$
\begin{Bmatrix} M_1 \\ M_2 \\ M_3 \\ M_4 \\ M_5 \end{Bmatrix} =
\begin{bmatrix}
0 & 0 & 0 & 0 & 0 \\
0 & 0 & 0 & 0 & 0 \\
0 & -L & 0 & 0 & 0 \\
0 & -\dfrac{L}{2} & 0 & \dfrac{L}{2} & 0 \\
0 & 0 & 0 & 0 & 0
\end{bmatrix}
\begin{Bmatrix} P_1 \\ P_2 \\ P_3 \\ P_4 \\ P_5 \end{Bmatrix}
$$

influence ordinates for M_4

influence ordinates for M_3

The influence ordinates for M_3 and M_4 are given by the entries of the third and fourth rows, and are used to plot the influence lines shown in parts (b) and (c).

6.2. Influence Lines for Trusses

Influence lines for trusses can be constructed by any of the methods presented in Sec. 6.1; the use of free bodies is common and will be demonstrated here for further exercise in truss analysis and to gain added insight into the nature of influence lines.

In analyzing ideal trusses it must be remembered that loads can only be applied to the panel points. Moving traffic loads on truss bridges usually bear on a bridge deck placed on longitudinal stringers, which may be assumed to transmit the loads by simple beam action to cross beams, which in turn are framed into the truss panel points, as shown in Fig. 6.1. Thus, any load acting on the bridge deck will be transmitted linearly to the adjacent truss

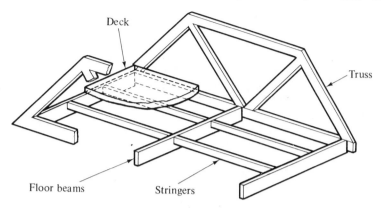

Fig. 6.1 Truss-bridge deck framing.

joints. For this reason, the influence line segments between truss joints will always be straight lines, and often unit loads are placed only on panel points, the influence ordinates at these points are calculated, and connected by straight lines.

In the usual types of simple-span bridge trusses, influence lines for top and bottom chords will not show reversal of sense; in such trusses, the top chord will always be in compression and the bottom chord in tension, due to gravity loads. Truss diagonals, whose function is mainly to resist shear, will usually show reversal of force as the load moves past, similar to the action represented by influence lines for beam shear, such as that of Ex. 6.1(c).

The units of influence ordinates for truss member forces will be kips/kip of applied load.

Example 6.4: For the truss shown in part (a) of the figure, calculate and draw influence lines for forces in members U_1U_2 and U_1L_2.

Solution: This truss was already considered in Ex. 4.2, where appropriate sections and equilibrium conditions for the members were considered.

For member U_1U_2 the truss is divided into two parts by the section shown in part (b). The unit load is placed at joint L_1, the reaction at L_5 is determined as $R = \frac{1}{5}$, and the desired member force is found by summing moments of the shaded portion of the truss about joint L_2:

$$\sum \overset{+}{M_{L2}} = 0: \qquad -\tfrac{1}{5} \cdot 90 \text{ ft} - .95\overline{U_1U_2} \cdot 40 \text{ ft} = 0; \qquad \overline{U_1U_2} = -.474 \text{ kip/kip}$$

Next, the unit load is placed at joint L_2, and a similar operation is performed:

$$\sum \overset{+}{M_{L2}} = 0: \qquad -\tfrac{2}{5} \cdot 90 \text{ ft} - .95\overline{U_1U_2} \cdot 40 \text{ ft} = 0; \qquad \overline{U_1U_2} = -.948 \text{ kip/kip}$$

From similar calculations for unit loads at the other joints, it will be seen that the influence line varies linearly from a maximum at L_2 to zero at L_5 as shown in part (c).

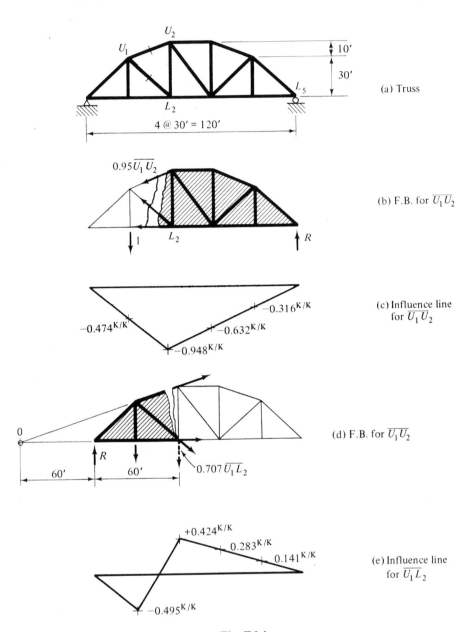

(a) Truss

(b) F.B. for $\overline{U_1 U_2}$

(c) Influence line for $\overline{U_1 U_2}$

(d) F.B. for $\overline{U_1 U_2}$

(e) Influence line for $\overline{U_1 L_2}$

Fig. E6.4

For the web member U_1L_2, reference to Ex. 4.2 shows a convenient moment equation; proceeding as before, referring to part (d), and expressing equilibrium of the shaded portion of the truss, we set for the unit load at L_1,

$$\sum \overset{+}{M_0^!} = 0: \qquad -\tfrac{4}{5} \cdot 60 \text{ ft} + 1 \cdot 90 \text{ ft} + .707 \overline{U_1L_2} \cdot 120 \text{ ft} = 0;$$
$$\overline{U_1L_2} = -.495 \text{ kip/kip}$$

For the unit load at L_2,

$$\sum \overset{+}{M_0^!} = 0: \qquad -\tfrac{3}{5} \cdot 60 \text{ ft} + .707 \overline{U_1L_2} \cdot 120 \text{ ft} = 0; \qquad \overline{U_1L_2} = .424 \text{ kip/kip}$$

Similar calculations yield the influence line shown in part (e).

The shape of the influence lines derived in Ex. 6.4 can be visualized by the theorem of Mueller–Breslau discussed in Sec. 6.1; the virtual distortion associated with axial force is a unit axial elongation of the member whose influence line is desired. These influence lines may be looked at as the deformed shape of the bottom chord of the truss (where the external, real unit load is acting) due to such virtual member distortions. Usually, this method is not favored for quantitative computation because of the relative obscureness of truss deformations.

If the appropriate member force transformation matrix of the truss is available, the influence ordinates can be easily found as the terms in one row of the [b] matrix of the truss.

6.3. Use of Influence Lines to Compute Forces Due to Arbitrary Loads

With the influence line for a certain force F drawn, we can find the value of this action due to any given loading condition, such as the one shown in Fig. 6.2(a). The influence line for the force whose value is desired has been drawn as in Fig. 6.2(b). The influence ordinate at any point x is denoted by y.

The effect of the concentrated load P_1 on the desired force F equals P_1 times the effect of a unit load, y_1:

$$F = P_1 y_1$$

The resultant of the distributed load $w(x)$ acting over the element of length dx is $w \cdot dx$. The contribution of this elementary load to the force F is $(w \cdot dx) \cdot (y)$, and that of the entire distributed load between two points A and B is

$$F = \int_A^B wy \, dx$$

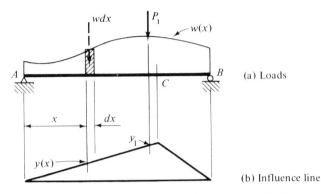

(a) Loads

(b) Influence line

Fig. 6.2 Use of influence line for determination of action.

This integral can always be evaluated, but another and sometimes very time-saving method is to replace the variable $w(x)$ by its average value w_{ave}, and take it outside the integral:

$$F = w_{ave} \int_A^B y \, dx$$

The integral $\int_A^B y \, dx$ can be interpreted geometrically as the area under the influence line between A and B, so that the value of the function due to this distributed load equals the average load intensity multiplied by the area under the portion of the influence line over which it acts. For uniform loads, this is very convenient. For linearly varying loads and influence lines, it can be shown that the average load w_{ave} is the value of the load intensity at the centroid of the area under the influence line.

The total value of the desired force due to the entire load is then

$$F = \int wy \, dx + \Sigma P \cdot y \tag{6.3}$$

Note that depending on the sign of the influence ordinate y, the contributions of the various forces can be either additive or canceling.

Example 6.5: By the use of influence lines, calculate
a. The shear at the quarter point, and
b. The moment at the quarter point
of the simple-span beam shown in part (a) of the figure due to a uniform dead load of w kips/ft and the concentrated midspan load of value wL kips.

Solution: a. The influence line for shear at the quarter-point B is drawn in part (b). Since the distributed load w is uniform, it can be taken outside the integral of Eq. (6.3) and multiplied by the total area under the influence line,

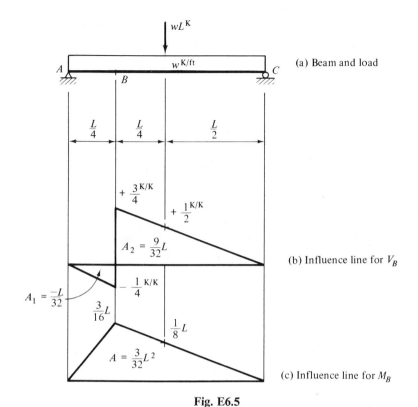

Fig. E6.5

which is computed as $\frac{9}{32}L - \frac{1}{32}L = \frac{1}{4}L$; the midspan load is multiplied by the midspan influence ordinate, so Eq. (6.3) yields

$$V_B = w\left(\frac{L}{4}\right) + (wL)\frac{1}{2} = \frac{3}{4}wL$$

Note that uniform load between points B and C only would lead to a larger value of shear, indicating that partial loading can sometimes be more critical than full loading.

b. In a similar fashion, the moment influence line is drawn in part (c); Eq. (6.3) for the specified loading and this influence line results in

$$M_B = w\frac{3}{32}L^2 + (wL)\frac{L}{8} = \frac{7}{32}wL^2$$

6.4. Use of Influence Lines to Compute Critical Loading Conditions

With moving or variable loads on a structure it becomes important to determine that loading condition which causes the critical value of the forces to be used in design. This is done by the use of influence lines. A number of theo-

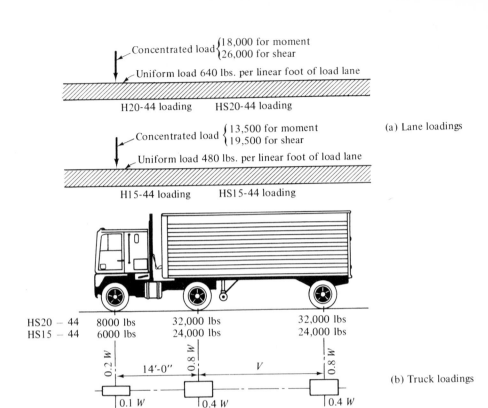

Concentrated load $\begin{cases} 18,000 \text{ for moment} \\ 26,000 \text{ for shear} \end{cases}$

Uniform load 640 lbs. per linear foot of load lane

H20-44 loading   HS20-44 loading

Concentrated load $\begin{cases} 13,500 \text{ for moment} \\ 19,500 \text{ for shear} \end{cases}$

Uniform load 480 lbs. per linear foot of load lane

H15-44 loading   HS15-44 loading

(a) Lane loadings

	8000 lbs	32,000 lbs	32,000 lbs
HS20 − 44	8000 lbs	32,000 lbs	32,000 lbs
HS15 − 44	6000 lbs	24,000 lbs	24,000 lbs

$0.2\ W$ 14'-0'' $0.8\ W$ V $0.8\ W$

$0.1\ W$ $0.4\ W$ $0.4\ W$

$0.1\ W$ $0.4\ W$ $0.4\ W$

(b) Truck loadings

W = Combined weight on the first two axles which is the same as for the corresponding H truck.

V = Variable spacing — 14 feet to 30 feet inclusive. Spacing to be used is that which produces maximum stresses.

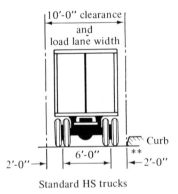

10'-0'' clearance
and
load lane width

Curb

2'-0'' →   6'-0''   **
←2'-0''

Standard HS trucks

Fig. 6.3 Highway bridge loadings (AASHO specifications). (From *Standard Specifications for Highway Bridges*, 10th ed., American Association of State Highway Officials, Washington, D.C., 1969.)

retical criteria are available for the calculation of the critical load position. Here, however, we shall only indicate some simple cases in which this determination can be done by common sense or a simple trial-and-error procedure. As a general rule, the heaviest concentrated load should be positioned at the point of maximum influence ordinate, and the remaining loads clustered around this point. Any variable distributed loads should be applied over the portion of the structure with the influence ordinate of appropriate sign. In cases of sign reversal of the influence diagram, both positive and negative critical values must be determined.

As an example of the way in which bridge loadings are specified, Fig. 6.3 shows typical design loadings taken from the 1969 edition of *Standard Specifications for Highway Bridges* of the American Association of State Highway Officials (AASHO). According to these provisions, a critical loading condition can be due to either of two cases: the *truck loading*, consisting of a single truck–trailer combination per lane of roadway, or the *lane loading*, a combination of uniform and concentrated loads, whichever produces maximum design forces when in its critical position on the bridge.

Obviously, both of these loadings are idealizations; no bridge is going to be in distress under the load of a single truck. The lane loading appears somewhat more plausible; it can be interpreted as a row of vehicles held up by a traffic jam with one extra-heavy vehicle in its midst. The development of design loadings of this type requires a great deal of field measurements, comparative studies, and probabilistic analyses.

The use of such specification design loads will be illustrated in the next example.

Example 6.6: Compute the critical forces in member U_1L_2 of the two-lane highway truss bridge shown in part (a) of the figure due to HS20–44 highway loading according to the AASHO specifications.

Solution: The influence line for member U_1L_2 is constructed in part (b). Maximum influence ordinates and positive and negative areas under the influence line are shown. It shows the possibility of either tensile or compressive member force, depending on the location of the load.

We first consider the truck loading; maximum member forces will result from a short trailer of 14-ft wheelbase, so the loading will be that of the truck of part (c); the truck can come from either direction.

We now place the truck so as to cause maximum tensile member force. Two possible critical truck positions are shown in part (d), each with a 32-kip axle load at the point of maximum influence ordinate. We could now calculate the member force due to each of the two load positions (or any other possible critical locations), and pick the largest of them. However, we can see by inspection that, although the contribution of the 32-kip loads is the same for both cases, the 8-kip load is placed under a smaller influence ordinate in position 1 than in position 2; therefore, the latter is critical:

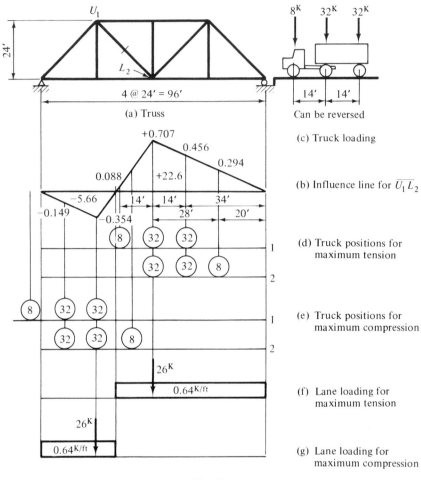

Fig. E6.6

$$\overline{U_1 L_2}_{(max)} = 32(.707 + .456) + 8(.294) = \underline{39.7 \text{ kips}}$$

In a similar fashion, the truck is placed over the region of negative influence ordinates to determine the critical compressive force. Again, a 32-kip force is located at the peak of the influence line; in position 1 the 8-kip force is off the truss; in position 2 it is in a region of positive influence ordinate and would thus diminish the compression in the member. Position 1 thus controls:

$$U_1 L_{2(min)} = 32(-.354 - .149) = \underline{-16.0 \text{ kips}}$$

Next, we have to evaluate the effect of the equivalent lane loading. The shape of the influence line, with its reversal of sense within the span, resembles that of an influence line for beam shear, so we use the 26-kip concentrated load. Parts (f) and (g) show the placement of this load for maximum tensile and compressive member forces. These are calculated by Eq. (6.3), using the previously calculated areas under the parts of the influence line, as

$$\overline{U_1 L_2}_{(max)} = +22.6 \cdot .64 \text{ kip/ft} + .707 \cdot 26 \text{ kips} = +32.9 \text{ kips}$$

and $\overline{U_1 L_2}_{(min)} = -5.66 \cdot .64 \text{ kip/ft} - .354 \cdot 26 \text{ kips} = -12.8 \text{ kips}$

We see that the forces due to the truck loading calculated earlier are critical. To these must be added the effect of dead and other loads, with due respect to sign, in order to obtain the extreme design forces for the member.

6.5. Envelopes of Forces

The outlined method enables the determination of the extreme values of one force, say, the moment at one section. This process now has to be repeated for every section of the structure (in practice, a finite number of sections is considered, such as $^1/_{10}$th points along each span of a bridge girder), so that all sections can be designed for the critical positive and negative moment values which can possibly arise under any possible location of the design load. The curve that connects the extreme moment values at all sections is called the *moment envelope*. For a complete design of bridge structures, both moment and shear envelopes are necessary. Repetitive work of this type is best assigned to the computer, but a simple example will be presented here.

Example 6.7: Compute the envelope of moments for the simple-span bridge shown in part (a) of the figure. The design live load consists of the truck shown in part (a), and the dead load is 2 kips/ft.

Solution: Because of the symmetry of the bridge, only half the span needs to be designed. The left half is split into four 10-ft segments, as shown in part (b), and the moment influence lines for their end points are calculated by any method and plotted as in part (c). Next, the critical truck positions are determined for each section, as shown in part (d). The critical live-load moments are found by the calculations shown in Table 6.1, and the dead-load moments are calculated and superposed, leading to the moment envelope ordinates at the bottom of this table. Finally, the moment envelope is plotted as shown in part (e); the span can now be designed to resist the critical moments at all sections.

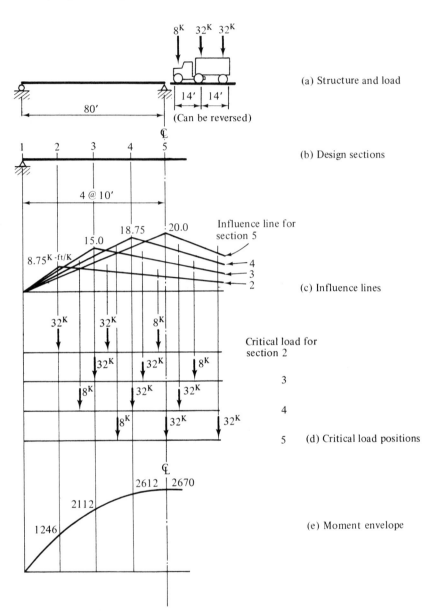

(a) Structure and load

(Can be reversed)

(b) Design sections

(c) Influence lines

Influence line for
section 5

Critical load for
section 2

(d) Critical load positions

(e) Moment envelope

Fig. E6.7

TABLE 6.1

AXLE	AXLE LOAD P	SECTION							
		2		**3**		**4**		**5**	
		y	P·y	y	P·y	y	P·y	y	P·y
1	8 kips	5.25	42.0	8.00	64.0	10.00	80.0	13.00	104.0
2	32 kips	7.00	224.0	11.50	368.0	18.75	600.0	20.00	640.0
3	32 kips	8.75	280.0	15.00	480.0	13.50	432.0	13.00	416.0
Live load total			546.0		912.0		1112.0		1160.0
Dead load	w	A	A·w	A	A·w	A	A·w	A	A·w
	2 kips/ft	350 ft²	700.0	600	1200.0	750	1500.0	800	1600.0
Dead + live load total			1246.0		2112.0		2612.0		2760.0

PROBLEMS

1. Calculate and draw
    a. Influence line for M_C.
    b. Influence line for V_B.

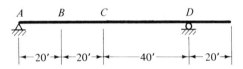

Prob. 1

2. Draw influence lines and calculate the maximum values for
    a. Reaction at 1.
    b. Shear at 2.
    c. Moment at 2.

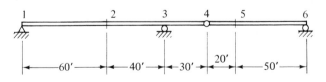

Prob. 2

3. Using the theorem of virtual displacements, draw directly for the beam of Prob. 2 the

  a. Influence line for reaction at 3.
  b. Influence line for shear in pin at 4.
  c. Influence line for moment at 3.

4. For the articulated beam of Prob. 2, write the force transformation matrix [*b*], of order (6 × 6), relating the moments at points 1 to 6 to the loads at these points. Identify the rows of this matrix as influence ordinates, and verify the influence lines already obtained in Probs. 2c and 3c.

5. Compute the critical midspan moment of a single beam of length *L* due to
  a. An HS20–44 truck loading.
  b. An HS20–44 equivalent lane loading.
  Plot the moment versus span length on the same graph for both cases, and draw conclusions regarding the range of span length for which one or the other of these loadings might be critical.

6. a. Draw the influence line for M_A.
  b. Calculate the largest values of positive and negative moment at point *A* due to the truck loading shown.

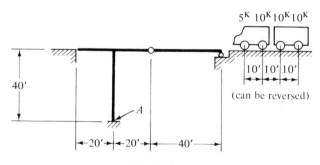

Prob. 6

7. Find the critical values of shear and moment at point *B*. Dead load of the structure is .5 kip/ft. Live load is as shown and can be reversed.

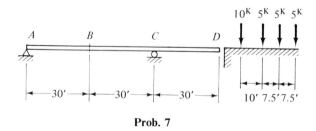

Prob. 7

8. The bridge shown is two lane, so that each truss is carrying the traffic load of one lane.
  a. Calculate and draw influence lines for members L_1L_2 and U_2L_3.

b. Calculate critical positive and negative forces in members L_1L_2 and U_2L_3 due to
   1. Dead load of 2 kips/ft of truss.
   2. HS20–44 highway loading.

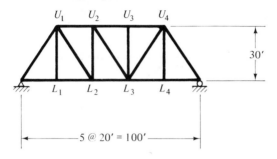

Prob. 8

9. a. Draw influence lines for forces in members U_1L_1, U_1U_2, and U_1L_2 of the bascule bridge shown.
   b. Calculate critical values of positive and negative live load forces in members U_1L_1, U_1U_2, and U_1L_2 due to the moving load shown.

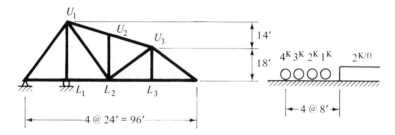

Prob. 9

10. The two-lane highway truss bridge weighs 3 kips/ft/truss, and is subjected to

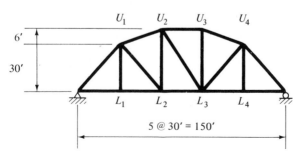

Prob. 10

an HS20–44 highway loading. Calculate critical design forces in members U_1U_2, U_1L_2, and U_2L_2.

11. The three-hinged arch carries a one-lane roadway supported as shown and is to be designed for an HS15–44 loading. Find the critical value of the axial force in member AB due to the live loading only.

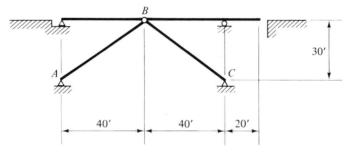

Prob. 11

12. Compute maximum tensile and compressive values of the member force in L_1U_2 due to the moving load shown; load can be reversed.

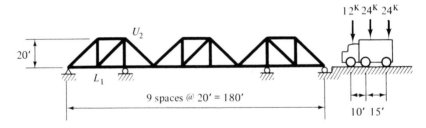

Prob. 12

13. It is required to find the critical member forces to be used in the design of members A, B, and C of the highway bridge shown. Live loading is HS20–44.
   a. Estimate the dead weight of the bridge truss and roadway structure. For dead weight of the truss, use the formula

$$w = \frac{L\sqrt{p}}{20} + \frac{L(b-16)}{10} + 50$$

where w = weight in pounds/foot of each truss, including floor beams, but not stringers and slab
L = length of span in feet
b = width of roadway in feet
p = live load/linear foot of each truss; use 1,200 lb/ft for two-lane bridge and HS20 loading.

Note that the weight of slab, curb, and stringers is to be calculated separately. Assume all dead weight tributary to the bottom panel points.

b. Draw influence lines for the forces in members *A*, *B*, and *C*.

c. Using the influence lines, calculate maximum and minimum design forces due to dead and live loads in members *A*, *B*, and *C*. Use either truck or equivalent lane loading, whichever is more critical.

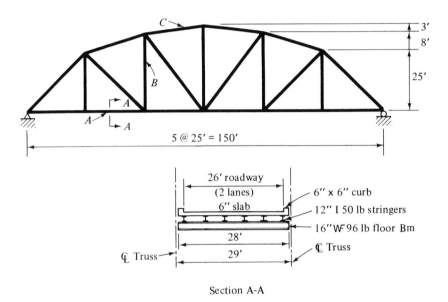

Section A-A

Prob. 13

14. Compute and draw the positive and negative moment envelopes for the left-hand half of the symmetrical articulated bridge span shown. Loading is HS20–44. Use sections spaced on 10-ft centers to compute the critical values.

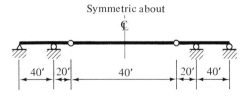

Prob. 14

15. Prove that when both influence line and distributed load are linear the integral $\int w \cdot y \cdot dx$ can be found by multiplying the load intensity at the centroidal point of the area under the influence line by that area.

Deformations of Statically Determinate Structures

Knowledge of the deformations of structures under load is necessary for two main purposes:

1. *To determine the stiffness and deformation of structures in order to prevent excessive deflections under loads.*

2. *To analyze statically indeterminate structures. In such cases, equations of deformation are required to supplement the equilibrium equations.*

Before deformations can be found, the internal member forces must be known. It is therefore necessary to perform a statical analysis of the structure first.

In this part of the book, deformations will be determined by two methods: first, by studying the geometry of the deformed structure, and second, by application of the theorem of virtual forces. Finally, the method of virtual forces will be cast in matrix form to obtain a numerical method suitable for computer analysis.

We shall here restrict ourselves to the case of small deformations, an assumption adequate for most engineering purposes. The study of large deformations leads to nonlinear problems, which are usually difficult to solve. This aspect, which may become important for the solution of complicated buckling problems, will be left to advanced texts.

The ability to sketch the deformed structure under load is all-important to the designer or analyst, and it can be claimed that he must be able to draw a qualitative picture of the structural deformations in order to get a real feel for the response and behavior of the structure, and to develop judgment as to the worth of any particular concept or arrangement of components. For this reason, even prior to any formal analysis, the designer should try to sketch the deformed shape of the structure, to be compared later with quantitative results.

Furthermore, at a time when analyses of large structures will routinely be performed by computer with vast amounts of numerical results being spewed out of a "black box," the danger is great that all judgment about the validity, or any possible errors in these solutions, might be lost. By comparing computer-calculated deformations with previously visualized deformed shapes, it may be possible on the one hand to detect errors, and on the other to gain confidence in the correctness of computer solutions.

CHAPTER 7

Geometrical Approach

The geometrical approach for computing deformations can give the best insight into the physical behavior of a structure under load, and will thus be covered first. In the following sections the geometrical method will be applied first to plane flexural structures, since for these types the meaning can be grasped most easily. Following this, Sec. 7.5 introduces the concept of discretization of deformations for further understanding of the nature of geometrical relations, and to pave the way for matrix formulation of deformations. Section 7.6, finally, shows an example of the geometrical approach for truss deformations.

For more complicated structures, such as three-dimensional configurations, the geometry of deformations is often too involved to be grasped easily, and other methods become preferable.

7.1. Geometrical Relations of Flexural Structures

We begin by defining some terms used in describing the geometry of the deformed structure.

1. *Curvature* Φ. The curvature is defined as the change of slope of the deformed neutral axis per unit length along the member, as shown in Fig. 7.1.

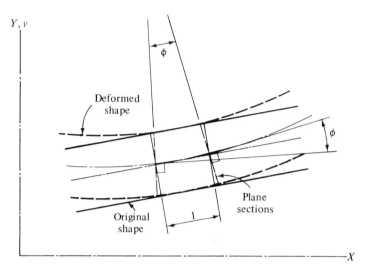

Fig. 7.1 Curvature.

$$\Phi = \frac{d}{dx}\left(\frac{dv}{dx}\right) = \frac{d^2v}{dx^2} \qquad (7.1)$$

where v is the deflection of any point parallel to the Y axis, that is, at right angles to the member axis. One method of determining v is by two successive integrations; this double-integration method is covered in courses on mechanics of deformable bodies, and will not be pursued here.

Since, according to technical beam theory, transverse planes remain normal to the bent neutral axis, the curvature can, by mutually perpendicular lines, as shown in Fig. 7.1, also be interpreted as the relative inclination of two plane transverse sections unity apart.

2. *Angle change and tangential deviation.* We consider a portion of an initially straight beam with end points A and B, as shown in Fig. 7.2. The angle between the tangents to the deformed beam axis at points A and B is

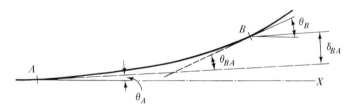

Fig. 7.2 Angle change and tangential deviation.

called the *angle change* between points A and B, and is denoted by the symbol

$$\theta_{BA} = \theta_B - \theta_A \tag{7.2}$$

The *tangential deviation* δ_{BA} is defined as the normal distance from point B on the deformed axis to the tangent drawn to the deformed axis at A, as shown in Fig. 7.2.

To determine the relation between the curvature Φ and the angle change and tangential deviation, we consider Fig. 7.3, which shows the beam segment AB subjected only to one kink of intensity Φ per unit length, extending over an infinitesimal length ds. Due to this single small kink, of magnitude $\Phi \cdot ds$, the small angle change is

$$d\theta_{BA} = \Phi \, ds$$

Due to all contributions between points A and B, the total angle change is

$$\theta_{BA} = \int_A^B d\theta_{BA} = \int_A^B \Phi \, ds \tag{7.3}$$

The infinitesimal tangential deviation at B due to the small kink is

$$d(\delta_{BA}) = (\Phi \, ds)s$$

where s must be measured from the point of tangential deviation, that is, B. The total tangential deviation is found by integrating the individual contributions between A and B:

$$\delta_{BA} = \int_A^B d(\delta_{BA}) = \int_A^B \Phi s \, ds \tag{7.4}$$

A convenient sign convention is the following: slopes such as θ_A and θ_B are positive if they represent an increase of the deflection v with increasing coordinate x. Angle changes such as θ_{BA} are positive in accordance with Eq. (7.2), that is, if they represent an increase of slope with increasing x. Tangential

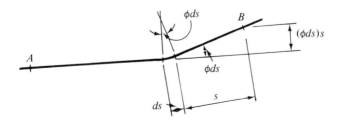

Fig. 7.3 Effect of infinitesimal kink.

deviations such as δ_{BA} are positive if they are in the positive sense of the deflection v as we move from the point of tangency A to the point of deviation B in the positive sense of x.

While the above sign conventions become necessary in complicated problems, they should not preoccupy the reader to the exclusion of a physical feel for the geometry of the deformed structure. The following examples show how various deformations can be computed by considering the geometry of a beam with known curvature.

Example 7.1: The simply supported beam AB, of length L, is subjected to a varying curvature

$$\Phi = ax$$

Calculate the resulting end slopes θ_A and θ_B.

Solution: From the sketch of the deformed shape, it is seen that

$$\theta_A = -\frac{\delta_{BA}}{L}$$

where, according to Eq. (7.4),

$$\delta_{BA} = \int_{x=0}^{L} \Phi(L - x)\, dx = a \int_0^L (Lx - x^2)\, dx = \frac{aL^2}{6}$$

The end rotation is

$$\theta_A = -\frac{\delta_{BA}}{L} = \underline{-\frac{aL^2}{6}}$$

The end slope θ_B will be determined by Eq. (7.2):

$$\theta_B = \theta_A + \theta_{BA}$$

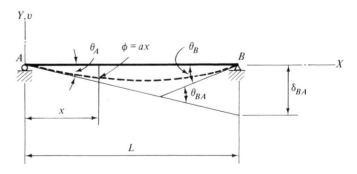

Fig. E7.1

where θ_A is already known and, according to Eq. (7.3),

$$\theta_{BA} = \int_{x=0}^{L} \Phi \, dx = a \int_{0}^{L} x \, dx = \frac{aL^2}{2}$$

so that

$$\theta_B = -\frac{aL^2}{6} + \frac{aL^2}{2} = \frac{aL^2}{3}$$

Example 7.2: For the beam of Ex. 7.1, write the equation of the deflected shape.

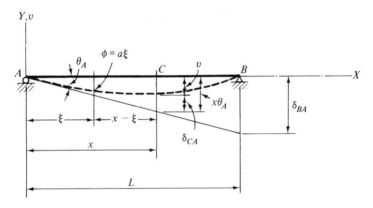

Fig. E7.2

Solution: From the geometry of the deflected shape shown in the sketch,

$$v = x\theta_A + \delta_{CA}$$

where θ_A has been determined as in Ex. 7.1:

$$\theta_A = -\frac{aL^2}{6}$$

and δ_{CA} is the tangential deviation from the point in question, C, to the tangent drawn at A, and is thus determined according to Eq. (7.4) as

$$\delta_{CA} = \int_{\xi=0}^{x} \Phi(x - \xi) \, d\xi = \int_{0}^{x} (a\xi)(x - \xi) \, d\xi = \frac{ax^3}{6}$$

so that the deflection at a point x from the left end is

$$v = -\frac{aL^2}{6} \cdot x + \frac{ax^3}{6} = \frac{aL^3}{6}\left[-\left(\frac{x}{L}\right) + \left(\frac{x}{L}\right)^3\right]$$

The following observations can now be made:

1. Any deformation quantity can be determined from the angle changes and tangential deviations.

2. All deformations must be visualized in terms of the geometry of the deformed shape of the beam, preferably by drawing a clear sketch.

3. No reference has been made to elastic behavior. Only geometrical relations have been used, and the method is therefore valid no matter what the cause of the curvature, that is, elastic or inelastic bending, temperature of creep effects, and the like.

In case the curvatures are due to the bending of an elastic beam of stiffness EI, the bending moment M and the resulting curvature are related by

$$\Phi = \frac{M}{EI} \tag{7.5}$$

and the geometrical relations, Eqs. (7.3) and (7.4), are expressed in terms of the moment as

$$\theta_{BA} = \int_A^B \frac{M\,ds}{EI} \tag{7.6}$$

and

$$\delta_{BA} = \int_A^B \frac{Ms\,ds}{EI} \tag{7.7}$$

For statically determinate structures, the method then consists of the following steps:

1. Analyze the structure and write moment equations.

2. From the moments determine the curvature $\Phi = M/EI$ as a function of position.
   From here, proceed as already shown in Exs. 7.1 and 7.2:

3. Draw the deflected shape of the structure, and determine the geometric relations in terms of angle changes and tangential deviations.

4. Calculate the required deformations by Eqs. (7.6) and (7.7).

Example 7.3: Calculate the deflection at point A of the elastic beam of bending stiffnesses shown in part (a) of the figure, subject to the concentrated load on the overhanging end.

Solution: The structure is analyzed first and its moment diagram and equations determined, as shown in part (b). The curvature diagram and the curvature

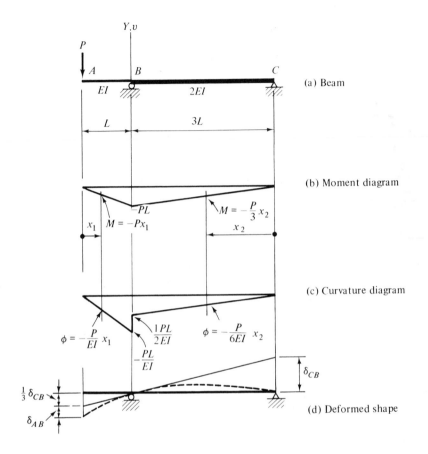

(a) Beam

(b) Moment diagram

(c) Curvature diagram

(d) Deformed shape

Fig. E7.3

equations are then found according to Eq. (7.5) by dividing the moment at any point by the appropriate bending stiffness EI. Next, from the geometry of the deformed beam of part (d), drawn in accordance with the curvature of part (c), we find the desired deflection as

$$v_A = \tfrac{1}{3}\delta_{CB} + \delta_{AB}$$

where

$$\delta_{CB} = \int_{x_2=0}^{3L} \left(-\frac{P}{6EI}x_2\right)(x_2)\,dx = -\frac{3PL^3}{2EI}$$

and

$$\delta_{AB} = \int_{x_1=0}^{L} \left(-\frac{P}{EI}x_1\right)(x_1)\,dx = -\frac{PL^3}{3EI}$$

so that

$$v = -\left(\frac{1}{3} \times \frac{3PL^3}{2EI} + \frac{PL^3}{3EI}\right) = \underline{-\frac{5}{6}\frac{PL^3}{EI}}$$

7.2. Curvature-Area Method

We consider a portion AB of a beam for which the curvature is known or has been calculated, and plot this curvature. This graph, as shown in Fig. 7.4, is called the *curvature diagram*. The previously derived geometrical equations (7.3) and (7.4) can now be interpreted in terms of the properties of the curvature diagram in the following fashion:

$$\theta_{BA} = \int_A^B \Phi \, ds = \text{area under curvature diagram between } A \text{ and } B$$

$$\delta_{BA} = \int_A^B \Phi s \, ds = \text{static moment of area under curvature diagram between } A \text{ and } B \text{ with respect to } B$$

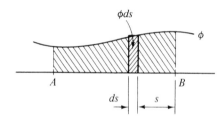

Fig. 7.4 Curvature-area method.

These relations are called the first and second curvature-area theorems. In cases of sufficiently simple curvature diagrams, known area properties may be used to determine the needed angle changes or deviations; for more complicated cases, analytical or numerical integrations are required.

Example 7.4: The simply supported beam AB, of length L, is subjected to the curvature variation $\Phi = ax$; determine the end slopes. (This problem is identical to Ex. 7.1.)

Solution: The curvature diagram is sketched; its area properties are expressed in terms of the resultant area A and its centroidal distance $\bar{x}$, as shown. Next, the deflected curve is drawn, and from its geometry we determine

$$\theta_A = \frac{\delta_{BA}}{L} = -\frac{A \cdot \bar{x}}{L} = -\frac{(\frac{1}{2}aL^2)(L/3)}{L} = \underline{-\frac{aL^2}{6}}$$

The end slope θ_B is found by Eq. (7.2):

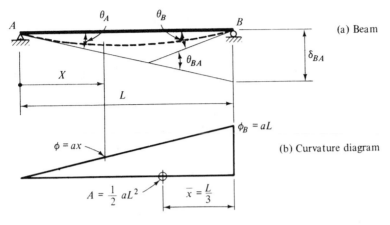

Fig. E7.4

$$\theta_B = \theta_A + \theta_{BA} = -\frac{aL^2}{6} + A = -\frac{aL^2}{6} + \frac{1}{2}aL^2 = \frac{aL^2}{3}$$

Example 7.5: The simply supported beam of part (a) of the figure subjected to a temperature gradient ΔT through its depth. The coefficient of thermal expansion is α. Determine the maximum deflection.

Solution: We first determine the curvature due to the differential strains $\alpha \, \Delta T$, which vary linearly across the beam depth, as shown in part (b), and consequently cause a curvature

$$\Phi = \frac{\alpha \, \Delta T}{d}$$

This curvature is uniform along the beam and is portrayed by the curvature diagram shown in part (c), with appropriate area properties indicated. The symmetrical deformed shape of part (a) drawn in accordance with the curvature diagram indicates that

$$\Delta_{max} = \delta_{AB} = A \cdot \bar{x} = \left(\frac{\alpha \, \Delta T}{d} \cdot \frac{L}{2}\right)\frac{L}{4} = \frac{\alpha \, \Delta T L^2}{8d}$$

In case the curvature is due to the action of a bending moment M acting on an elastic beam of stiffness EI, the curvature diagram is obtained by dividing the moments by the stiffness EI of the beam. This plot is sometimes called the M/EI diagram, and in much of the literature the curvature-area theorems are referred to as the *moment-area theorems*. Here, this notation is not favored, because it obscures the geometric nature of the relations, as well as their generality.

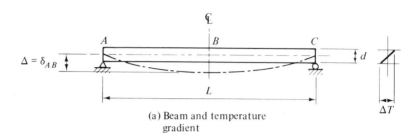

(a) Beam and temperature
gradient

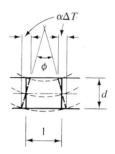

(b) Curvature due to
temperature gradient

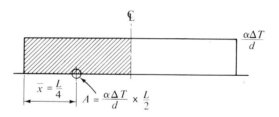

(c) Curvature diagram

Fig. E7.5

Example 7.6: Calculate Δ_A of Ex. 7.3 by the curvature-area method.
Solution:

$$\Delta_A = \tfrac{1}{3}\delta_{CA} + \delta_{AB}$$

where

$$\delta_{CA} = \left(\frac{3}{4}\frac{PL^2}{EI}\right)2L = \frac{3}{2}\frac{PL^3}{EI}$$

$$\delta_{AB} = \left(\frac{1}{2}\frac{PL^2}{EI}\right)\frac{2L}{3} = \frac{1}{3}\frac{PL^3}{EI}$$

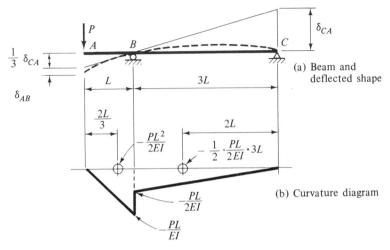

Fig. E7.6

so that

$$\Delta_A = \frac{5}{6}\frac{PL^3}{EI}$$

7.3. Deformations of Rigid Frames by Integration of Curvatures

Deformations of statically determinate rigid frames can be found by the geometrical method using the following steps:

1. Determine the curvature of all members. If these curvatures are due to elastic bending, they can be represented by the M/EI diagram.

2. Carefully draw the deformed shape of the structure in accordance with the curvatures, and express the desired deformations in terms of appropriate tangential deviations and angle changes.

3. Calculate the required tangential deviations and angle changes by integration of the curvatures, or using the properties of the curvature diagram.

4. Substitute the values found in step 3 into the relations of step 2, and determine the desired results.

Example 7.7: For the rigid frame shown, determine
a. The horizontal deflection at *B*.
b. The horizontal deflection at *D*.

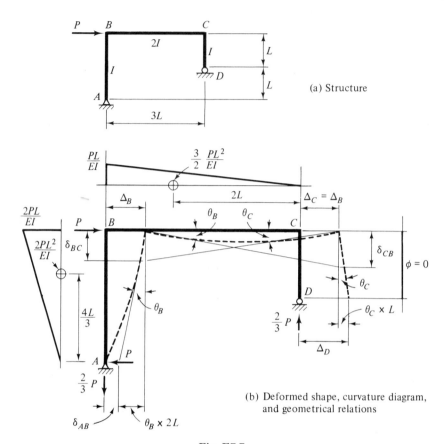

(a) Structure

(b) Deformed shape, curvature diagram, and geometrical relations

Fig. E7.7

Solution: We determine the moments by statics, and draw the corresponding curvature diagram, as shown in the figure. The curvature diagram enables us to draw the appropriate deformed shape of the structure. To help in further calculations, we indicate the resultant curvature areas and their locations, as in part (b).

a. $\Delta_B = \delta_{AB} + \theta_B \cdot 2L$, where

$$\theta_B = \frac{\delta_{CB}}{3L} = \frac{1}{3L}\left(\frac{3}{2}\frac{PL^2}{EI} \cdot 2L\right) = \frac{PL^2}{EI}$$

and

$$\delta_{AB} = \frac{2PL^2}{EI} \cdot \frac{4L}{3} = \frac{8}{3}\frac{PL^3}{EI}$$

so that

$$\Delta_B = \left(\frac{8}{3} + 2\right)\frac{PL^3}{EI} = \frac{14}{3}\frac{PL^3}{EI}$$

b. $\Delta_D = \Delta_C + \theta_C \cdot L$.

Neglecting axial deformations we set

$$\Delta_C = \Delta_B = \frac{14}{3}\frac{PL^3}{EI}$$

and

$$\theta_C = \frac{\delta_{BC}}{3L} = \frac{1}{3L}\left(\frac{3}{2}\frac{PL^2}{EI}\cdot L\right) = \frac{1}{2}\frac{PL^2}{EI}$$

so that

$$\Delta_D = \left(\frac{14}{3}+\frac{1}{2}\right)\frac{PL^3}{EI} = \frac{31}{6}\frac{PL^3}{EI}$$

7.4. General Arch Equations

The general arch equations constitute an extension of the concept of integration of curvatures to the case of plane curved members of arbitrary shape. They are presented here to show how the geometric method can be applied to more complicated structures than beams.

We consider the fixed-ended cantilever of arbitrary plane shape shown in Fig. 7.5. The structure is subjected to a curvature Φ due to any cause, such as elastic bending under loads, temperature effects, and so on. It is required to calculate the displacements u and v along the X and Y axes, and the rotation θ of point B.

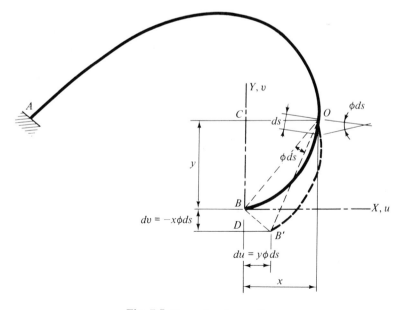

Fig. 7.5 General arch equations.

We place the origin of the X and Y axes at point B. Due to the curvature Φ over an infinitesimal length ds located at (x, y), we get rigid-body rotation of the dashed portion of the structure shown in Fig. 7.5 through a small angle $\Phi\, ds$; this rotation causes the small rotation $d\theta$ and the small displacements du and dv at point B given by

$$d\theta = \Phi\, ds$$
$$du = (\Phi\, ds)y$$
$$dv = -(\Phi\, ds)x$$

These relations can be proved by the similarity of triangles OCB and BDB' in Fig. 7.5.

Due to all portions of the structure between A and B subjected to curvature, the total rotation and displacements of point B will be

$$\theta = \int_A^B \Phi\, ds \tag{7.8}$$

$$u = \int_A^B \Phi y\, ds$$

$$v = -\int_A^B \Phi x\, ds \tag{7.10}$$

Equations (7.8), (7.9), and (7.10) are called the *general arch equations*. For structures composed of straight members, these equations can be interpreted as the area and static moments of the area under the curvature diagram between the fixed end and the point in question with respect to the X and Y axes through this point. In this sense they can be considered as an extension of the curvature-area method to curved planar members.

> **Example 7.8:** Calculate horizontal and vertical deflections at B of the quarter-circular arch shown. The constant stiffness is EI, and radius R.
>
> **Solution:** We compute the moment variation in terms of polar coordinates and divide by the bending stiffness EI to obtain the curvature:
>
> $$M = -PR(1 - \sin\theta)$$
> $$\Phi = -\frac{PR}{EI}(1 - \sin\theta)$$
>
> Applying Eqs. (7.9) and (7.10), we find the deflections at point B:
>
> $$u = \int_{\theta=0}^{\pi/2} \Phi y\, ds = \int_0^{\pi/2} -\frac{PR}{EI}(1 - \sin\theta)[-R(1 - \sin\theta)]R\, d\theta$$
> $$= \left(\frac{3}{4}\pi - 2\right)\frac{PR^3}{EI}$$

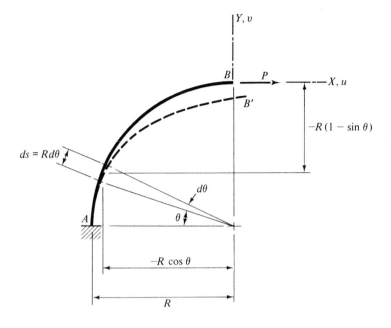

Fig. E7.8

$$v = -\int \Phi x \, ds = -\int_0^{\pi/2} -\frac{PR}{EI}(1 - \sin\theta)[-R(\cos\theta)]R \, d\theta$$

$$= -\frac{1}{2}\frac{PR^3}{EI}$$

Example 7.9: Member *BC* of the frame shown is subject to a temperature gradient ΔT (hot inside, cold outside). The coefficient of thermal expansion is α. Find the deflection at *C*.

Solution: The curvature diagram due to the temperature distortion of member *BC* is drawn, and the static moments of its area about the *X* and *Y* axes through point *C* are computed:

$$\Phi = \frac{\alpha \Delta T}{d}$$

$$u = \int \Phi y \, ds = \left(\frac{\alpha \Delta T \sqrt{2} L}{d}\right)\left(-\frac{L}{2}\right) = -.707\frac{\alpha \Delta T L^2}{d}$$

$$v = +\int \Phi x \, ds = \left(\frac{\alpha \Delta T \sqrt{2} L}{d}\right)\left(-\frac{L}{2}\right) = +.707\frac{\alpha \Delta T L^2}{d}$$

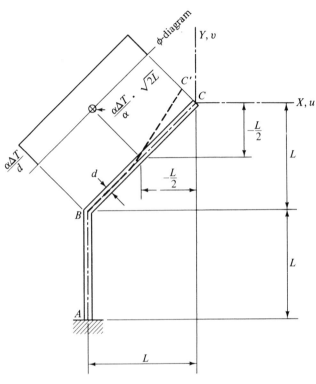

Fig. E7.9

7.5. Discretization and Matrix Calculation of Deformations

Just as in statical calculations, it is sometimes convenient to discretize the deformations and proceed by numerical or graphical calculations, thus avoiding formal integrations. When used with judgment, this also makes it possible to arrive at quick approximations, but it should be realized that, in general, with each approximation the accuracy becomes less satisfactory; thus, if the loads are first discretized to get approximate moments, as in Sec. 3.6, and this is followed by a further discretization of curvatures, extreme caution is required in the interpretation of the results. Nevertheless, the method is presented here because it may lead to a better understanding of the geometrical relations of the preceding sections, and show additional uses of the matrix method of bookkeeping.

To be specific, we consider the prismatic simple beam of Fig. 7.6(a), subjected to equal end moments M so that constant curvature $\Phi = M/EI$ results along the entire beam. If we now consider the total angle change

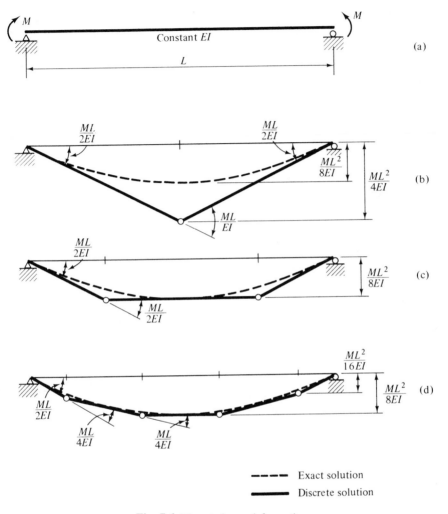

Fig. 7.6 Discrete beam deformations.

between beam ends, $\theta_{BA} = \int_A^B \Phi \, ds = ML/EI$, lumped as a concentrated kink at midspan, the resulting beam deformations will be those shown by the deflected shape in Fig. 7.6(b), consisting of two straight segments with concentrated angle change. The geometry of this configuration is simple, so the midspan deflection can be calculated immediately as $(ML/2EI) \cdot (L/2) = ML^2/4EI$; this is 100 per cent above the exact midspan deflection of $ML^2/8EI$.

We next consider the angle change over each half of the beam, of value $ML/2EI$, lumped as a concentrated kink at the quarter-point, leading to the deflected shape of Fig. 7.6(c), in which the midspan deflection has the exact value, but the quarter-point deflections are too large by 33 per cent.

With more refined discretization, the piecewise-linear deformed shape will constitute a better and better fit to the exact deflection curve, as shown by the very good approximation of Fig. 7.6(d), in which the angle changes occurring over quarter-spans have been lumped at the midpoint of each element. In each case, we note that the deviation between exact and approximate deflection is a maximum at the concentrated element midpoint kinks and vanishes at the element boundaries. Thus, if a displacement at a specified point is desired, the beam curvatures should be lumped in such a fashion that this point falls between segments. It may be noted that this procedure was already used in the curvature-area method of Sec. 7.2.

In cases of more general loading, which causes nonuniform moments along each beam element, a good approximation usually results if the moment value at element midpoint is used; in such cases one cannot, of course, count on obtaining exact deflection values at element boundaries.

To proceed in a somewhat more general sense, we consider the relation between unit values of the kinks at points 1 to 3 of the simple beam shown in Fig. 7.7(a) and the resulting deflections at these points. Figures 7.7(b) and (c) show the deformations resulting from a single unit kink at points 1 and 2, respectively. The effects of a unit kink at point 3 are not shown, since by symmetry of the structure they can be obtained from Fig. 7.7(b) by reversing the order of points 1 to 3. The deflections d_{ij} at point i due to a unit kink at point j can be written as the elements of the $[d]$ matrix:

$$[d] = \frac{L}{16} \begin{bmatrix} 3 & 2 & 1 \\ 2 & 4 & 2 \\ 1 & 2 & 3 \end{bmatrix} \tag{7.11}$$

Due to kinks at point j of magnitude θ_j, stored in the matrix $\{\theta\}$, the deflections Δ_i at points i are contained in the $\{\Delta\}$ matrix:

$$\{\Delta\} = [d]\{\theta\}$$

It is interesting to observe that the $[d]$ matrix of deflections due to unit kinks is identical with the transpose of the force transformation matrix $[b]$ of moments due to unit loads. This $[b]$ matrix has been presented in Eq. (3.13). The general relation $[d] = [b]^T$ was derived by the theorem of virtual work in Sec. 5.4.

The kink θ_j at point j due to elastic bending can be approximated by considering the total angle change occurring over the beam segment bounded by center lines between adjacent discrete points, of length l and stiffness EI:

$$\theta_j = \int_0^l \frac{M}{EI}\, ds \approx M_j \cdot \frac{l}{EI}$$

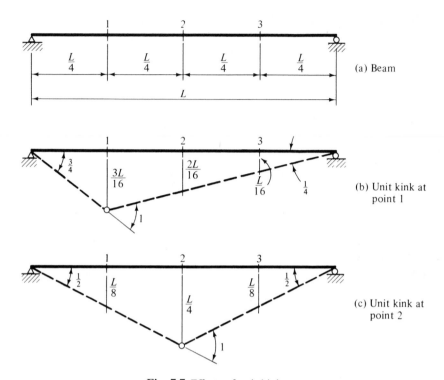

Fig. 7.7 Effects of unit kinks.

where the average moment may be assumed to be at point j, that is, M_j. For a given loading consisting of concentrated loads P, the average moments M_j are given by the force transformation matrix

$$\{M\} = [b]\{P\}$$

The concentrated angle changes $\{\theta\}$ for the case of constant spacing l of the discrete points and constant EI can now be related to the internal moments and the applied loads by the relation

$$\{\theta\} = \frac{l}{EI}\{M\} = \frac{l}{EI}[b]\{P\}$$

The deflections Δ_i are now contained in the matrix

$$\{\Delta\} = [d]\frac{l}{EI}[b]\{P\}$$

Since, according to earlier discussion, $[d] = [b]^T$, we finally obtain

$$\{\Delta\} = [b]^T \frac{l}{EI}[b]\{P\} \tag{7.12}$$

A similar expression will be derived by the theorem of virtual forces in Sec. 9.5.

The matrices $\{P\}$ and $\{\Delta\}$ are due to one loading condition. If several, say l, loading conditions are to be analyzed, we provide l columns in the $[P]$ and $[\Delta]$ matrices, one for each loading condition, so that the element P_{ij} (or Δ_{ij}) is the load (or deflection) at point i of loading condition j. Equation (7.12) thus becomes

$$[\Delta] = [b]^T \frac{l}{EI}[b][P] \tag{7.13}$$

We now consider a structure with i discrete points, subjected to i loading conditions, each consisting of a unit load applied at one of the i discrete points, beginning with the first and ending with the ith. For this case the $[P]$ matrix will be a unit matrix $[I]$, of order ($i \times i$), which can be deleted. The deformation at point i resulting from a unit load at point j is called the *structure flexibility*, or *compliance*, F_{ij}. These quantities are stored in the *structure flexibility matrix* $[F]$, of order ($i \times i$), given by

$$[F] = [b]^T \frac{l}{EI}[b] \tag{7.14}$$

Once this matrix, which is a structure characteristic, is obtained, the deflections due to any loading conditions can be obtained by

$$[\Delta] = [F][P] \tag{7.15}$$

Because of the formulation of discrete kinks in terms of the discrete average moments, this procedure furnishes exact results only for cases of loadings that lead to constant curvature over each beam segment; for other more common moment variations, the method may still yield useful, if approximate, results, as shown in the following problem.

Example 7.10: Calculate the deflections at points 1 to 3 of the prismatic simple beam of part (a) of the figure, due to the parabolically varying load, by consideration of discrete kinks at these points.

Solution: We could use Eq. (7.12) directly, but instead recall that the moments in this beam due to the indicated loading were already computed in Ex. 3.8:

$$\{M\} = [b]\{P\} = \frac{L}{16}\begin{bmatrix} 3 & 2 & 1 \\ 2 & 4 & 2 \\ 1 & 2 & 3 \end{bmatrix} \cdot \frac{w_0 L}{192}\begin{Bmatrix} 4 \\ 13 \\ 28 \end{Bmatrix} = \begin{Bmatrix} .0215 \\ .0378 \\ .0372 \end{Bmatrix} w_0 L^2$$

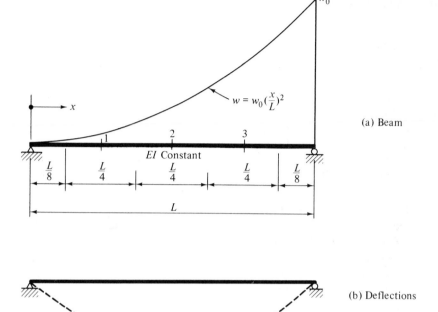

Fig. E7.10 Discrete deformations.

The kink at each point is approximately equal to the segmental angle change $M \cdot (L/4EI)$, and the $[d]$ matrix is the transpose of the above $[b]$ matrix; thus, Eq. (7.12) becomes

$$\{\Delta\} = [b]^T \frac{L}{4EI}[b]\{P\} = \frac{L^2}{64EI}\begin{bmatrix} 3 & 2 & 1 \\ 2 & 4 & 2 \\ 1 & 2 & 3 \end{bmatrix}\begin{Bmatrix} .0215 \\ .0378 \\ .0372 \end{Bmatrix}w_0 L^2$$

$$= \begin{Bmatrix} .00277 \\ .00419 \\ .00326 \end{Bmatrix}\frac{w_0 L^4}{EI}$$

Part (b) shows these deflections superimposed upon the exact deflection curve (to be obtained in Ex. 8.1). In spite of the crude division of the beam and the rapid variation of curvatures in this beam, all values are within 10 per cent of the exact deformations. This error could be radically decreased by finer division or more intelligent calculation of the discrete kinks.

7.6. A Geometric Method for Truss Deformations

The graphical method for computation of truss deflections devised by the Frenchman M. Williot and the German O. Mohr is a convenient way of obtaining all truss joint displacements at once. Its geometric significance is recognized by the construction of Fig. 7.8(a), which considers a two-member truss whose member AC is shortened by a given amount ΔL_1, and whose member BC is elongated by an amount ΔL_2 due to some cause.

To calculate the displacement of joint C due to these changes in member length, we visualize the following steps:

1. We remove the pin at joint C and let the members change lengths as specified without allowing them to rotate, as shown in Fig. 7.8(b).

2. In order to reconnect the members at joint C, we allow them to rotate about their fixed ends A and B, as shown by the dashed arcs of Fig. 7.8(c), until they intersect at the point marked by the circle, whereupon we can reinsert the pin into joint C in its displaced position. A vector drawn from the original to the displaced location C' of joint C gives us the displacement.

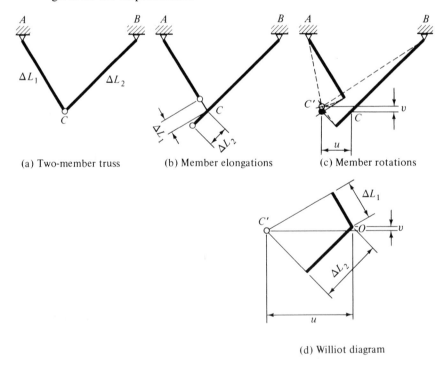

(a) Two-member truss   (b) Member elongations   (c) Member rotations

(d) Williot diagram

Fig. 7.8 Truss displacement by geometric method.

The member elongations ΔL are actually an order of magnitude smaller than the member lengths, so that the scale of the displacements becomes immeasurably small. We therefore resort to the linearization of the following steps:

3. Instead of the arcs shown dashed in Fig. 7.8(c), we draw their tangents shown solid in that figure. For small deflections, their intersection, shown by a solid circle, will be sufficiently close to that of the arcs.

4. The substitution of the tangents for the arcs allows us to draw all deformation quantities to a convenient scale, which is independent of that of the truss, as shown in Fig. 7.8(d); this figure is called the *Williot diagram.* Here, the vectors denoting the member elongations are laid off in the appropriate direction and sense from a common origin O, and the displacement vectors due to the member rotations (shown in fine line) are drawn as normals to be elongation vectors to their intersection C'. The actual joint displacement, or its components u and v, can now be scaled from the origin O.

The member elongations ΔL can be due to any cause; when they are due to elastic deformations resulting from applied loads, an analysis is necessary to determine the member forces S, from which the member elongations can be determined by usual methods, for instance, by the formula $\Delta L = SL/AE$ for a prismatic member of length L and elastic stiffness AE.

The general procedure is to start with any two joints of known location, and to determine the displaced location of the third joint of the triangular panel. In general, for each panel there will be four displacement vectors: two member elongation vectors and two normal vectors due to member rotation, as shown for instance in Fig. 7.8(d). In this way we progress from panel to panel, traversing the entire truss.

With these concepts, we now find the displacements for the truss shown in Fig. 7.9(a). Joint A is fixed in location, and support D prevents any rotation of member AD. With all member elongations computed, as shown in Table 7.1, we can now draw the Williot diagram, as shown in Fig. 7.9(b), as follows. We establish the origin O designating the undisplaced location of all joints, and proceed by laying off the elongation of member AD horizontally to the chosen scale to determine the displaced location D' of joint D.

Similarly, we draw the elongation vector of the unrotated member DC (whose lower end D has been displaced to D'), and of member AC, starting at point A', which coincides with the origin since it is fixed. These elongation vectors are shown in heavy line in Fig. 7.9(b); their sense can be visualized by positioning ourselves at the near member end and observing the sense of movement of the other end due to member lengthening or shortening. The normals from the far ends of the elongation vectors, representing the tangents

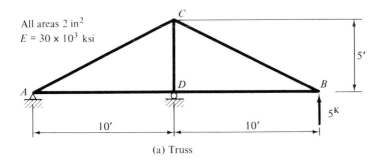

All areas 2 in^2
$E = 30 \times 10^3$ ksi

10'   10'

5'

5^K

(a) Truss

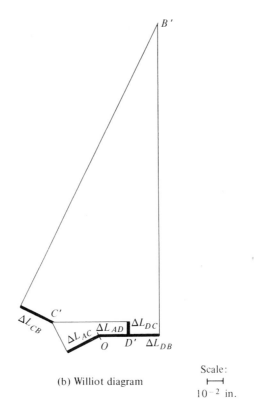

B'

C'

ΔL_{CB}

ΔL_{AC} ΔL_{AD} ΔL_{DC}

O D' ΔL_{DB}

Scale:

10^{-2} in.

(b) Williot diagram

Fig. 7.9 Williot diagram.

to the arcs described by the member rotations about D' and A', respectively, are carried to their intersection at point C', which locates the displaced joint C.

We next draw the elongation vectors of the unrotated members DB and CB, whose near ends are now displaced to points C' and D'; their sense is

TABLE 7.1. Member Forces and Elongations

MEMBER	L/AE (10^{-3} in./kip)	S (kips)	$\Delta L = \dfrac{SL}{AE}$ (10^{-3} in.)
AD	2.00	$+10.00$	$+20.0$
DB	2.00	$+10.00$	$+20.0$
AC	2.24	-11.20	-25.1
CB	2.24	-11.20	-25.1
CD	1.00	$+10.00$	$+10.0$

found by positioning ourselves at these ends and observing the far ends approaching or receding due to member lengthening or shortening. Again the normals at the far ends of the elongation vectors are drawn to their intersection, defining the displaced location B' of joint B.

The distances from the origin O (which defined the undeformed location of all joints) to the primed points give the required joint displacements to the chosen scale.

The truss of Fig. 7.9(a), which has one member constrained against rotation, is simpler than the general case represented by the truss of Fig. 7.10(a), in which all members are able to rotate. In this case, we consider a reference point that maintains its position and a reference member attached to it, which, for the time being, is assumed to maintain its direction; let us here use joint A as reference point and member AD as reference member. In effect, we substitute a support at D for the real support at B, which, for the time being, we disregard.

Proceeding now as before, we analyze the member forces due to the load on the truss of Fig. 7.10(a), and find that they, as well as the member elongations, happen to be identical to those of the previous truss of Fig. 7.9(a). Since the support conditions assumed temporarily are identical for the two trusses, it follows that the joint displacements are those given by the Williot diagram of Fig. 7.9(b). These truss deformations, based on the assumed reference member AD, are shown again to exaggerated scale in Fig. 7.10(b). [To firm up the ideas, it might be useful to match up the various components of joint displacements shown in this figure with those of the Williot diagram, Fig. 7.9(b).]

Because, contrary to fact, member AD was assumed fixed in direction, joint B has a vertical component of displacement that does not satisfy the given support conditions of Fig. 7.10(a). It therefore remains to rotate the entire truss as a rigid body into such a position that joint B comes back to rest on its support. This operation is accomplished by the *Mohr correction diagram*.

The Mohr correction diagram consists of a scaled view of the undeformed truss, shown by the dashed and dotted line in Fig. 7.10(b), whose joints are here defined by double-primed letters. Joint B'', representing the joint to be

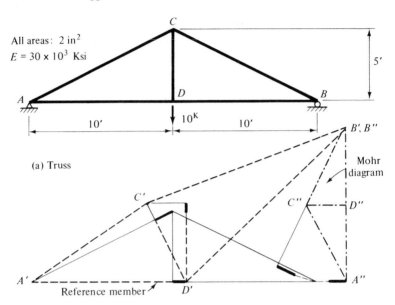

All areas: 2 in^2
$E = 30 \times 10^3$ Ksi

5'

10^K

(a) Truss

Mohr diagram

(b) Deformed truss (exaggerated) with correction diagram

Reference member

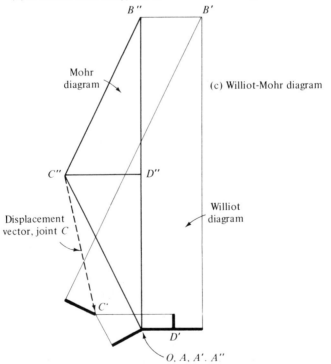

Mohr diagram

(c) Williot-Mohr diagram

Displacement vector, joint C

Williot diagram

O, A, A', A''

Fig. 7.10 Truss displacements by Williot–Mohr diagram.

placed on its specified support, coincides with B'; joint A'', representing the reference joint, is placed on the supporting plane of the truss. The correction diagram is so oriented that the line $B''A''$ forms a tangent line to the arc that is described by joint B as the truss is rotated into its proper position on support plane B (for small deflections, line $A'B'$ is almost parallel to line AB; therefore, this tangent line can be drawn normal to line AB through the center of rotation A).

We now claim that the vector drawn from any double-primed point of the Mohr diagram to the reference point A'' represents the displacement of that point due to the corrective rotation of the truss onto its support B. To prove this, we observe that

1. The line from any point on the Mohr diagram to point A'' is parallel to the translation of the actual point due to a corrective rotation about point A.

2. The line from any point on the Mohr diagram to point A'' is of length equal to the translation of the corresponding actual point due to the corrective translation. We shall prove this for joint C:
   Corrective angle of rotation:

$$\theta = \frac{B''A''}{BA}$$

By similar triangles,

$$\frac{C''A''}{CA} = \frac{B''A''}{BA} = \theta$$

Therefore,

$$C''A'' = CA\cdot\theta = \text{distance translated by joint } C \text{ due to}$$
$$\text{small rotation } \theta \text{ about point } A$$

We conclude that we can add the vectors connecting the double-primed points to reference point A'' to the previously determined displacement vectors to obtain the total joint translation.

To perform this superposition at a convenient scale, we add the Mohr correction diagram to the Williot diagram drawn previously in Fig. 7.9(b). Adhering to the previously defined orientation of the correction truss, we place its reference point A'' on the origin O, and expand its scale so that the supported joint B'' lies on a line through B' parallel to the permitted movement of this joint, that is, on a horizontal line for this case. The combined Williot–Mohr diagram is shown in Fig. 7.10(c).

The total displacement of any joint, say, joint C, now consists of two portions to be added:

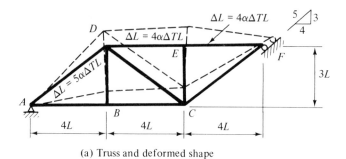

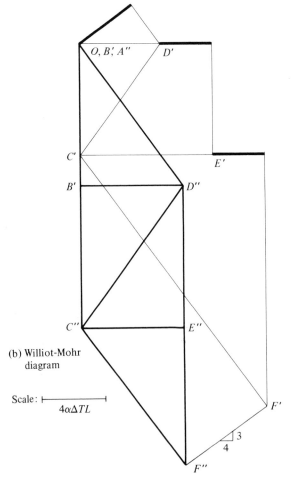

Joint	u	v
A	0	0
B	0	+ 10.7
C	0	+ 13.0
D	− 1.7	+ 10.7
E	+ 2.2	+ 13.0
F	+ 6.2	+ 4.6

All values × $\alpha\Delta TL$

(a) Truss and deformed shape

(b) Williot-Mohr
diagram

Scale: $\vdash\!\!\!\longrightarrow\!\!\!\dashv$
$4\alpha\Delta TL$

Fig. E7.11

1. The vector OC', representing the displacement due to member elongations of the truss with member AD held against rotation.

2. The vector $C''O$, representing the displacement due to the corrective
   rotation to reestablish the given support conditions.

The sum of these two vectors is given by the vector $C''C'$. We generalize now
to draw the following conclusion:

> The vector drawn from the double-primed to the single-primed
> points in the Williot–Mohr diagram represents the total displace-
> ment of the point due to truss deformations.

As in all graphical methods, a certain skill acquired by practice is necessary
to perform the construction efficiently; in particular, judgment must be
developed regarding choice of scale and position of the sketch on the paper.

Example 7.11. The top chord members AD, DE, and EF of the truss shown
in part (a) of the figure are subjected to a temperature increase ΔT. They are
of material of coefficient of thermal expansion α. Compute the truss deformations due to this effect.

Solution: The top chord members elongate by the amount indicated on these
members in part (a). All other members remain at their original length.

   We consider point A as reference point and member AB as reference
member, and construct the Williot diagram as shown in part (b). Beginning
at joints A and B (neither of which displaces, so A' and B' will coincide with
the origin O), we scale off ΔL_{AD}, and drop the normals to members AD and
BD to their intersection D'. Next, we drop normals to members DC and BC
(neither of which elongates) from D' and B' to their intersection C'. We
continue by establishing, successively, points E' and F' to complete the
Williot diagram.

   The prescribed displacement of joint F of the truss is along the supporting
plane at F shown in part (a). Had the correct rotation of the truss been taken
into account, then the displacement vector OF' would be oriented in this
direction. To superimpose the results of the rigid-body truss rotation necessary
to satisfy the support requirements, we add the Mohr correction diagram
shown in part (b). This consists of the truss turned through 90°, with reference
point A'' at the origin O, and point F'' on a line passing through F' and parallel
to the supporting plane. All joint displacements can now be scaled as vectors
from the double-primed points of the Mohr diagram to the single-primed
points of the Williot diagram. The components of the resulting displacements
are tabulated, and the deformed shape of the truss, to a smaller scale, is shown
as a dashed line in part (a).

PROBLEMS

   In all problems the deformed shape of the structure should be carefully drawn
using appropriate instruments, and the critical values indicated.

1. a. Calculate the transverse deflection v of the parabolically loaded beam of constant cross section as a function of x.

   b. Find the maximum deflection.

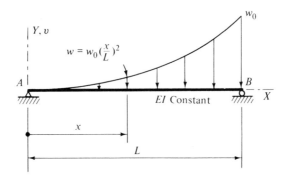

Prob. 1

2. a. Determine the deflection v as a function of position x due to the cranked-in moment M.

   b. Find the end rotations at A and B.

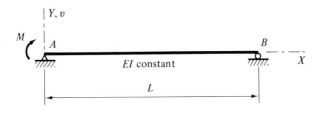

Prob. 2

3. The cantilever beam of coefficient of thermal expansion α is subjected to a temperature variation $T = \Delta T[1 + 2(y/d)][1 - (x/L)]$.

   a. Sketch the variation of temperature transversely and longitudinally in the beam.

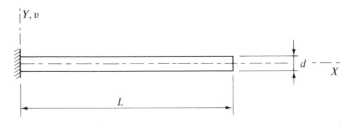

Prob. 3

b. Compute the deflection $v(x)$ of the beam, as well as its maximum displacement and slope.

4. The moment-curvature relation of the nonlinear cantilever beam of part (a) of the figure is shown in part (b).
  a. Draw the plot of curvature along the beam due to the cranked-in moment at B.
  b. Compute the deflection v along the beam and the maximum deflection.

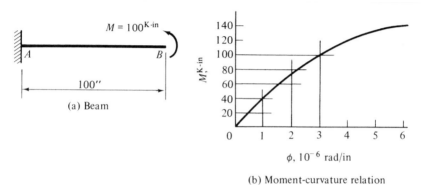

(a) Beam

(b) Moment-curvature relation

Prob. 4

5. The elastic beam is of rectangular cross section of constant width b and linearly decreasing depth, tapering to a point at the free end. The load intensity varies parabolically from zero at the free end to a maximum w_0 at the fixed end.
  a. Establish a suitable origin and write the moment equation.
  b. With the same origin, write the equation for variation of beam stiffness.
  c. Write the curvature equation, and compute the elastic curve v of the beam.

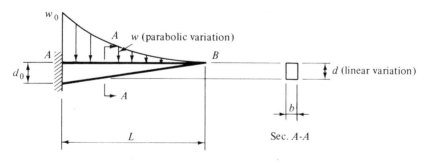

Prob. 5

6. For the simple beam of constant stiffness EI under the concentrated loads shown, calculate
  a. The end rotation at point A.

b. The equation of the elastic curve between points B and C, in terms of x, as shown.
c. The maximum deflection of the beam.

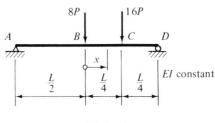

Prob. 6

7. a. Sketch the curvature diagram for beam AB under the self-equilibrating loads shown.
   b. Compute the rotations of the beam at ends A and B and the maximum deflection.

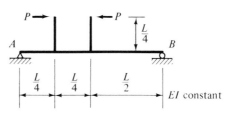

Prob. 7

8. Calculate the deflection under the concentrated load. Cross section:

Beam is 14 WF 43: $I = 429$ in.^{4} web thickness $= .28$ in.

$A = 12.65$ in.^{2} $\bar{x} = .70$ in.   (from channel back to centroid of section)

Channel is 12 C 20.7: $I = 3.9$ in.^{4}

$A = 6.03$ in.^{2}

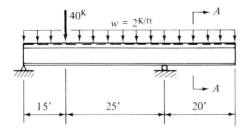

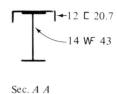

Sec. *A A*

Prob. 8

9. The beam shown is coverplated between points C and D as indicated in the section. It is subjected to a uniform load of 1 kip/ft over its entire length. Find the vertical deflection of the beam at B.

$$10 \text{ WF } 21: \quad I = 106.3 \text{ in.}$$
$$E_{st} = 30 \times 10^6 \text{ psi}$$

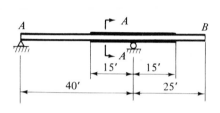

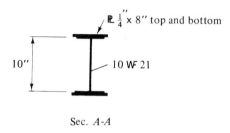

Sec. *A-A*

Prob. 9

10. Find the slope and deflection of the beam at point B due to the 100-kip load. Stiffness is E.

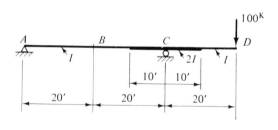

Prob. 10

11. The simple beam AB has suffered a kink $\theta = 0.1$ rad. Compute the maximum deflection due to this effect.

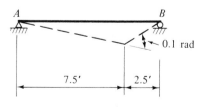

Prob. 11

12. The straight prismatic beam, of stiffness EI, contains an elastic spring of rotational stiffness EI/l kip-in./rad at the third point, as shown. Compute the deflection of the beam under the load P. (*Hint:* Consider deformations in the beam and spring separately, and use superposition.)

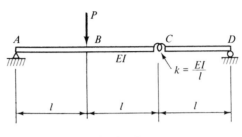

Prob. 12

13. Point C of the frame shown has suffered a kink of 0.1 rad.
    a. Draw the deformed shape of the frame due to this kink.
    b. Compute the rotation of end A and the displacements of points C and D due to this effect.

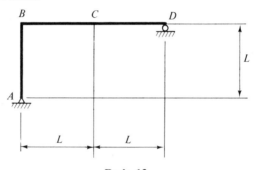

Prob. 13

14. Member BC, of depth d and coefficient of thermal expansion α, is subjected to a temperature gradient ΔT. Compute the maximum deflection of member BC and the displacement of point D.

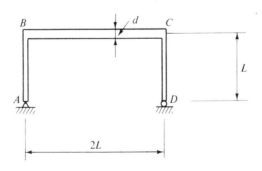

Prob. 14

15. a. Draw shear and moment diagrams for the structure and load shown.
  b. Find the maximum deflection of beam *BC* of the rigid frame. All members have same stiffness *EI*.

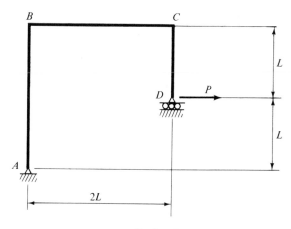

Prob. 15

16. Calculate the horizontal displacement and the rotation of point *C* of the rigid frame under the load shown. Consider bending distortions only. Flexural stiffnesses are as shown.

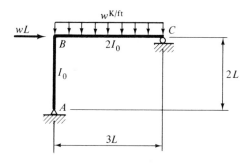

Prob. 16

17. Calculate the horizontal deflection of point *A* and the rotation of point *B* of the structure shown. Stiffness of the material is *E*.

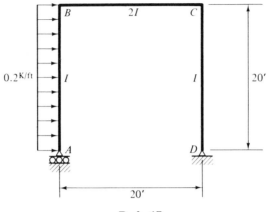

Prob. 17

18. Calculate the horizontal and vertical deflections of the free end of the semi-circular cantilever due to the applied vertical load *P*.

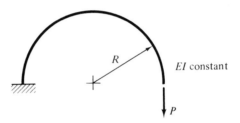

EI constant

Prob. 18

19. a. Establish suitable origins and write expressions for the variation of shear and moment in the structure due to the load shown.

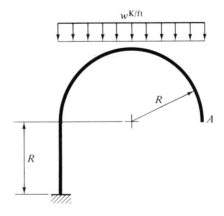

Prob. 19

b. Find the deflection components, and rotation of point *A*. *EI* is constant.

20. For the discrete sections 1 to 4 of the simple beam,
    a. Write the [*b*] matrix of internal moments due to unit loads, of order (4 × 4), and the [*d*] matrix of deflections due to unit kinks, of order (4 × 4).
    b. Write the (4 × 4) matrix [*F*], where F_{ij} is the approximate deflection at point *i* due to a unit load applied at point *j*.
    c. For $P_1 = P$, $P_2 = 2P$, $P_3 = 3P$, and $P_4 = 0$, compute the approximate deflections at points 1 to 4.
    d. Compare the results obtained in Part c with exact results, and draw conclusions.

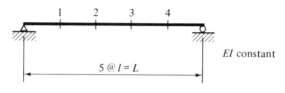

Prob. 20

21. The beam of Prob. 20 is loaded with a parabolically varying load as specified in Ex. 7.10.
    a. By discretization and matrix methods, find approximate deflections at points 1 to 4.
    b. By comparison with the results of Ex. 7.10 and exact results (see Ex. 8.1), try to draw conclusions about the convergence of the method with increasing number of segments.

22. By use of the [*F*] matrix found in Prob. 20, Part b, find the (4 × 3) matrix [**Δ**] of discrete beam deflections at points 1 to 4 due to the following three loading conditions:
    a. $P_1 = 1$ kip, $P_2 = P_3 = 2$ kips, $P_4 = 0$.
    b. $P_1 = 2$ kips, $P_2 = P_3 = 0$, $P_4 = 3$ kips.
    c. $P_1 = P_2 = 0$, $P_3 = P_4 = 2$ kips.

23. The prismatic beam shown is loaded with a sinusoidally varying load. Discretize

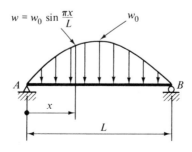

Prob. 23

the left-hand side of the beam and load into four equally spaced nodal points, and find

a. The moments $\{M\}$ at these points.
b. The flexibility matrix $[F]$ at these points.
c. The matrix $\{\Delta\}$ of deflections at these points.
d. Compare the results of the discretized solution with exact results $v = (w_0 L^4/\pi^4) \sin(\pi x/L)$.

24. Due to temperature effects, the member AC of the truss is shortened by an amount ΔL. Calculate the vertical deflection of joint C due to this effect.

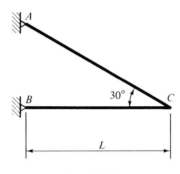

Prob. 24

25. Find the total deflection of all panel points of the truss shown. Tabulate your results in alphabetical order and give values in inches. Timber truss: All areas 20 in.^{2}; $E = 2{,}000$ ksi.

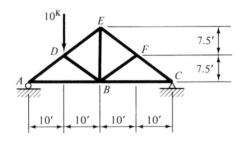

Prob. 25

26. Cross-sectional areas of all members are 2 in.^{2}; $E = 30{,}000$ ksi. Find the horizontal and vertical components of deflection of all joints. Tabulate the results in alphabetical order of joints.

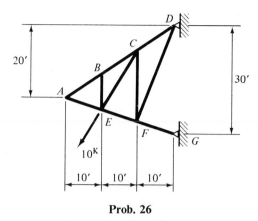

Prob. 26

CHAPTER 8

The Method of Virtual Forces

As was already pointed out in Chapter 5, the theorem of virtual work provides a unified approach to the solution of statical and geometrical problems, which, because of its systematic procedure, lends itself particularly well to the solution of lengthy or difficult problems. Here, we want to study the use of the theorem of virtual forces in the calculation of structural deformations more closely.

In the theorem of virtual forces, the equilibrium system P, σ in the virtual work equation

$$\text{EVW} = \text{IVW}$$

$$\sum P^{**} \cdot \Delta^* = \int_V \sigma^{**} \cdot \epsilon^* \cdot dV \tag{8.1}$$

will be arbitrary and can be chosen for convenience. Its use will be made clear in the following sections by application to various structures, and it will finally be used in Chapter 9 to derive a matrix method of calculating deformations.

8.1. Internal Virtual Work for Structural Members

It becomes convenient to express the internal virtual work $\int_V \sigma^{**}\epsilon^* \, dV$ for individual structural members in terms of their stress resultants and characteristic distortions. We shall carry this out for various types of distortions acting in a general flexural member.

1. *Axial extension ϵ^*.* The virtual stresses due to a virtual axial force n^{**} are

$$\sigma^{**} = \frac{n^{**}}{A}$$

Therefore [see Figs. 8.1(a) and (b)],

$$\text{IVW} = \int_V \sigma\epsilon \, dV = \int_L \left(\int_A \frac{n^{**}}{A} \, dA \right)(\epsilon^* \, ds) = \int_L n^{**}\epsilon^* \, ds \qquad (8.2)$$

n^{**} is here the virtual axial load; ϵ^* is the real axial strain.

2. *Flexural curvature Φ^*.* The virtual internal moment at any section results from the integration $m^{**} = \int_A \sigma^{**}y \, dA$ [Fig. 8.1(c)], and the curvature is, according to the plane section assumption [Fig. 8.1(d)], $\Phi^* = \epsilon^*/y$. By multiplying top and bottom of the internal virtual work integrand by y and rearranging, we write, under substitution of the above expressions,

$$\text{IVW} = \int_V \sigma^{**}\epsilon^* \, dV = \int_L \left(\int_A \sigma^{**}y \, dA \cdot \frac{\epsilon^*}{y} \right) ds = \int_L m^{**}\Phi^* \, ds$$

$$(8.3)$$

The last form, in which m^{**} is the virtual moment and Φ^* is the real curvature, is a convenient form of expressing the internal virtual work due to bending.

It should be observed that in this derivation the stress resultant $m^{**} = \int_A \sigma^{**}y \, dA$ was not based on any specific stress distribution (such as, for instance, the linear stress variation in elastic bending). Any stress field can be visualized that is in equilibrium with the moment m^{**}. The result embodied in Eq. (8.3) can be explained very simply by considering Fig. 8.1(e), which shows a beam segment of length ds, subjected to virtual moments m^{**} and subjected to real curvature Φ^*, which causes a real inclination of the ends of the segment $\Phi^* \, ds$ relative to each other. Thus, the virtual work done in the

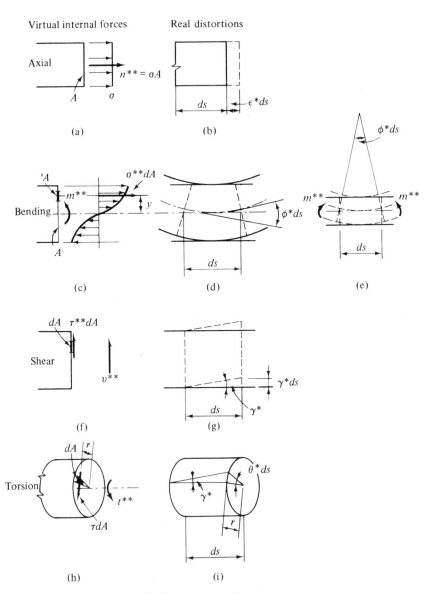

Fig. 8.1 Internal virtual work.

single element, obtained by multiplying the virtual moment by the real angle through which it rotates, is $m^{**}\Phi^* \, ds$, and the virtual work of the entire structure due to bending is $\int_L m^{**}\Phi^* \, ds$. The other results of this section can be interpreted in a similar fashion.

3. *Shear distortion γ^*.* The virtual internal shear force $v^{**} = \int_A \tau^{**} \, dA$ [Fig. 8.1(f)], and the corresponding transverse displacement per unit length under the assumption of uniform shear strain is γ^* [Fig. 8.1(g)]. Accordingly, over the length L of the member,

$$\text{IVW} = \int_V \tau^{**}\gamma^* \, dV = \int_L \left(\int_A \tau^{**} \, dA \, \gamma^* \right) ds = \int_L v^{**}\gamma^* \, ds \qquad (8.4)$$

v^{**} is here the virtual shear force, and γ^* is the real shear strain.

4. *Torsional twist θ^*.* The virtual torsional moment results from the integration of torsional shear stresses times their lever arms r [Fig. 8.1(h)], $t^{**} = \int_A \tau^{**} r \, dA$, and the usual strength of materials assumption that radii remain straight during twisting leads to the relations between shear strain γ^* and unit angle of twist θ^* [Fig. 8.1(i)]:

$$\gamma^* = \theta^* r$$

By proceeding in a fashion similar to that used in deriving the work in bending, we obtain

$$\text{IVW} = \int_V \tau^{**}\gamma^* \, dV = \int_L \left(\int_A \tau^{**} r \, dA \cdot \frac{\gamma^*}{r} \right) ds = \int_L t^{**}\theta^* \, ds \qquad (8.5)$$

t^{**} is here the virtual twisting moment, and θ^* is the real unit angle of twist.

In members subject to various types of distortions, the appropriate contributions to the internal virtual work can be superposed. For a general member in which the distortions associated with all six internal forces are important, the internal virtual work is

$$\text{IVW} = \int_L [n^{**}\epsilon^* + v_s^{**}\gamma_s^* + v_w^{**}\gamma_w^* + t^{**}\theta + m_s^{**}\Phi_s^* + m_w^{**}\Phi_w^*] \, ds \qquad (8.6)$$

where v_s^{**}, γ_s^*, m_s^{**}, and Φ_s^* represent shear force, shear distortion, bending moment, and curvature with respect to the strong bending axis, and v_w^{**}, γ_w^*, m_w^{**}, and Φ_w^* the corresponding quantities with respect to the weak bending axis.

8.2. Virtual Work Equation for Structural Members

The virtual forces represented by lower-case letters in Eq. (8.6) are due to a conveniently chosen virtual load. For the purpose of finding a certain displacement, it is convenient to select a concentrated virtual unit force (or moment) at the point and in the direction of the desired displacement, so that the virtual work equation (8.1) becomes

EVW = IVW

$$1^{**} \cdot \Delta^* = \int_L [n^{**}\epsilon^* + v_s^{**}\gamma_s^* + v_w^{**}\gamma_w^* + t^{**}\theta^* + m_s^{**}\Phi_s^* + m_w^{**}\Phi_w^*] \, ds$$

$$(8.7)$$

The double-starred virtual internal forces are due to the virtual unit load, and must be determined from analysis. The single-starred real distortions are calculated from whatever physical cause is specified, such as temperature distortion, creep, elastic or inelastic loads, or specified support settlements.

If several different displacements of the structure due to a given cause are required, a separate virtual unit load and a corresponding separate analysis are required for each separate displacement. As will be seen in the following sections, not all the internal virtual work terms are usually required. In planar cases the torsional and w-subscripted distortions vanish; in flexural structures axial and shear distortions are often sufficiently small so that they can be neglected; and in ideal trusses only axial distortions occur. Because of the use of the virtual unit load, the method is often called the *unit-load method* or the *dummy-load method*. Its use will be illustrated by several examples.

The real distortions ϵ^*, Φ^*, γ^*, and θ^* in Eq. (8.7) can be due to any cause; if, however, they arise from the action of real internal forces due to applied loads, such as the axial force N^*, the bending moment M^*, the shear force V^*, or the torsional moment T^* in the elastic range of the member, then the corresponding internal distortions can be calculated by elementary mechanics of materials methods:

$$\epsilon^* = \frac{N^*}{AE}; \qquad \Phi^* = \frac{M^*}{EI}; \qquad \gamma^* = \frac{V^*}{kAG}; \qquad \theta^* = \frac{T^*}{GC} \qquad (8.8)$$

where AE, EI, kAG, and GC are the axial, flexural, shear, and torsional member stiffnesses. In this case, the virtual work equation (8.7) becomes (neglecting one bending and one shear term)

$$\Delta = \int_L \frac{N^*n^{**}}{AE} \, ds + \int_L \frac{M^*m^{**}}{EI} \, ds + \int_L \frac{V^*v^{**}}{kAG} \, ds + \int_L \frac{T^*t^{**}}{GC} \, ds$$

$$(8.9)$$

The single-starred upper-case forces denote the real member forces due to the real applied loads, and the double-starred lower-case forces denote the virtual member forces due to the appropriately placed unit virtual load. Two different analyses are necessary to determine these forces, one for the real, the other for the virtual loading condition. It will also be necessary to express the internal member forces as functions of position along the members in order to perform the integrations indicated in Eq. (8.9).

In the following sections this method will be used to evaluate the deformations of various types of structures.

8.3. Application to Plane Flexural Structures

We shall now consider some cases of deformation of plane beams, frames, and arches in which flexural distortions are of prime importance. Neglecting then the axial and shear distortions, the virtual work equation (8.7) becomes

$$\Delta = \int_L m^{**}\Phi^* \, ds \qquad (8.10)$$

and, in case the real curvature is due to elastic bending, Eq. (8.9) becomes

$$\Delta = \int_L \frac{M^*m^{**}}{EI} \, ds \qquad (8.11)$$

In each of the following examples, the real curvature is calculated and expressed as a function of position along the structure. Then, for each displacement desired, a unit load is applied and a separate analysis performed to determine the virtual moments m^{**} as a function of position, whereby it is necessary to use the same coordinate system for the real and the virtual quantities. Finally, the indicated integrations are performed to obtain the required deformations.

> **Example 8.1:** For the simply supported, prismatic beam under the parabolically varying load of part (a) of the figure calculate
> a. The rotation at end *B*.
> b. An expression for the elastic curve, and the maximum deflection.
> Neglect effect of shear distortions.
>
> **Solution:** Since only the bending distortions will be considered, only the internal virtual work $\int_L \Phi^* m^{**} ds$ enters into our calculations. We recall that the variation of real moments due to the given load was already obtained in Ex. 3.2 as
>
> $$M^* = \frac{w_0 L^2}{12}\left[\left(\frac{\xi}{L}\right) - \left(\frac{\xi}{L}\right)^4\right]$$
>
> We substitute the variables ξ here to save the variable x for Part b of this problem.
> The real curvature, whose variation is shown in part (b) of the figure is,

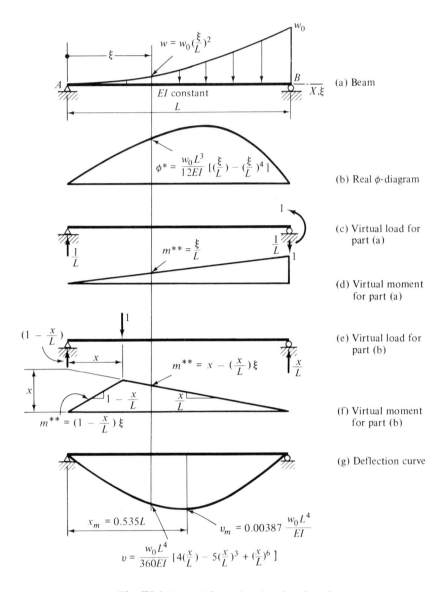

(a) Beam

(b) Real ϕ-diagram

(c) Virtual load for part (a)

(d) Virtual moment for part (a)

(e) Virtual load for part (b)

(f) Virtual moment for part (b)

(g) Deflection curve

Fig. E8.1 Beam deformations by virtual work.

for the constant bending stiffness EI of the beam,

$$\Phi^* = \frac{M^*}{EI} = \frac{w_0 L^2}{12EI}\left[\left(\frac{\xi}{L}\right) - \left(\frac{\xi}{L}\right)^4\right]$$

This real curvature Φ is the same for both parts of the problem; the virtual load, and the resulting internal virtual forces, will be different for the two parts:

a. To determine the end rotation at B, we visualize a virtual unit moment, as in part (c), applied at B, and draw the virtual moment diagram due to this moment in part (d). These virtual moments are given by

$$m^{**} = \frac{\xi}{L}$$

where we make sure to calculate the real curvature and the virtual moments in terms of the same coordinate ξ.

Equating the external and internal virtual work due to the real and virtual loads, we obtain

$$\text{EVW} = \text{IVW}$$

$$1 \cdot \theta_B = \int_{\xi=0}^{L} \frac{w_0 L^2}{12 EI} \left[\left(\frac{\xi}{L} \right) - \left(\frac{\xi}{L} \right)^4 \right] \left[\frac{\xi}{L} \right] d\xi$$

from which the desired rotation is

$$\theta_B = \frac{1}{72} \frac{w_0 L^4}{EI}$$

b. Corresponding to the desired deflection at an arbitrary point x, we visualize a virtual unit load, as in part (e), draw the virtual moment diagram in part (f), and write the moment equations, using the same position coordinate as before:

$$m^{**} = \frac{L-x}{L} \xi = \left[1 - \left(\frac{x}{L} \right) \right] \xi \qquad \text{for } 0 \leq \xi \leq x$$

$$m^{**} = x - \left(\frac{x}{L} \right) \xi \qquad\qquad \text{for } x \leq \xi \leq L$$

Note that the discontinuity at the point of applied unit load requires the writing of the virtual moment m in two distinct expressions, and, accordingly, the integration for the internal virtual work will require two parts, from $\xi = 0$ to x, and from x to L. Again equating external and internal virtual work, using the same real curvature as before, we obtain

$$\text{EVW} = \text{IVW}$$

$$1^{**} \cdot v^* = \int_{\xi=0}^{x} \frac{w_0 L^2}{12 EI} \left[\left(\frac{\xi}{L} \right) - \left(\frac{\xi}{L} \right)^4 \right] \left[1 - \left(\frac{x}{L} \right) \right] \xi \, d\xi$$

$$+ \int_{\xi=x}^{L} \frac{w_0 L^2}{12 EI} \left[\left(\frac{\xi}{L} \right) - \left(\frac{\xi}{L} \right)^4 \right] \left[x - \left(\frac{x}{L} \right) \xi \right] d\xi$$

Performing the integration and simplifying, we obtain the deflection at any point x as

$$v = \frac{w_0 L^4}{360EI}\left[4\left(\frac{x}{L}\right) - 5\left(\frac{x}{L}\right)^3 + \left(\frac{x}{L}\right)^6\right]$$

Positive values indicate deflections in the sense of the virtual unit load, that is, downward. The deflection curve is plotted in part (g).

The maximum deflection can be obtained by equating the derivative of the deflection function, with respect to (x/L), to zero:

$$\frac{dv}{d(x/L)} = \frac{w_0 L^4}{360EI} \cdot \frac{d}{d(x/L)}\left[4\left(\frac{x}{L}\right) - 5\left(\frac{x}{L}\right)^3 + \left(\frac{x}{L}\right)^6\right] = 0$$

or

$$4 - 15\left(\frac{x_m}{L}\right)^2 + 6\left(\frac{x_m}{L}\right)^5 = 0$$

By trial and error, we find $(x_m/L) = .535$. Substituting this value for the location of the point of maximum deflection into the general deflection equation, we obtain

$$v_{max} = \frac{w_0 L^4}{360EI}[2.14 - .77 + .02] = .00387\frac{w_0 L^4}{EI}$$

Example 8.2: Find the horizontal sway of point C of the unsymmetric frame shown in part (a) the figure due to
a. The uniform vertical load shown.
b. A uniform temperature gradient ΔT on beam BC only; this beam is of depth d and has a coefficient of thermal expansion α.

Solution:
a. The statically determinate structure is analyzed and the real moments M^* are calculated and divided by the flexural stiffness EI to obtain the real curvatures; these are drawn alongside the members as shown.

Similarly, a virtual unit load is applied horizontally at joint C, another analysis is performed, and the virtual moments m^{**} are also drawn in the figure. Choosing appropriate origins for each member, the expressions for real curvatures and virtual moments are written. We note that had the origin for member BC been taken at any point other than C, the virtual moment equation would contain two terms, thus doubling the labor of integration.

The virtual moments on member AB will not cause internal virtual work because they do not rotate through any real curvature; we therefore need consider only the virtual work in member BC and equate this to the external virtual work:

$$1 \cdot \Delta = \int_{x=0}^{2L}\left[\frac{w}{EI}\left(Lx - \frac{x^2}{2}\right)\left(\frac{x}{2}\right)\right]dx = \frac{wL^4}{3EI}$$

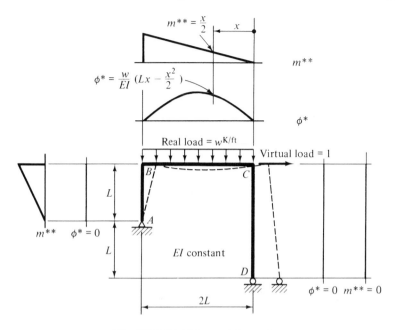

Fig. E8.2 Rigid-frame deformation.

This concludes part (a) of the problem. The deformed shape of the frame is also shown by the dashed line in the figure.

b. As discussed before (see Ex. 7.5), the specified temperature gradient causes a real constant curvature along BC of

$$\Phi^* = \frac{\alpha \, \Delta T}{d}$$

In the other members, no real curvature occurs. The virtual moments are identical to those of Part a of the problem; thus, the virtual work equation becomes

$$1 \cdot \Delta = \int_{x=0}^{2L} \left(\frac{\alpha \, \Delta T}{d}\right)\left(\frac{x}{2}\right) dx = \underline{\frac{\alpha \, \Delta T \cdot L^2}{d}}$$

Note that this deflection is independent of the structure stiffness; it depends only on the geometry imposed by the specified curvature.

Example 8.3: Calculate the vertical deflection at the crown of the three-hinged semicircular arch shown in part (a) of the figure due to its dead weight, of intensity w kips/ft. Consider only flexural distortions.

Solution: This structure and loading were already considered in Ex. 3.3, where the moment was calculated in terms of the variable angle θ; dividing this

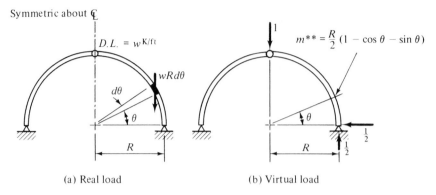

(a) Real load   (b) Virtual load

Fig. E8.3 Deformation of three-hinged arch.

moment by the constant flexural stiffness *EI* leads to the real curvature:

$$\Phi^* = \frac{wR^2}{EI}\left[\frac{\pi}{2}(1 - \sin\theta) + \left(\theta - \frac{\pi}{2}\right)\cos\theta\right]$$

The virtual unit load, at the point and in the direction of the desired deflection, is shown in part (b). It causes the virtual moment

$$m^{**} = \frac{R}{2}[1 - \cos\theta - \sin\theta], \qquad 0 \le \theta \le \frac{\pi}{2}$$

This expression is valid only for $0 \le \theta \le \pi/2$ because of the discontinuity at the crown. However, since both real curvature and virtual moment are symmetric about $\theta = \pi/2$, we can compute the internal virtual work over half the arch and double it to find the total amount; accordingly, the virtual work equation becomes, with $ds = R \cdot d\theta$,

$$1 \cdot \Delta = 2 \int_{\theta=0}^{\pi/2} \frac{wR^2}{EI}\left[\frac{\pi}{2}(1 - \sin\theta) + \left(\theta - \frac{\pi}{2}\right)\cos\theta\right]\frac{R}{2}$$
$$\cdot [1 - \cos\theta - \sin\theta] R \, d\theta$$
$$= \frac{wR^4}{16EI}[7\pi^2 - 6\pi - 44] = .375\frac{wR^4}{EI}$$

8.4. Application to Three-Dimensional Structures

In space structures consisting of flexural members, torsional moments and consequent twisting must be considered. In planar structures loaded normal to their plane, bending about one axis also occurs; however, in more complicated structures bending about both principal cross-sectional axes may take place, so that, again neglecting axial and shear distortions, the virtual work equation for deformations of elastic structures due to loads becomes

$$\Delta = \int_L \frac{M_s^* m_s^{**}}{(EI)_s} \, ds + \int_L \frac{M_w^* m_w^{**}}{(EI)_w} \, ds + \int_L \frac{T^* t^{**}}{GC} \, ds \qquad (8.12)$$

where the subscripts s and w refer to the strong and weak bending axes of the cross section.

As in earlier applications, the deformations due to axial elongations and due to shear distortions in the principal planes are neglected in Eq. (8.12), but can be included without particular difficulties in the rare cases when they become important.

The torsional stiffness GC, defined as the ratio of the applied torque T to the resulting unit angle of twist θ, deserves a further word. In texts on elementary mechanics of deformable bodies the case of torsion members of circular cross section is covered; in this case, the torsional stiffness $T/\theta = GJ$, where J is the polar moment of inertia of the cross section, and G is the shear modulus of the material.

Cases of more general cross section are analyzed by St. Venant's theory of torsion, which forms a part of the theory of elasticity.* From this approach it is found that in all cases the torsional stiffness can be represented in the form $T/\theta = GC$, where C represents a torsional stiffness factor that depends only on the properties of the cross section. Table 8.1, for instance, gives values of C for rectangular elastic cross sections of various proportions.

TABLE 8.1. Torsional Stiffness Factor for
Rectangular Sections; $C = kb^3 d$

$\frac{d}{b}$	k
1.0	.141
1.2	.166
1.5	.196
2.0	.229
2.5	.249
3.0	.263
4.0	.281
5.0	.291
10.0	.312
∞	.333

In the case of thin-walled closed sections, such as box beams and cell-like structures widely used in aerospace design, the torsional stiffness factor is

*S. P. Timoshenko and J. N. Goodier, *Theory of Elasticity*, 3rd ed., McGraw-Hill Book Company, New York, 1970.

given by the expression

$$C = \frac{4A^2}{\oint \frac{ds}{t}} \tag{8.13}$$

where A is the area bounded by the centerline of the cell walls, t is the wall thickness, and the integration is performed around the cell perimeter.

Example 8.4: Calculate the vertical deflection of the free end of the semi-circular cantilever bow girder due to its dead weight w kips/ft. The cross section is shown in Sec. A-A of the figure. Consider bending and torsional distortions. Poisson's ratio μ of the material is .3.

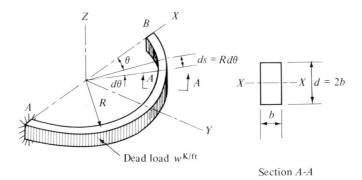

Fig. E8.4 Circular bow girder.

Solution: As a preliminary step, we consider the cross-sectional stiffness of the section. Bending occurs about the horizontal axis X-X, with respect to which the moment of inertia $I = \frac{1}{12}bd^3 = \frac{1}{12}b(2b)^3$. The flexural stiffness is therefore $EI = .667Eb^4$.

The torsional stiffness factor can be obtained, according to Table 8.1, from the formula $C = kb^3d$, where b is the shorter side of the section, d the longer, and k is a constant, which according to Table 8.1 for $d/b = 2$ has the value $k = .229$. The torsional stiffness factor is thus $C = .229b^3(2b) = .458b^4$. The shear modulus $G = E/[2(1 + \mu)] = .385E$, so the torsional stiffness becomes $GC = .176Eb^4$. Because we shall later on factor out the stiffness EI, it is convenient to express the torsional as a ratio of the flexural stiffness: $GC = (.176/.667)EI = .265EI$. Considering flexural and torsional distortions, the virtual work equation is in this case, with $ds = R\,d\theta$,

$$1 \cdot \Delta = \int_{\theta=0}^{\pi} \frac{M^* m^{**}}{EI} R\,d\theta + \int_{\theta=0}^{\pi} \frac{T^* t^{**}}{GC} R\,d\theta$$

where, from Ex. 3.5, the real moments are

$$M = wR^2(1 - \cos \theta)$$
$$T = -wR^2(\theta - \sin \theta)$$

The virtual moments due to a unit virtual downward load at point *B* are (using the same coordinates and convention as in Ex. 3.5)

$$m^{**} = R \sin \theta$$
$$t^{**} = -R(1 - \cos \theta)$$

The virtual work equation now becomes

$$\Delta = \frac{wR^2}{EI} \int_0^\pi (1 - \cos \theta)(R \sin \theta) R \, d\theta$$

$$+ \frac{wR^2}{GC} \int_0^\pi (\theta - \sin \theta) R(1 - \cos \theta) R \, d\theta$$

$$= \frac{wR^4}{EI} \int_0^\pi \Big[(\sin \theta - \sin \theta \cos \theta)$$

$$+ \frac{1}{.265} (\theta - \theta \cos \theta - \sin \theta + \sin \theta \cos \theta) \Big] d\theta$$

In performing the indicated integrations, Wallis's formula for definite integrals is helpful:

$$\int_0^\pi \sin^2 \theta \, d\theta = \frac{\pi}{2}; \qquad \int_0^\pi \cos^2 \theta \, d\theta = \frac{\pi}{2}; \qquad \int_0^\pi \sin \theta \cos \theta \, d\theta = 0$$

Completing the integration, we obtain

$$\Delta = \frac{wR^4}{EI} [2.00 + 18.56] = 20.56 \frac{wR^4}{EI}$$

The second term within the brackets, representing the effect of torsion, is vastly greater than the first term, which contains the flexural effects; it is obvious that in such cases torsion must be considered.

Example 8.5: Compute the vertical deflection at the free end *A* of the helicoidal cantilever girder due to the load *P* applied at point *A*. The cross section is shown. Poisson's ratio $\mu = .3$.

Solution: In this space structure, bending about two principal axes as well as twisting must be considered, so the virtual work equation (8.12) is applicable. The cross-sectional properties are identical to those of Ex. 8.4, so $GC = .265EI_s$, and $EI_w = \frac{1}{12}b^3d = .25EI_s$.

The real internal bending and twisting moments due to the load *P* were

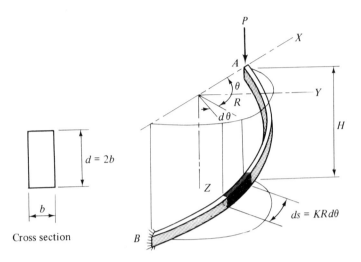

Fig. E8.5 Helicoidal girder.

computed in Ex. 3.6:

$$M_s^* = PR \sin \theta; \qquad M_w^* = -\frac{PH}{\pi K}(1 - \cos \theta); \qquad T^* = -\frac{PR}{K}(1 - \cos \theta)$$

where $K = [1 + (H/\pi R)^2]^{1/2}$ defines the steepness of the structure. The virtual unit load for the desired deflection is identical to the real load, except that $P = 1$; the resulting virtual moments are therefore identical to the real moments, with $P = 1$:

$$m_s^{**} = R \sin \theta; \qquad m_w^{**} = -\frac{H}{\pi K}(1 - \cos \theta); \qquad t^{**} = -\frac{R}{K}(1 - \cos \theta)$$

The element length for this inclined member is, referring to Ex. 3.6,

$$ds = \sqrt{(dx)^2 + (dy)^2 + (dz)^2} = \sqrt{(R \sin \theta)^2 + (R \cos \theta)^2 + \left(\frac{H}{\pi}\right)^2} \cdot d\theta$$

$$= \sqrt{1 + \left(\frac{H}{\pi R}\right)^2} \cdot R \, d\theta \equiv KR \, d\theta$$

The virtual work equation now becomes

$$\Delta = \int_{\theta=0}^{\pi} \left[\left(\frac{PR}{EI_s} \sin \theta\right)(R \sin \theta) + \frac{PH}{\pi K EI_w}(1 - \cos \theta)\frac{H}{\pi K}(1 - \cos \theta)\right.$$

$$\left. + \frac{PR}{KGC}(1 - \cos \theta)\frac{R}{K}(1 - \cos \theta)\right] KR \, d\theta$$

$$= \frac{PR^3}{EI_s} \int_0^{\pi} \left[K \sin^2 \theta + \frac{(H/R)^2}{.25\pi^2 K}(1 - \cos \theta)^2 + \frac{1}{.265K}(1 - \cos \theta)^2\right] d\theta$$

$$= \frac{PR^3}{EI_s}\left[1.57K + \frac{1}{K}\left(1.92\left(\frac{H}{R}\right)^2 + 17.80\right)\right].$$

The first term represents the effects of strong-axis bending, the second those of weak-axis bending, and the last those of torsion. It is interesting to observe that with increasing steepness of the helix, the torsional effects lose their importance in comparison with the bending effects. Trends of this type should be recognized by the designer for a complete understanding of the structure.

8.5. Application to Trusses

The deformations of ideal trusses are due only to member elongations. No bending or shear takes place in the members. Furthermore, when all loads are applied at the joints, each member will be subject to a constant axial force or strain along its length. If, as is usually the case, the members are of constant cross section A, then the virtual internal work in one member, using the appropriate virtual load, becomes

$$\text{IVW} = s^{**}\epsilon^* \int_L ds = s^{**}\epsilon^* L = s^{**}\, \Delta L^*$$

where ΔL^* is the real member elongation, and s^{**} is the virtual member force.

For all m members of the truss

$$\text{IVW} = \sum_m s^{**}\, \Delta L^*$$

where $\Delta L^* = \epsilon^* L =$ real elongation of each member. Due to a unit virtual load, the virtual work equation becomes

$$\Delta^* = \sum_m s^{**}\, \Delta L^*$$

and, in case the real elongations ΔK^* are caused by real member forces S^* acting in the elastic truss, Eq. (7.9) becomes

$$\Delta = \sum S^* s^{**} \frac{L}{AE}$$

The following examples illustrate an efficient longhand computational scheme for solving truss deflections by virtual work.

Example 8.6: For the simple truss shown, determine the vertical deflection of joint D due to
a. A horizontal load of 10 kips at joint B.
b. A difference of temperature of the top chord only, from 100°F in summer to -20°F in winter; the coefficient of thermal expansion is 6.5×10^{-6} in./in./°F.

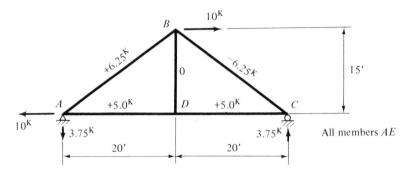

(a) Real load and forces, loading condition 1

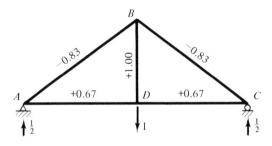

(b) Virtual load and forces

Fig. E8.6 Truss deflections.

Solution: To determine the real member elongations, it is first necessary to analyze the structure. Part (a) of the figure shows the real loading condition 1, as well as the member forces resulting from analysis. With these forces and the member properties, the real elongations can be calculated.

Loading condition 2, due to temperature change, leads to real member elongations that can be computed as $\Delta L^* = \alpha \Delta TL$, member by member. The virtual member forces are obtained by analysis due to the virtual unit load applied vertically at joint D, as shown in part (b); these virtual forces are also shown.

We next compute the internal virtual work in the truss, member by member, for each loading condition. The suggested tabular procedure provides column 2 for the flexibility L/AE for each member; here, because the value of AE is identical for all members, this is factored out, as well as the multiplier 12 to take care of units. Column 3 contains the real member forces due to loading condition 1, and column 4 the real member elongations $\Delta L^* = S^*L/AE$ due to loading condition 1. Similarly, column 5 contains the real elongations due to the total temperature change of 120°F in the two top chord members specified by loading condition 2. The virtual member forces s already calculated in part (b) are stored in column 6, and the virtual internal work

TABLE E8.6

1	2	3	4	5	6	7	8
MEMBER	$\left(\frac{L}{AE}\right) \times \frac{AE}{12}$, (IN.)	S_1^*, kips, LD. COND. 1	$\Delta L^* \cdot \frac{AE}{12}$, LD. COND. 1 (IN.)	ΔL^*, LD. COND. 2 (IN.)	s^{**}, kips/kip	$s^{**} \Delta L^* \frac{AE}{12}$, LD. COND. 1	$s^{**} \Delta L^*$, LD. COND. 2
AB	25	+6.25	+156.25	+.234	−.83	−130.0	−.194
BC	25	−6.25	−156.25	+.234	−.83	+130.0	−.194
AD	20	+5.00	+100.00	0	+.67	+66.7	0
DC	20	+5.00	+100.00	0	+.67	+66.7	0
DB	15	0	0	0	+1.00	0	0

$$\Delta_1 = +133.3 \times \frac{12}{AE}$$

$$= \frac{1600}{AE} \text{ in.}$$

$$-.388 \text{ in.} = \Delta_2$$

$\Delta L*S**$ for each member due to two loading conditions is contained in columns 7 and 8. Adding the individual member contributions in each of columns 7 and 8 sums the entire virtual work stored in the truss. By the theorem of virtual forces, this sum is equal to the desired deflections.

PROBLEMS

In all problems, document your analyses for real and virtual loads with separate calculations and figures, carefully labeling each one.

1. Due to the cranked-in end moment M, calculate:
   a. Rotation of end A.
   b. Rotation of end B.

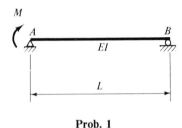

Prob. 1

2. For the beam and load of Prob. 1, compute the elastic curve. Select the origin carefully.

3. The elastic beam contains a spring as shown.
   a. Compute the deflection under the load P.
   b. Compute the end rotation at D.

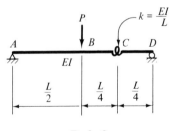

Prob. 3

4. The articulated prismatic beam is subjected to a uniform load w. Compute the rotation of the hinge at point B. (*Hint:* Consider a virtual load consisting of two equal and opposite unit moments applied at the hinge.)

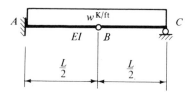

Prob. 4

5. a. Compute the deflection at C due to a unit load at B.
   b. Compute the deflection of B due to a unit load at C.
   c. Draw conclusions from the results of Parts a and b of the problem.

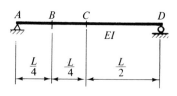

Prob. 5

6. Calculate the deflections $v(x)$ of the nonprismatic beam due to the cranked-in end moments at points A and B.

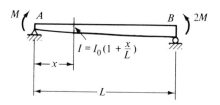

Prob. 6

7. The simply supported steel beam is prismatic, of moment of inertia I and area A. The ratio of modulus of rigidity G to modulus of elasticity E is 2: 5.
   a. Compute the midspan deflection, considering flexural and shear distortion.
   b. Compute the ratio of the deflection due to shear, Δ_s, and that due to bending, Δ_B, in terms of the ratio of beam properties $EI/(G \cdot AL^2)$.
   c. For a W 16 $\times$ 36 steel beam ($I = 446.3$ in.^{4}; $A = 10.59$ in.^{2}) plot Δ_s/Δ_B as a function of the length of the beam, and draw engineering conclusions.

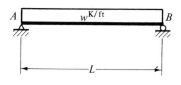

Prob. 7

8. Calculate the horizontal displacement and the rotation of point C of the rigid frame under the load shown. Consider bending distortions only. Flexural stiffnesses are as shown.

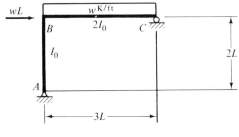

Prob. 8

9. Frame ABC is made of reinforced concrete, the column AB is 12×12 in. in cross section, and the beam BC is 12×16 in. (16-in. dimension vertical). E of concrete is 2×10^6 psi. Find the deflection of point C due to the load shown.

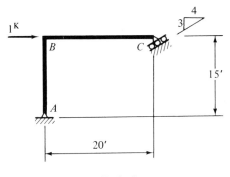

Prob. 9

10. Given:

$$E = 30 \times 10^6 \text{ ksi}$$

$$I = 200 \text{ in.}^4$$

Find:

a. The horizontal deflection of point C.
b. The vertical deflection of point E.

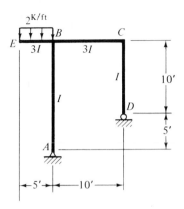

Prob. 10

11. Calculate the deflection of point D of the structure due to the loading shown. Inclined members AB and CD are of W 16 × 36 beams: $I = 446\ in^4$. Beam BC is of W 16 × 36 coverplated with $\frac{1}{4}$ × 8 in. plates top and bottom. $E = 30{,}000$ ksi.

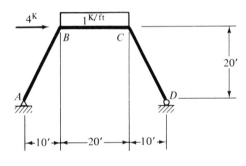

Prob. 11

12. Compute the horizontal and vertical displacement and the rotation of point B due to the applied load P.

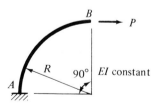

Prob. 12

13. Calculate the horizontal deflection of point *B* due to the cranked-in moment
M. The arch is of constant stiffness.

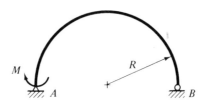

Prob. 13

14. a. Determine the vertical deflection of point *B*.
   b. Determine the rotation of the hinge at *B*. (*Hint:* Apply a virtual load at *B*
   consisting of equal and opposite unit moments.) Consider only effects of
   flexural distortions.

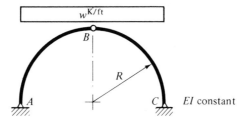

Prob. 14

15. The three-hinged arch of Prob. 14 is made of a W 14 × 87 steel beam of
$I = 966.9$ in.^{4}, $A_{\text{web}} = 5.88$ in.^{2}, $E = 30,000$ ksi, and $G = 12,000$ ksi. Compute
the portion of the vertical deflection of the crown *B* due to axial distortions of
the member, compare it to that due to flexural distortions already computed
in Prob. 14a, and discuss these results from an engineering viewpoint.

16. The bent cantilever girder lies in the horizontal plane. It is of constant cross
section, 2.5 in. wide by 4 in. deep. Material is steel.
   a. Calculate the vertical deflection at *C* due to the dead weight of the structure.

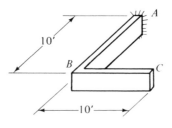

Prob. 16

b. Due to sunshine, the top fiber heats more than the bottom fibers by 100°F. Calculate the vertical deflection of point C due to this effect.

17. The semicircular bow girder on three supports is of constant cross section, $A = 10$ in.^{2}; I (about horizontal axis) $= 100$ in.^{4}, $C = 50$ in.^{4}, and $E = 2G$. Compute the twist of section B due to the dead weight W of the structure.

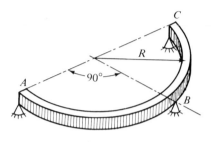

Prob. 17

18. For the helicoidal girder discussed in Ex. 8.5, compute
a. The vertical deflection at point A due to a unit vertical load at any point θ.
b. By integration of the effects of the loads of Part a of the problem, determine the vertical deflection of point A due to the dead load of the structure, of intensity w kips/ft of girder.

19. Compute the horizontal and vertical components of displacement of point B. Timber truss: All areas 20 in.^{2}. $E = 2,000$ ksi.

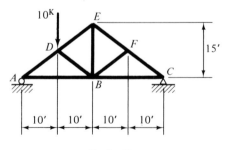

Prob. 19

20. The truss shown is of steel ($E = 30,000$ ksi). Chord members have a cross-sectional area of 6 in.^{2}, web members of 3 in.^{2}. Find the horizontal and vertical deflections of point A due to the load shown.

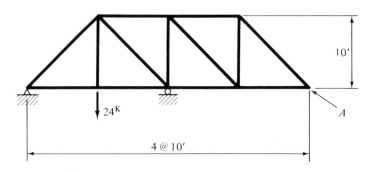

Prob. 20

21. Find the horizontal and vertical components of displacement of the crown *B* of the arch. All members have identical area *A* and stiffness *E*.

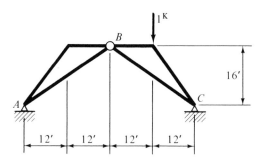

Prob. 21

CHAPTER 9

Flexibility Coefficients and the Matrix Calculation of Deformations

The concept of *flexibility* of a structure is an important building block in structural theory. In the following sections the term will be defined for both individual elements and for structures, and the theorem of virtual forces will be expressed in matrix form and used for the calculation of structural deformations.

9.1. Nodal Numbering

As a framework for efficient bookkeeping of our calculations, we introduce the concept of *nodes*. These are numbered generalized coordinates attached to the structure in accordance with the quantities to be defined. For each force and each deformation to be considered, such a node must be introduced and numbered.

For instance, in the beam shown in Fig. 9.1 we wish to determine the rotations at both member ends due to cranked-in end moments. In this case, the two numbered nodes shown allow us to label clearly the moments as well as the rotations in their positive sense.

Similarly, in the propped cantilever beam shown in Fig. 9.2, reactions at the supported points A and C, as well as loads and resulting displacement at the intermediate point B, are to be labeled. Accordingly, we introduce the nodal numbering system shown, which allows us to designate clearly the

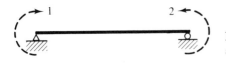

Fig. 9.1 Nodes for beam end moments and rotations.

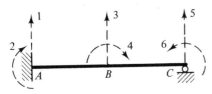

Fig. 9.2 Nodes for propped cantilever beam.

forces $\begin{Bmatrix} X_1 \\ \vdots \\ X_6 \end{Bmatrix}$ and deformations $\begin{Bmatrix} \Delta_1 \\ \vdots \\ \Delta_6 \end{Bmatrix}$ at the involved points in their positive

sense. It is interesting to observe that for any node, either the nodal force or the nodal displacement is unknown, but never both. So, for the nodes indicated on the beam in Fig. 9.2, nodes 1, 2, and 5 contain unknown reactive forces but known displacements (equal to zero), and nodes 3, 4, and 6 contain known forces (the loads) but unknown displacements. This fact will become clearer in the sections dealing with matrix analysis of statically indeterminate structures.

9.2. Virtual Work Equation in Matrix Form

We desire the displacement Δ_{ij} at node i due to an applied load at node j. The theorem of virtual forces, Eq. (8.1), will then appear as

$$1^{**} \cdot \Delta_{ij}^* = \int_V \sigma(x)_i^{**} \epsilon(x)_j^* \, dV \tag{9.1}$$

where $\sigma(x)_i$ denotes the internal virtual forces due to a virtual unit load at node i, and $\epsilon(x)_j$ denotes the real internal distortions due to the real load at node j. Both $\sigma(x)$ and $\epsilon(x)$ may be functions of position.

The subscripts of Eq. (9.1) indicate that it may be considered one element of a matrix. Indeed, if we desire displacements at several nodes i due to loads at several nodes j, we can compute all of these quantities in matrix form:

$$
\begin{bmatrix}
\Delta_{11} & \Delta_{12} & \cdots & \Delta_{1j} & \cdots \\
\Delta_{21} & \Delta_{22} & \cdots & \Delta_{2j} & \cdots \\
\cdot & & & & \\
\cdot & & & & \\
\Delta_{i1} & \Delta_{i2} & \cdots & \Delta_{ij} & \cdots \\
\cdot & & & & \\
\cdot & & & &
\end{bmatrix}
=
\begin{bmatrix}
\int_V \sigma_1 \epsilon_1 \, dV & \int_V \sigma_1 \epsilon_2 \, dV & \cdots & \int_V \sigma_1 \epsilon_j \, dV & \cdots \\
\int_V \sigma_2 \epsilon_1 \, dV & \int_V \sigma_2 \epsilon_2 \, dV & \cdots & \int_V \sigma_2 \epsilon_j \, dV & \cdots \\
\cdot & \cdot & & \cdot & \\
\cdot & \cdot & & \cdot & \\
\int_V \sigma_i \epsilon_1 \, dV & \int_V \sigma_i \epsilon_2 \, dV & \cdots & \int_V \sigma_i \epsilon_j \, dV & \cdots \\
\cdot & \cdot & & \cdot & \\
\cdot & \cdot & & \cdot &
\end{bmatrix}
$$

The integral can, according to usual matrix notation, be taken outside the matrix, and the equation rewritten as

$$[\Delta] = \int_V [\sigma_1 \quad \sigma_2 \quad \cdots \quad \sigma_j \quad \cdots]^T [\epsilon_1 \quad \epsilon_2 \quad \cdots \quad \epsilon_i \quad \cdots]\, dV$$

or

$$[\Delta] = \int_V [\sigma(x)^{**}]^T [\epsilon(x)^*]\, dV \tag{9.2}$$

Flexibilities f_{ij} are the displacements Δ_{ij} at node i due to unit loads at node j; thus, Eq. (9.2) becomes

$$[f] = \int_V [\sigma(x)]^T [\epsilon(x)]\, dV \tag{9.3}$$

where $\epsilon(x)_j$ are the internal distortions due to a *unit* load at node j, and $\sigma(x)_i$ are the internal forces due to a unit load at node i.

The quantities $\epsilon(x)$ and $\sigma(x)$ can represent strain and stress, or distortions and corresponding stress resultants, such as the curvature $\Phi(x)$ and the moment $M(x)$, obtained by integration of the stresses over the beam cross section. For this case, for instance, Eq. (9.3) becomes

$$[f] = \int_L [M(x)]^T [\Phi(x)]\, dl = \int_L [M(x)]^T \left[\frac{M(x)}{EI}\right] dl \tag{9.4}$$

The matrix $[f]$ is called the *flexibility matrix*.

We note that now both the real curvature $\phi_j^* = M_j^*/EI$, and the virtual moments, M_i^{**}, are due to similar causes, namely, a unit load at nodes i and j, respectively; it becomes therefore much less important for calculation purposes (although not for understanding) to distinguish between real and virtual moments, and we shall at times delete the asterisk system of distinguishing between real and virtual forces.

Example 9.1: By use of Eq. (9.4), determine the flexibility matrix for the beam with the nodal numbering shown in part (a) of the figure.

Solution: The moments $M(x)_1$ and $M(x)_2$ are shown and their functions written in parts (b) and (c). They are inserted in the appropriate slots of the matrices of Eq. (9.4), and the indicated operations are performed:

$$[f] = \begin{bmatrix} f_{11} & f_{12} \\ f_{21} & f_{22} \end{bmatrix} = \int_L \left[\left(1 - \frac{x}{L}\right) \quad \left(\frac{x}{L}\right)\right]^T \cdot \frac{1}{EI}\left[\left(1 - \frac{x}{L}\right) \quad \left(\frac{x}{L}\right)\right] dx$$

$$= \frac{1}{EI} \int_L \begin{bmatrix} \left(1 - \dfrac{x}{L}\right) \\ \left(\dfrac{x}{L}\right) \end{bmatrix} \left[\left(1 - \frac{x}{L}\right) \quad \left(\frac{x}{L}\right)\right] dx$$

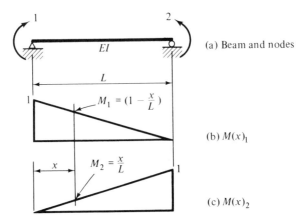

Fig. E9.1

$$= \frac{1}{EI} \begin{bmatrix} \int_L \left(1 - \frac{x}{L}\right)^2 dx & \int_L \left(1 - \frac{x}{L}\right)\left(\frac{x}{L}\right) dx \\ \int_L \left(\frac{x}{L}\right)\left(1 - \frac{x}{L}\right) dx & \int_L \left(\frac{x}{L}\right)^2 dx \end{bmatrix}$$

$$= \frac{L}{6EI} \begin{bmatrix} 2 & 1 \\ 1 & 2 \end{bmatrix}$$

9.3. Element Flexibility Matrix

The flexibility f_{ij} of a structural element k has been defined as the deformation of node i due to a unit force applied at node j. For instance, in the simple case of the straight prismatic element of Fig. 9.3, the element flexibility is

$$f_{11}^k = \frac{\Delta_1}{P_1} = \frac{L}{AE} \qquad (9.5)$$

Note that the flexibility enables us to write the relation between an applied force P_1 and the resulting deformation Δ_1:

$$\Delta_1 = f_{11}^k P_1$$

The rotational end flexibilities of the beam element k shown in Fig. 9.4(a) can be expressed in terms of the nodal system shown. There is a total of four flexibilities f_{ij}, $(i, j = 1, 2)$, that is, the rotations at ends $i = 1, 2$ each due to unit moments at ends $j = 1, 2$, as shown in Figs. 9.4(b) and (c). These four quantities are stored in the element flexibility matrix already found in Ex. 9.1:

$$[f^k] = \begin{bmatrix} f_{11} & f_{12} \\ f_{21} & f_{22} \end{bmatrix} = \frac{L}{6EI} \begin{bmatrix} 2 & 1 \\ 1 & 2 \end{bmatrix} \qquad (9.6)$$

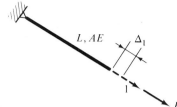

Fig. 9.3 Axial-element flexibility.

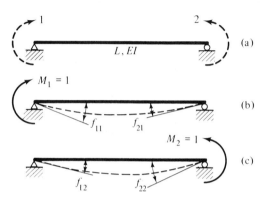

Fig. 9.4 Rotational-beam end flexibilities.

The first subscript of each term of this matrix, corresponding to the row number, designates the deformation node, and the second, corresponding to the column number, records the node of the unit force that causes it. That is, the first subscript denotes the effect, the second the cause.

In general, all relevant member nodes are specified, and the corresponding flexibilities defined accordingly. For instance, in a straight prismatic member shown in Fig. 9.5, which might be part of a space structure, the distortions due to bending about one axis, torsion, and axial force might be important. With the member restrained as shown, there are four nodal quantities, which are labeled, and the flexibilities computed and inserted in the $[f^k]$ matrix, which relates the displacements $\{\Delta\}$ to the corresponding forces $\{S\}$:

$$
\begin{Bmatrix} \Delta_1 \\ \Delta_2 \\ \Delta_3 \\ \Delta_4 \end{Bmatrix} =
\begin{bmatrix}
\dfrac{L}{3EI} & \dfrac{L}{6EI} & 0 & 0 \\
\dfrac{L}{6EI} & \dfrac{L}{3EI} & 0 & 0 \\
0 & 0 & \dfrac{L}{AE} & 0 \\
0 & 0 & 0 & \dfrac{L}{GC}
\end{bmatrix}
\begin{Bmatrix} S_1 \\ S_2 \\ S_3 \\ S_4 \end{Bmatrix}
\tag{9.7}
$$

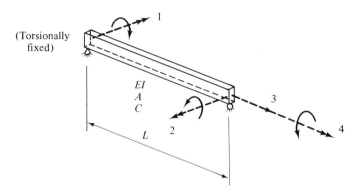

Fig. 9.5 Nodes for element flexibilities of Eq. (9.3).

The zero elements simply denote that the bending deformations are not affected by the torsional moment or axial force, and vice versa.

We note that all member flexibility matrices discussed so far are symmetric; that this must always be so for elastic structures can be deduced by the theorem of virtual forces. We consider, for instance, a flexibility f_{12} of a member whose deformations are due to bending. The flexibility is then calculated by Eq. (9.4):

$$f_{12} = \int_L m_1^{**} \frac{m_2^*}{EI} \, ds$$

The real moments m_2^* are due to a single real unit load at node 2; the virtual moments due to a virtual unit load at node 1 are m_1^{**}.

Similarly, the flexibility f_{21} of the same structure is due to a real unit load at node 1, which causes the real moments m_1^*, while the virtual unit load at node 2 causes the moments m_2^{**}:

$$f_{21} = \int_L m_2^{**} \frac{m_1^*}{EI} \, ds$$

It is obvious that $f_{12} = f_{21}$, or generally,

$$f_{ij} = f_{ji} \tag{9.8}$$

thus showing the symmetry of the $[f]$ matrix.

9.4. Structure Flexibility

The *structure flexibility* F_{ij} represents the displacement at node i of a structure due to an applied unit force at node j, all other forces equal to zero. In the statically determinate structure of Fig. 9.6, for instance, flexibility F_{12} repre-

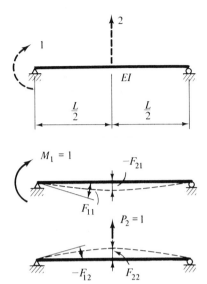

Fig. 9.6 Structure flexibilities.

sents the end rotation in the sense of node 1 due to a unit value of the midspan load in the sense of node 2. For the two numbered nodes of this structure, there are two displacements due to each of two unit loads, for a total of four flexibilities.

These flexibilities can be summarized in the *structure flexibility matrix* [*F*], defined as

$$\begin{Bmatrix} \theta_1 \\ \Delta_2 \end{Bmatrix} = \begin{bmatrix} F_{11} & F_{12} \\ F_{21} & F_{22} \end{bmatrix} \begin{Bmatrix} M_1 \\ P_2 \end{Bmatrix}$$

or
$$\{\Delta\} = [F]\{X\} \tag{9.9}$$

The individual structure flexibilities can be calculated by any of the methods of deformation analysis; for the case of the system of Fig. 9.6, the flexibility matrix is computed as

$$[F] = \frac{L}{48EI} \begin{bmatrix} 16 & -3L \\ -3L & L^2 \end{bmatrix}$$

The structure flexibility matrix establishes the relation between any combination of applied loads and resulting deformations. If the deformations due to several, say *l*, loading conditions are to be computed, the [Δ] and [*P*] matrices contain *l* columns, one for each loading condition, whereas the [*F*] matrix, which depends only on the properties of the structure, remains the same. It is only necessary to have specified all nodes that are required to define the loads and displacements involved.

Example 9.2: For the beam of Fig. 9.6, whose flexibility matrix is given in Sec. 9.3, calculate the displacements at nodes 1 and 2 due to the following three loading conditions:

a. A downward load of 10 kips at node 2.
b. A clockwise moment of $5L$ kip-ft at node 1.
c. A downward load of 10 kips and a clockwise moment of $5L$ kip-ft at node 1 applied jointly.

Solution: The $[P]$ and $[\Delta]$ matrices consist of two rows corresponding to the two nodes and three columns corresponding to the loading conditions:

$$[\Delta] = [F][X] = \frac{L}{48EI} \begin{bmatrix} 16 & -3L \\ -3L & L^2 \end{bmatrix} \begin{bmatrix} 0 & 5L & 5L \\ -10 & 0 & -10 \end{bmatrix}$$

$$= \frac{L}{48EI} \begin{bmatrix} 30L & 80L & 110L \\ -10L^2 & -15L^2 & -25L^2 \end{bmatrix}$$

9.5. Beam Deformations by the Matrix Method

The application of the virtual work matrix, Eq. (9.1), for the determination of displacements of an entire structure becomes unhandy because of the discontinuities in the internal forces and distortions. For such purposes, the method outlined in the following, which decomposes the structure into individual elements, each with continuous functions, becomes preferable.

To explain the matrix calculation of displacements and flexibilities, we shall use the example of a statically determinate beam. We begin by considering the single displacement Δ_{ij} at node i due to a single concentrated load P_j at node j, to be obtained by the method of virtual forces.

To compute the internal virtual work in the beam, it will be convenient to divide the beam into individual segments, each of constant stiffness EI and with linear variation of moments, so that the element flexibilities discussed in Sec. 9.3. can be used. We now have to calculate the internal moments acting on the end sections k of these elements. The unit virtual load is applied to node i, and the resulting internal virtual moments at the sections k will be summarized in the force transformation matrix $\{b_i^{**}\} = \{m_{ki}^{**}\}$ consisting of one single column.

The real moments at the sections k of the beam due to the applied load P_j will be called $\{M_{kj}^*\}$, and can be represented by means of the force transformation matrix $\{b_j^*\}$ due to a unit load at j, which also consists of a single column:

$$\{M_{kj}^*\} = \{b_j^*\}P_j$$

The subscripts of $\{b_i^{**}\}$ and $\{b_j^*\}$ designate the location of the loads that cause the virtual and real internal moments.

We now consider the internal virtual work done in one single prismatic

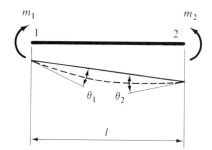

Fig. 9.7 Beam element with
end moments.

beam element of length l, subjected only to end moments, as shown in Fig.
9.7. This virtual work is given by the real end rotations θ_1^* and θ_2^* (measured
with respect to the chord), multiplied by the corresponding virtual end
moments m_1^{**} and m_2^{**}; this operation can be performed in matrix form:

$$\text{IVW} = m_1^{**}\theta_1^* + m_2^{**}\theta_2^* = [m_1^{**} \quad m_2^{**}]\begin{Bmatrix}\theta_1^*\\\theta_2^*\end{Bmatrix}$$

$$= \{m_i^{**}\}^T\{\theta^*\} = \{b_i^{**}\}^T\{\theta^*\}$$

We now find the real end rotations θ_1^* and θ_2^* from the real moments M_1^* and
M_2^* by recalling the rotational end flexibilities for such a beam element derived
in Eq. (9.6):

$$\{\theta^*\} = \begin{Bmatrix}\theta_1^*\\\theta_2^*\end{Bmatrix} = \frac{l}{6EI}\begin{bmatrix}2 & 1\\1 & 2\end{bmatrix}\begin{Bmatrix}M_1^*\\M_2^*\end{Bmatrix} = [f^k]\{M_j^*\} = [f^k]\{b_j^*\}P_j$$

The internal virtual work in one element is then

$$\text{IVW} = [m_1^{**} \quad m_2^{**}][f^k]\begin{Bmatrix}M_1^*\\M_2^*\end{Bmatrix}$$

$$= \{b_i^{**}\}^T[f^k]\{b_j^*\}P_j$$

For the virtual work in the entire beam, we add the contributions from
all n elements of the structure on the right-hand side, and equate it to the
external virtual work on the left-hand side:

$$1 \cdot \Delta_{ij} = \sum_{k=1}^{n} \{b_i^{**}\}^T[f]\{b_j^*\}P_j$$

$$= [[M_{1i}^1 M_{2i}^1][M_{1i}^2 M_{2i}^2]\cdots]\begin{bmatrix}\frac{l_1}{6EI_1}\begin{bmatrix}2 & 1\\1 & 2\end{bmatrix} & & 0 & \\ & & \frac{l_2}{6EI_2}\begin{bmatrix}2 & 1\\1 & 2\end{bmatrix} & \\ 0 & & & \ddots\end{bmatrix}\begin{Bmatrix}M_{1j}^1\\M_{2j}^1\\M_{1j}^2\\M_{2j}^2\\\vdots\end{Bmatrix}P_j$$

or

$$\underset{(1 \times 1)}{\Delta_{ij}} = \underset{(1 \times 2n)}{\{b_i^{**}\}^T} \quad \underset{(2n \times 2n)}{[f]} \quad \underset{(2n \times 1)}{\{b_j^*\}} \quad \underset{(1 \times 1)}{P_j} \qquad (9.10)$$

The quantities in parentheses denote the order of the matrices. The $[f]$ matrix is the composite of the n individual element flexibility matrices $[f^k]$:

$$[f] = \begin{bmatrix} [f^1] & 0 & \\ 0 & [f^2] & \\ & & \ddots \end{bmatrix}$$

Example 9.3: Calculate the rotation $\theta_A (= \Delta_{12})$ at node 1 due to a force of -5 kips at node 2 of the nonprismatic beam shown in part (a) of the figure.

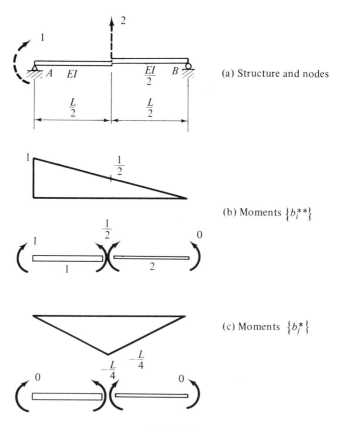

(a) Structure and nodes

(b) Moments $\{b_i^{**}\}$

(c) Moments $\{b_j^*\}$

Fig. E9.3

Solution: We attach a nodal system 1, 2 to the structure, as shown, corresponding to the desired displacement and to the real load. We also split the beam into two elements, 1 and 2, each of constant stiffness $(EI)_k$ and with linear variation of moments under all loading conditions, real and virtual.

We next calculate the $\{b_i^{**}\} = \{b_1^{**}\}$ matrix of virtual element end moments due to a virtual unit moment at node 1; it may be useful to draw the moment diagram for this, as shown in part (b).

$$\{b_1^{**}\} = \left\{ \begin{matrix} 1 \\ \frac{1}{2} \\ \frac{1}{2} \\ 0 \end{matrix} \right\} = \frac{1}{2} \left\{ \begin{matrix} 2 \\ 1 \\ 1 \\ 0 \end{matrix} \right\}$$

Similarly, we compute the $\{b_j^*\} = \{b_2^*\}$ matrix due to a real unit load at node 2; the free bodies and moment diagram are shown in part (c), according to which the force transformation matrix becomes

$$\{b_2^*\} = \left\{ \begin{matrix} 0 \\ -\frac{1}{4}L \\ -\frac{1}{4}L \\ 0 \end{matrix} \right\} = \frac{L}{4} \left\{ \begin{matrix} 0 \\ -1 \\ -1 \\ 0 \end{matrix} \right\}$$

The flexibility matrix $[f]$ for the two beam elements is composed according to Eq. (9.6), where each of the individual element flexibility matrices must contain the prismatic element properties and length:

$$\begin{bmatrix} \dfrac{L/2}{6EI} \begin{bmatrix} 2 & 1 \\ 1 & 2 \end{bmatrix} & 0 \\ 0 & \dfrac{L/2}{6(EI/2)} \begin{bmatrix} 2 & 1 \\ 1 & 2 \end{bmatrix} \end{bmatrix} = \frac{L}{12EI} \begin{bmatrix} 2 & 1 & & 0 \\ 1 & 2 & & \\ & & 4 & 2 \\ 0 & & 2 & 4 \end{bmatrix}$$

Inserting these matrices into Eq. (9.6), we obtain

$$\Delta = \{b_i^{**}\}^T [f][b_j^*]P$$

$$= \frac{L^2}{96EI} [2 \quad 1 \quad 1 \quad 0] \begin{bmatrix} 2 & 1 & & 0 \\ 1 & 2 & & \\ & & 4 & 2 \\ 0 & & 2 & 4 \end{bmatrix} \begin{bmatrix} 0 \\ -1 \\ -1 \\ 0 \end{bmatrix} (-5)$$

$$= -\frac{5}{12} \frac{L^2}{EI}$$

We next proceed to the determination of deformations Δ_{ij} at a single node i, due to several, say l, loading conditions with applied forces acting on m nodal points j. For each additional loading condition, we add one column to the $[P_j]$ matrix and to the matrix of the resulting deformations $[\Delta]$, for a total of l columns. Since the applied loads are acting on m nodes, we have to expand the $[P]$ matrix to m rows and the $[b^*]$ matrix by a corresponding number of columns in order to allow for the internal moments due to unit loads at the m nodes. Accordingly, the matrix equation (9.10) is expanded to the order indicated by the quantities in parentheses:

$$
\underset{(1 \times l)}{[\Delta]} = \underset{(1 \times 2n)}{[b^{**}_i]^T} \quad \underset{(2n \times 2n)}{[f]} \quad \underset{(2n \times m)}{[b^*_j]} \quad \underset{(m \times l)}{[P]} \tag{9.11}
$$

Note that the $[b^{**}_i]$ matrix, which is due to the single virtual unit load at point i, does not change, nor does the element flexibility matrix $[f]$, which depends only on the structural properties of the beam.

Example 9.4: For the beam of Ex. 9.3, shown again in the figure, calculate the displacement at node 1 (the rotation θ_A) due to three different loading conditions:
a. $M_1 = 3L$ kip-ft; $P_2 = 0$.
b. $M_1 = 0$; $P_2 = -5$ kips.
c. $M_1 = 3L$ kip-ft; $P_2 = -5$ kips.

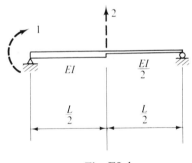

Fig. E9.4

Solution: The $[f]$ matrix is the same as already derived in Ex. 9.3, and since the same deformation as in that problem is required, so is the $\{b^{**}_i\} = \{b^{**}_1\}$ matrix.

The load matrix $[P]$, however, will now contain three columns, one for each loading condition, and two rows, to accommodate the nodes at which the loads are acting:

$$
[P] = \begin{bmatrix} 3L & 0 & 3L \\ 0 & -5 & -5 \end{bmatrix}
$$

The $[b^*_j]$ matrix must be expanded to two columns to account for moments at the element ends due to unit loads at nodes 1 and 2:

$$[b_j] = \begin{bmatrix} 1 & 0 \\ \frac{1}{2} & -\frac{1}{4}L \\ \frac{1}{2} & -\frac{1}{4}L \\ 0 & 0 \end{bmatrix} = \frac{1}{4} \begin{bmatrix} 4 & 0 \\ 2 & -L \\ 2 & -L \\ 0 & 0 \end{bmatrix}$$

Note that this matrix represents a composite of the $\{b_i\}$ and $\{b_j\}$ matrices of Ex. 9.3.

The $[\Delta]$ matrix will now contain three columns for the displacements due to three loading conditions. Inserting these matrices into Eq. (9.11), we obtain

$$[\Delta] = \{b_i^{**}\}^T [f][b_j^*][P]$$

$$= \frac{L}{96EI} [2 \quad 1 \quad 1 \quad 0] \begin{bmatrix} 2 & 1 & & \\ 1 & 2 & & 0 \\ & & 4 & 2 \\ 0 & & 2 & 4 \end{bmatrix} \begin{bmatrix} 4 & 0 \\ 2 & -L \\ 2 & -L \\ 0 & 0 \end{bmatrix} \begin{bmatrix} 3L & 0 & 3L \\ 0 & -5 & -5 \end{bmatrix}$$

$$= \begin{bmatrix} \dfrac{9}{8} & -\dfrac{5}{12} & \dfrac{17}{24} \end{bmatrix} \dfrac{L^2}{EI}$$

Note that the displacement due to the third loading condition can also be obtained by superposition of the first two.

We are now ready to proceed to the general case in which we seek to determine the displacements Δ_{ij} at d nodes i due to l different loading conditions P_j, consisting of forces applied at m nodes j. For each of these additional displacements we add one row to the $[\Delta]$ matrix and one column to the $[b_i^{**}]$ matrix to account for the additional virtual unit loads corresponding to each deformation, for a total of d rows and columns, respectively. The virtual work equation (9.7) is thus expanded to the order indicated by the quantities in parentheses:

$$\begin{array}{ccccc} [\Delta] & = & [b_i^{**}]^T & [f] & [b_j^*] & [P] \\ (d \times l) & & (d \times 2n) & (2n \times 2n) & (2n \times m) & (m \times l) \end{array} \qquad (9.12)$$

Each column of the $[\Delta]$ matrix now contains the d deformations due to one of l loading conditions.

Example 9.5: For the beam of Exs. 9.3 and 9.4, compute the rotation $\theta_A (= \Delta_1)$ and the deflection $\Delta_B (= \Delta_2)$ due to each of the three loading conditions of Ex. 9.4.

Solution: The $[f]$, $[b_j^*]$, and $[P]$ matrices of Ex. 9.4 are valid, but the $[b_i^{**}]$ matrix must now be expanded to two columns to allow the internal moments due to virtual unit loads at nodes 1 and 2 to be stored:

$$[b_i^{**}] = \tfrac{L}{4}\begin{bmatrix} 4 & 0 \\ 2 & -L \\ 2 & -L \\ 0 & 0 \end{bmatrix}$$

Note that this is identical to the previously computed $[b_j^*]$ matrix because the virtual unit loads happen to be located at the same nodes as the real applied loads. The $[\Delta]$ matrix must be expanded to two rows to accommodate two different displacements. Substituting into Eq. (9.12),

$$[\Delta] = [b_i^{**}]^T[f][b_j^*][P]$$

$$= \frac{L}{192EI}\begin{bmatrix} 4 & 2 & 2 & 0 \\ 0 & -L & -L & 0 \end{bmatrix}\begin{bmatrix} 2 & 1 & & \\ 1 & 2 & & 0 \\ \hline & & 4 & 2 \\ 0 & & 2 & 4 \end{bmatrix}\begin{bmatrix} 4 & 0 \\ 2 & -L \\ 2 & -L \\ 0 & 0 \end{bmatrix}\begin{bmatrix} 3L & 0 & 3L \\ 0 & -5 & -5 \end{bmatrix}$$

or,

$$\begin{Bmatrix} \Delta_1 \\ \Delta_2 \end{Bmatrix} = \frac{L^2}{EI}\begin{bmatrix} \tfrac{9}{8} & -\tfrac{5}{12} & \tfrac{17}{24} \\ -\tfrac{1}{4}L & -\tfrac{5}{32}L & -\tfrac{13}{32}L \end{bmatrix}$$

9.6. Structure Flexibility Matrix

To calculate the structure flexibility matrix by the methods of Sec. 9.5, we attach an appropriate system of d nodes to the structure. The structure flexibility matrix will now contain d displacements at nodes i due to each of d unit loads at nodes j, applied one at a time.

The virtual unit loads as well as the real applied loads are now acting at the same points, so that the $[b_i]$ and the $[b_j]$ matrices become identical:

$$[b_i^{**}] = [b_j^*] \equiv [b]$$

The loads $[P]$, which are the cause of the deformations, become unity and are applied one at a time:

$$[P] = \begin{bmatrix} 1 & 0 & 0 & \cdots \\ 0 & 1 & 0 & \cdots \\ 0 & 0 & 1 & \cdots \\ \cdot & \cdot & \cdot & \cdots \\ \cdot & \cdot & \cdot & \cdots \\ \cdot & \cdot & \cdot & \cdots \end{bmatrix} = [I]$$

The structure flexibility matrix $[F]$ is then given by Eq. (9.12):

$$\underset{(d \times d)}{[F]} = [b]^T[f][b][I] = \underset{(d \times 2n)}{[b]^T}\ \underset{(2n \times 2n)}{[f]}\ \underset{(2n \times d)}{[b]} \qquad (9.13)$$

Thus, the calculation of the $[F]$ matrix requires the force transformation matrix $[b]$ and the element flexibility matrix $[f]$, as well as the triple matrix multiplication indicated by Eq. (9.13). The orders indicated for Eq. (9.13) are for planar beam elements.

That the structure flexibility matrix must always be symmetric can be proved by some simple matrix operations; to write the transpose of the $[F]$ matrix, we set

$$[F]^T = [b]^T[f]^T[b]$$

Since by Eq. (9.4) the element flexibility matrix $[f]$ is symmetric, $[f]^T = [f]$, and

$$[F]^T = [b]^T[f][b] = [F]$$

or $$F_{ij} = F_{ji} \qquad (9.14)$$

Expressed in words, this important relation states: the displacement at node i due to a unit load at node j is equal to the displacement at node j due to a unit load at node i. It is called *Maxwell's law of reciprocal deflections.*

Example 9.6: Calculate the flexibility matrix $[F]$ for nodes 1 and 2 of the beam shown.

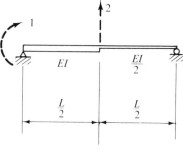

Fig. E9.6

Solution: This beam was already discussed in Ex. 9.5, where the $[b]$ and $[f]$ matrices were already calculated. We now recall that flexibilities are displacements due to unit loads acting at the specified nodes, one at a time; thus, there are two loading conditions:

$$[P] = \begin{bmatrix} 1 & 0 \\ 0 & 1 \end{bmatrix} = [I]$$

Thus, we arrive at Eq. (9.13):

$$[F] = [b]^T[f][b]$$

$$= \frac{L}{192EI}\begin{bmatrix} 72 & -16L \\ -16L & 6L^2 \end{bmatrix} = \frac{L}{EI}\begin{bmatrix} \frac{3}{8} & -\frac{1}{12}L \\ -\frac{1}{12}L & \frac{1}{32}L^2 \end{bmatrix}$$

9.7. Flexibility Matrices for General Statically Determinate Structures

Equation (9.13) is generally applicable for elastic structures; it is only necessary to define the individual elements appropriately and to decide which types of distortions are of importance in their effect on the structural deformations. Accordingly, the element flexibility matrices $[f^k]$ are defined, and the associated internal forces $[b]$ acting on these elements due to unit nodal forces are calculated and substituted into Eq. (9.13). If each of n elements of the structure has p element nodes, and there are d structure nodes to be considered, then the $[f]$ matrix will be of order ($pn \times pn$), and the $[b]$ matrix will be of order ($pn \times d$).

The procedure is best explained by means of some examples.

Example 9.7: Calculate the flexibility matrix for the truss shown (which was already discussed in Ex. 4.3) for the nodes indicated.

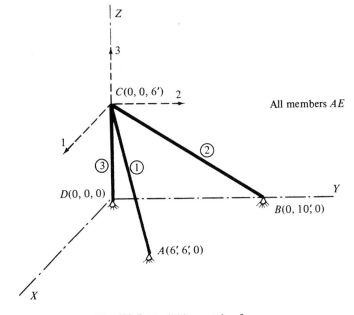

Fig. E9.7 Flexibility matrix of truss.

Solution: Only axial elongations are assumed to contribute to truss deflections, and, accordingly, we need only to determine the virtual internal work due to member elongations. Each member of the truss forms an element, each with only one nodal deformation, corresponding to which, according to Eq. (9.5), the flexibility is $[f^k] = L/AE$, and the force is the axial member force due to a unit applied nodal load. For three members and three structure nodes, the

[b] matrix is of order (3 × 3) and can be taken from the results of Ex. 4.3:

$$[b] = \begin{bmatrix} -1.732 & 0 & 0 \\ +1.167 & -1.167 & 0 \\ +.398 & +.600 & +1.000 \end{bmatrix}$$

The [f] matrix, for three elements of one node each, is also of order (3 × 3):

$$[f] = \begin{bmatrix} \left(\dfrac{L}{AE}\right)^1 & & \\ & \left(\dfrac{L}{AE}\right)^2 & \\ & & \left(\dfrac{L}{AE}\right)^3 \end{bmatrix} = \frac{1}{AE}\begin{bmatrix} 10.4 & 0 & 0 \\ 0 & 11.67 & 0 \\ 0 & 0 & 6.0 \end{bmatrix}$$

Substituting these matrices into Eq. (9.13), we obtain

$$[F] = [b]^T[f][b]$$

$$= \begin{bmatrix} -1.732 & +1.167 & +.398 \\ 0 & -1.167 & +.600 \\ 0 & 0 & +1.000 \end{bmatrix} \frac{1}{AE}\begin{bmatrix} 10.4 & 0 & 0 \\ 0 & 11.67 & 0 \\ 0 & 0 & 6.0 \end{bmatrix}$$

$$\begin{bmatrix} -1.732 & 0 & 0 \\ +1.167 & -1.167 & 0 \\ +.398 & +.600 & +1.000 \end{bmatrix}$$

$$= \frac{1}{AE}\begin{bmatrix} 48.04 & & \text{sym} \\ -14.46 & 18.05 & \\ 2.39 & 3.60 & 6.00 \end{bmatrix}$$

The next example considers a case of a flexural structure in space in which bending and torsional distortions are of importance.

Example 9.8: Calculate the flexibility matrix for nodes 1, 2, and 3 of the pipe structure shown in part (a) of the figure. Consider effects of flexural and torsional distortions. Assume Poisson's ratio of .5, leading to $G = \frac{1}{2}E$.

Solution: The structure is decomposed into two elements as shown in part (b); since bending about two cross-sectional axes and torsion are involved, five element nodes are introduced in each element, as indicated in part (b), which, in order, define bending effects about the horizontal and the vertical element axes, and torsion, analogous to Eq. (9.7); bending stiffnesses about two axes and torsional stiffness must be calculated on the basis of the cross-sectional properties; for the circular section of the structure,

$$(EI)_x = (EI)_y = EI; \qquad GC = GJ = \frac{E}{2}\cdot 2I = EI$$

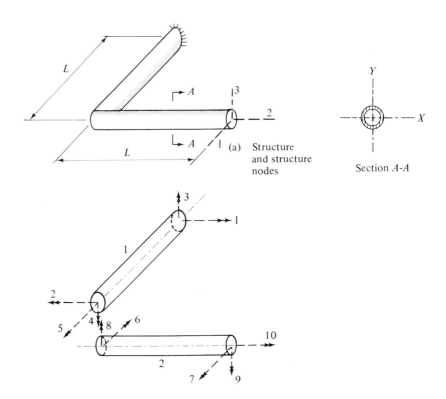

(a)   Structure
and structure
nodes

Section *A-A*

(b) Elements and element nodes

Fig. E9.8

so all stiffness values are identical, and $[f^k]$ becomes, referring to Eq. (9.7),

$$[f^k] = \frac{L}{6EI}\begin{bmatrix} 2 & 1 & & & \\ 1 & 2 & & & \\ & & 2 & 1 & \\ & & 1 & 2 & \\ & & & & 6 \end{bmatrix}$$

The two elements constituting this structure are identical, so that the $[f]$ matrix, of order (10×10), is a diagonal arrangement of the two matrices $[f^1] = [f^2]$:

$$[f] = \begin{bmatrix} [f^1] & \\ & [f^2] \end{bmatrix}$$

The $[b]$ matrix, of order (10×3), is calculated by statics; each of the three columns contains the internal element forces, in accordance with the number-ing of part (b), due to a unit load along each of the three nodes (to save space, we show here the transpose of the $[b]$ matrix):

$$[b]^T = \begin{bmatrix} 0 & 0 & L & L & 0 & 0 & 0 & L & 0 & 0 \\ 0 & 0 & -L & 0 & 0 & 0 & 0 & 0 & 0 & 0 \\ L & 0 & 0 & 0 & L & L & 0 & 0 & 0 & 0 \end{bmatrix}$$

Because of the many zeros in this matrix, the operations of Eq. (9.13) are easily performed by hand, leading to the structure flexibility matrix

$$[F] = [b]^T[f][b] = \frac{L}{6EI} \begin{bmatrix} 8L^2 & & \text{sym} \\ -3L^2 & 2L^2 & \\ 0 & 0 & 10L^2 \end{bmatrix}$$

The zero terms in this flexibility matrix indicate that there is no relation between in-plane and out-of-plane effects; that is, they are "decoupled."

9.8. Computer Calculation of the Structure Flexibility Matrix

Once the matrices $[b]$ and $[f]$ are known, the operations of Eq. (9.13) for the structure flexibility matrix,

$$[F] = [b]^T[f][b]$$

are simply programmed. They consist of the following steps:

1. Take the transpose of $[b]$.
2. Multiply $[b]^T[f] = [C]$.
3. Multiply $[C][b] = [b]^T[f][b] = [F]$.

After the matrices $[b]$ and $[f]$ have been fed into computer storage, this triple matrix multiplication is performed by subroutine TRMULT, which is listed in Appendix A. This subroutine in turn calls on two additional subrou-tines, TRANSM, which performs steps 1 and 2, and MULT, by means of which step 3 is carried out.

The reader is encouraged to try this routine on a suitably simple problem, such as Ex. 9.8.

PROBLEMS

1. For the plane prismatic cantilever beam of cross-sectional properties A, I, and modulus of elasticity E, compute the (3×3) flexibility matrix $[f]$ containing the elements f_{ij} denoting the deformation of node i due to a unit force in the sense of node j. Consider flexural and axial distortions of the beam.

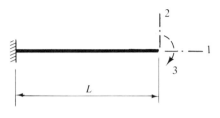

Prob. 1

2. The inclined beam is of flexural stiffness *EI*. Compute the (2 × 2) flexibility matrix [*f*], neglecting axial distortions.

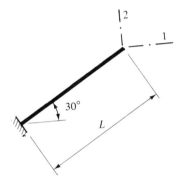

Prob. 2

3. The inclined cantilever beam of Prob. 2 has a cross-sectional area *A* and shear modulus *G*. Compute the (2 × 2) flexibility matrix [*f*], considering flexural, axial, and shear distortions.

4. The tapering I-section beam has a cross section that for analysis purposes can be considered as consisting of two equal thin flanges, each of area $A/2$ and of mean depth *d*, varying linearly from d_0 at end *A* to $2d_0$ at end *B*. The modulus is *E*. Compute the (2 × 2) flexibility matrix [*f*], considering flexural distortions only. (*Hint:* Integrations can be simplified by taking the origin at point *O*.)

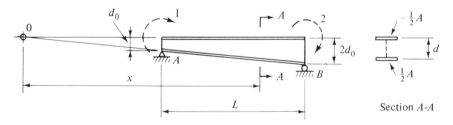

Prob. 4

5. Calculate the (3 × 3) flexibility matrix [*f*] for the quarter-circular prismatic member of bending stiffness *EI*. Neglect axial or shear distortions.

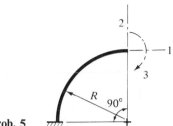

Prob. 5

6. Compute the flexibility matrix [*f*], of order (2 × 2), for the prismatic semi-circular member of stiffness *EI*. Consider only flexural distortions.

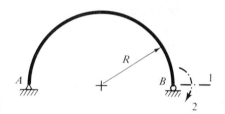

Prob. 6

7. Compute the (2 × 2) structure flexibility matrix [*F*] for the haunched simple beam. Consider only flexural distortions.

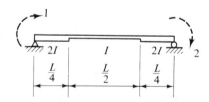

Prob. 7

8. Compute the (3 × 3) structure flexibility matrix [*F*] for the prismatic simple

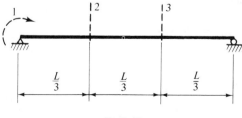

Prob. 8

beam by considering an appropriate number of elements and computing the virtual work by matrix methods. Consider flexural distortions only.

9. Compute the (3×3) flexibility matrix $[F]$, considering axial and flexural distortions.

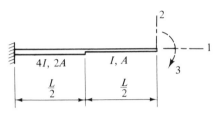

Prob. 9

10. Compute the (3×3) flexibility matrix $[F]$ for the structure shown. Consider only flexural distortions. (*Hint:* Use the results of Probs. 1 and 5.)

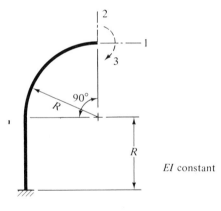

Prob. 10

11. The haunched beam is of I cross section, as shown. For analysis purposes only the thin flanges must be considered; the effect of the web may be neglected.

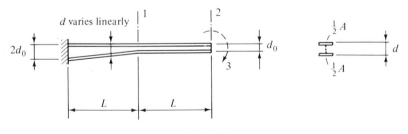

Prob. 11

Compute the (3 × 3) flexibility matrix [F], considering only flexural distortions. (*Hint:* Use the results of Prob. 4.)

12. Calculate the flexibility matrix [F] relating the deflections 1 and 2 to the applied loads 1 and 2. For all members the cross-sectional area is A and the modulus of elasticity is E.

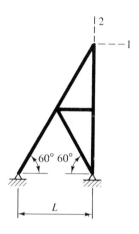

Prob. 12

13. The cantilever bent girder is of constant cross section, of area $A = 10$ in., moment of inertia (about both axes) $I = 100$ in.^{4}, torsional stiffness factor $C = 50$ in.^{4}, and modulus of elasticity $E = 2 \times$ shear modulus G. Compute the flexibility matrix [F], of order (3 × 3), considering axial, flexural, and torsional distortions.

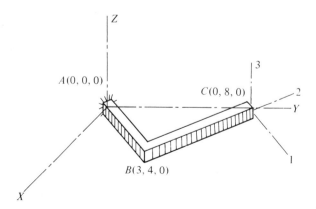

Prob. 13

IV

Statically
Indeterminate Analysis
by the Force Method

The structures treated so far have been statically determinate; as discussed in Sec. 2.4, a vaster class of structures, including most of those common in modern construction practice, is *statically indeterminate; that is, the equations of statics are insufficient for the solution of the unknown forces. The ability to distinguish between the two types is essential, and has been covered in Sec. 2.4.*

Two basic methods are available for the solution of statically indeterminate structures: the force *or* flexibility *method, and the* displacement *or* stiffness *method. In Part IV, we shall consider the force method, which, as the name implies, considers the forces as unknown. It is the oldest of the methods of analysis, the easiest to grasp, and for many problem types the most effective. As presented here, it will also serve to derive some of the building blocks that, in Part V, will be used in the displacement method.*

The force method will be introduced in Chapter 10 in the classical manner, and then represented in matrix formulation. It is then applied to a variety of structures to gain facility and to indicate its universality. In Chapter 11 the force method is used to develop the concepts of fixed-end forces and member stiffness in preparation for their use in later work. Some special variants and applications of the force method are finally covered in Chapter 12.

CHAPTER 10

The Force Method

This chapter contains in Sec. 10.1 the classical formulation of the *force method*, also called the *flexibility method* or the *method of consistent deformations*. The particular version called the *three-moment equation*, which is applicable to continuous beams, is also indicated here. The method is cast in simple matrix form in Sec. 10.2, and applied to the solution of several statically indeterminate structures in Sec. 10.3. A more general matrix development of the method is presented in Sec. 10.4, and this in turn is extended to different loading cases in Sec. 10.5.

10.1. Method of Consistent Deformations

We consider a structure of degree r of static indeterminacy; that is, the number of unknown forces exceeds the number of available equilibrium equations by r. The equations of statics must be supplemented by a number of equations of geometry equal to the degree of redundancy, r, of the structure. The r unknown or *redundant* forces are the unknowns in these equations, which express the consistency of the deformations with the specified support and continuity conditions. Hence, the method in its classical formulation is also called the *method of consistent deformations*.

The method will be outlined step by step in terms of the two-times statically indeterminate continuous beam of Fig. 10.1(a):

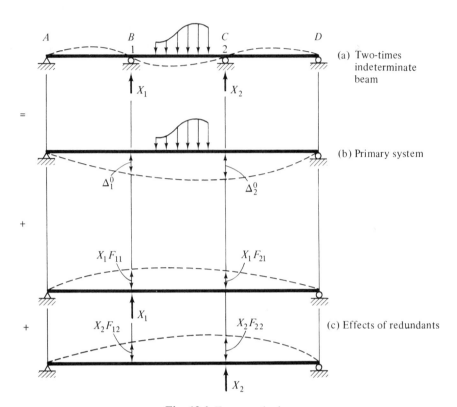

Fig. 10.1 Force method.

1. Reduce the structure to static determinacy by removing r forces (the redundants), here, the two intermediate supports labeled 1 and 2. The r points at which forces have been removed are called the *cuts* or *releases*. The resulting statically determinate structure is called the *primary structure*, here, the simple beam shown in Fig. 10.1(b). Note that the conditions of geometry of the actual structure are violated at the r cuts. The aim of the subsequent calculations is to determine those values of the r redundants necessary to restore the specified conditions of geometry at the cuts.

2. Calculate the deformations Δ_i° at cuts i of the primary structure due to the applied loads. In general, r deformations Δ_i° will have to be computed for each loading condition.

3. Calculate the deformations at the cuts i of the primary structure due to unit values of the redundants at j, shown in Fig. 10.1(c). These are the flexibilities F_{ij}, of total number $r \times r$. Note that the flexibilities are properties of the structure, not of the loading, and will therefore be the same irrespective of the loading condition. By the principle of

superposition, the deformation at point i due to the actual value of the redundant X_j at point j will be $X_j F_{ij}$.

4. The total deformations at the r cuts are obtained by adding the deformations for each cut i due to the applied loads, Δ_i°, and due to all redundants, $X_j F_{ij}$; compatibility with actual support conditions is restored by equating this sum to the specified deformations, which, for the problem illustrated, are zero. For the case of Fig. 10.1, $r = 2$:

$$\begin{aligned}
\sum \Delta_1 = 0: \quad & \Delta_1^\circ + X_1 F_{11} + X_2 F_{12} = 0 \\
\sum \Delta_2 = 0: \quad & \Delta_2^\circ + X_1 F_{21} + X_2 F_{22} = 0
\end{aligned} \tag{10.1}$$

These r *continuity equations* have to be solved for the r unknown redundants X_j. Once the redundants are found, all other forces can be determined by statics.

In the case of specified support displacements Δ_1 and Δ_2 at the points of redundancy, these values must be substituted for the zeros on the right-hand side of the equations. We note that in the case of statically indeterminate structures (as contrasted with determinate structures) support displacements can cause internal forces, which must be considered in design.

Thus, in general formulation, the conditions of geometry for an r-times statically indeterminate structure will be

$$\begin{aligned}
X_1 F_{11} + X_2 F_{12} + \cdots + X_r F_{1r} + \Delta_1^\circ &= \Delta_1 \\
X_1 F_{21} + X_2 F_{22} + \cdots + X_r F_{2r} + \Delta_2^\circ &= \Delta_2 \\
&\vdots \\
X_1 F_{r1} + X_2 F_{r2} + \cdots + X_r F_{rr} + \Delta_r^\circ &= \Delta_r
\end{aligned} \tag{10.2}$$

The displacements Δ_i° can also be due to given temperature changes, prestressing, or other causes that lead to strains within the structure. The values Δ_i° must then be calculated as the relative displacements of the redundant points of the primary structure; an example of such calculations is shown in Example 10.2c.

Example 10.1: Analyze the continuous beam shown in part (a) of the figure. Draw shear and moment diagrams due to the concentrated load P.

Solution: With four unknown vertical reactions, but only two independent equilibrium equations for a coplanar, parallel force system, this structure is indeterminate to the second degree. We must therefore remove two of the unknown reactions to reduce the beam to its primary system, as shown in part (b). We apply then, in sequence, the applied load P, the redundant X_1, and the redundant X_2 to this primary structure, as shown in parts (b) through (d).

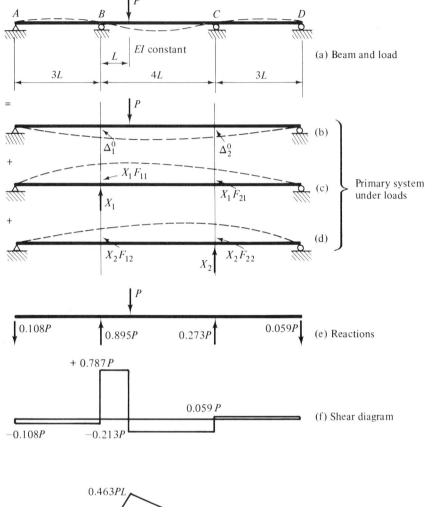

(a) Beam and load

(b)

(c) Primary system under loads

(d)

(e) Reactions

(f) Shear diagram

(g) Moment diagram

Fig. E10.1

Each of these forces will cause the displacements at the points of redundancy indicated in the figures. The conditions of geometry of the actual structure require that the sum of these displacements at both points of redundancy be zero, as expressed by Eq. (10.1). Prior to solving these equations for X_1 and X_2, we must compute the values of the displacements Δ_1° and Δ_2° and the structure flexibilities F_{11}, F_{12}, F_{21}, and F_{22} by any of the means discussed in Chapters 7 through 9. Without going into detailed calculations at this time, we obtain, assuming upward deflections positive,

$$\Delta_1^\circ = -16.50 \frac{PL^3}{EI}; \quad \Delta_2^\circ = -15.00 \frac{PL^3}{EI}$$

$$F_{11} = F_{22} = 14.70 \frac{L^3}{EI}; \quad F_{12} = F_{21} = 12.30 \frac{L^3}{EI}$$

and the equations of geometry become

$$14.70 \frac{L^3}{EI} X_1 + 12.30 \frac{L^3}{EI} X_2 = 16.50 P \cdot \frac{L^3}{EI}$$

$$12.30 \frac{L^3}{EI} X_1 + 14.70 \frac{L^3}{EI} X_2 = 15.00 P \cdot \frac{L^3}{EI}$$

Solving for the unknown redundants, we obtain

$$X_1 = .895P; \quad X_2 = .273P$$

The indeterminate portion of the solution is now finished, and the rest of the forces can be obtained by statics. For instance, we can consider the redundant reactions along with the specified load acting as in part (e) and find the remaining reactions at points A and D as shown. The shear and moment diagrams are then drawn as in parts (f) and (g). This completes the problem. Note that the inflection points of the moment diagram verify the deformed shape of the continuous beam indicated in part (a).

We note that in the set of simultaneous equations of geometry the bending stiffness EI is common to all terms and can be canceled out prior to solving. This is a usual feature of such problems, and for this reason the stiffnesses should always be left in symbolic form in the deformation calculations, resulting in some savings of computational effort.

An important consideration in solving statically indeterminate structures is the choice of the primary system. In general, the closer the structural action of the primary system resembles that of the real structure, the better-conditioned will be the resulting set of simultaneous continuity equations, and the less the round-off errors of the numerical solution. In this sense, the action of the redundants may be considered as "corrections"—the slighter the correction, the better. Another consideration is the ready availability of flexibilities, so that decomposition of the structure into standard components with known flexibilities (such as the rotational end flexibilities of prismatic members) should be considered. In this light, the force method of analysis of a continuous beam of $r + 1$ spans might be undertaken as shown in Fig. 10.2; the beam is cut over each interior support, releasing the r support moments, which become the redundant actions; the requirement of slope continuity of the real beam must now be satisfied by adding the slope changes (or "kinks")

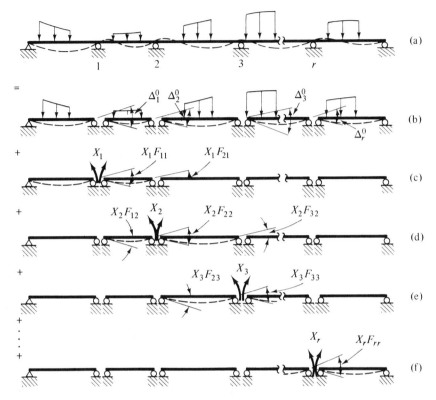

Fig. 10.2 Force-method analysis of continuous beam.

over the supports due to all effects on the primary system, and equating their sum to zero.

Figures 10.2(b) through (f) show the primary system under the applied load and the redundant moments applied one at a time. With the redundants and deformations as defined in these figures, the r continuity conditions become

$$\sum \theta_1 = 0: \quad X_1 F_{11} + X_2 F_{12} + 0 \quad\quad + 0 \quad\quad + 0 + \cdots + 0 \quad\quad = -\Delta_1^\circ$$

$$\sum \theta_2 = 0: \quad X_1 F_{21} + X_2 F_{22} + X_3 F_{23} + 0 \quad\quad + 0 + \cdots + 0 \quad\quad = -\Delta_2^\circ$$

$$\sum \theta_3 = 0: \quad 0 \quad\quad + X_2 F_{32} + X_3 F_{33} + X_4 F_{34} + 0 + \cdots + 0 \quad\quad = -\Delta_3^\circ$$

$$\vdots$$

$$\sum \theta_r = 0: \quad 0 \quad\quad + 0 \quad\quad + 0 \quad\quad + 0 \quad\quad + 0 + \cdots + X_r F_{rr} = -\Delta_r^\circ$$

In this case, the flexibilities are the differences (with due respect to signs) of the appropriate end rotations due to cranked-in moments; if the members happen to be prismatic, these flexibilities can be written in a unified manner for all redundant nodes:

$$F_{ij} = \left(\frac{L}{3EI}\right)_i + \left(\frac{L}{3EI}\right)_{i+1} \qquad \text{for } i = j$$

$$F_{ij} = \left(\frac{L}{6EI}\right)_i \qquad \text{for } i = j - 1 \qquad\qquad (10.3)$$

$$F_{ij} = \left(\frac{L}{6EI}\right)_{i+1} \qquad \text{for } i = j + 1$$

The Δ°'s are the differences of the end rotations of the simple spans due to the specified loads.

It is also noted that, in contrast to the action of the redundants chosen in Fig. 10.1, in this case the effect of each redundant is felt only in the adjacent beam spans, so that each continuity equation contains only three redundants, all other terms being zero. For this reason this version of the force method is called the *three-moment equation*. When expressed in matrix form, it leads to a *tri-diagonal matrix*, which can be solved very efficiently.

We observe that the labor of analysis can be vastly reduced by the appropriate choice of the primary system.

Example 10.2: Analyze the prismatic two-span beam of part (a) of the figure due to the following effects:

a. A uniform load w over the entire beam.
b. A support settlement Δ of point B.
c. A temperature variation ΔT between top and bottom of the beam; the depth of beam is d and the coefficient of thermal expansion is α.

Solution:

a. The beam is statically indeterminate to the second degree; it is decomposed into the primary system by removing the moments M_1 and M_2, which become the redundants. Parts (b), (c), and (d) show the relative slope changes, which are set equal to zero:

$$\sum \theta_1 = 0: \qquad \Delta_1^\circ + M_1 F_{11} + M_2 F_{12} = 0$$
$$\sum \theta_2 = 0: \qquad \Delta_2^\circ + M_1 F_{21} + M_2 F_{22} = 0$$

where the deformations are calculated for part (a) by any method as

$$\Delta_1^\circ = 2 \cdot \frac{wL^3}{24EI} = \frac{1}{12}\frac{wL^3}{EI}; \qquad \Delta_2^\circ = \frac{1}{24}\frac{wL^3}{EI}$$

$$F_{11} = 2 \cdot \frac{L}{3EI}; \qquad F_{22} = \frac{L}{3EI}; \qquad F_{12} = F_{21} = \frac{L}{6EI}$$

so that, placing the effects of the load, Δ_i°, on the right-hand side, and simplifying,

$$\frac{2}{3} M_1 + \frac{1}{6} M_2 = -\frac{1}{12} wL^2$$

$$\frac{1}{6} M_1 + \frac{1}{3} M_2 = -\frac{1}{24} wL^2$$

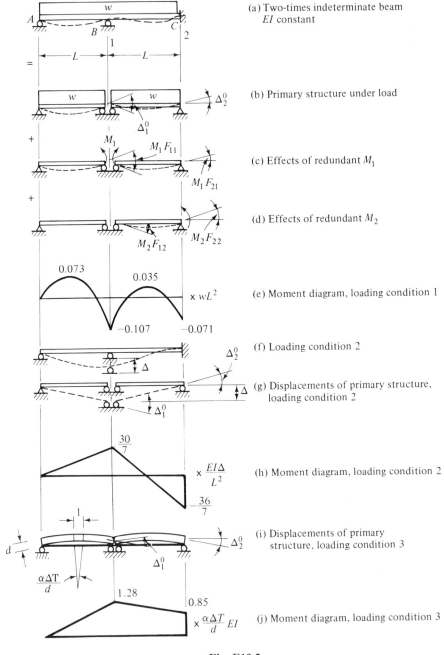

(a) Two-times indeterminate beam
 EI constant

(b) Primary structure under load

(c) Effects of redundant M_1

(d) Effects of redundant M_2

(e) Moment diagram, loading condition 1

(f) Loading condition 2

(g) Displacements of primary structure, loading condition 2

(h) Moment diagram, loading condition 2

(i) Displacements of primary structure, loading condition 3

(j) Moment diagram, loading condition 3

Fig. E10.2

from which

$$M_1 = -.107wL^2 \quad \text{and} \quad M_2 = -.071wL^2$$

The moment diagram is shown in part (e). It shows the general curvatures and inflection points, which confirm the anticipated deformations shown in part (a).

b. Part (f) shows the anticipated deformations of the continuous beam due to the specified support displacement; the curvatures of the deflection curve indicate the presence of moments in the beam.

The same primary system as in part (a) of the example is selected; part (g) shows the deformation of the primary system due to the support settlement, from which the kinks at supports 1 and 2 are calculated as

$$\Delta_1^\circ = -2\frac{\Delta}{L}; \qquad \Delta_2^\circ = +\frac{\Delta}{L}$$

The kinks due to the redundants are as before, so that the conditions of geometry become

$$\frac{2}{3}M_1 + \frac{1}{6}M_2 = 2\frac{EI}{L^2}\Delta$$

$$\frac{1}{6}M_1 + \frac{1}{3}M_2 = -\frac{EI}{L^2}\Delta$$

from which

$$M_1 = \frac{30}{7}\frac{EI}{L^2}\Delta; \qquad M_2 = -\frac{36}{7}\frac{EI}{L^2}\Delta$$

The moment diagram is shown in part (h); note that the moments are a function of the beam stiffness. Its inflection point confirms the validity of the deflected shape of part (f).

c. The curvatures resulting from temperature gradients in beams were discussed in Ex. 7.5 and found to be $\Phi = \alpha\,\Delta T/d$. Again, the same primary system as before is chosen, and the deformations due to the temperature differences are shown in part (i). The relative angle changes at supports 1 and 2, using for instance, a geometric method such as curvature area, are found to be

$$\Delta_1^\circ = -\frac{\alpha\,\Delta T}{d}L; \qquad \Delta_2^\circ = -\frac{1}{2}\frac{\alpha\,\Delta T}{d}L$$

Substituting these into the two equations of geometry, we obtain

$$\Sigma\,\Delta_1 = 0: \qquad \frac{2}{3}\frac{L}{EI}\cdot M_1 + \frac{1}{6}\frac{L}{EI}\cdot M_2 = \frac{\alpha\,\Delta T}{d}L$$

$$\frac{1}{6}\frac{L}{EI}\cdot M_1 + \frac{1}{3}\frac{L}{EI}\cdot M_2 = \frac{1}{2}\frac{\alpha\,\Delta T}{d}L$$

from which

$$M_1 = 1.28\frac{\alpha\,\Delta T}{d}EI; \qquad M_2 = .85\frac{\alpha\,\Delta T}{d}EI$$

The resultant moment diagram is shown in part (j). Since the curvature of the primary structure was in this case not associated with any moments, the moment diagram is proportional only to the portion of the curvature due to the redundants. We note that only the right-hand side of the set of equations, containing the load terms, varied for the three parts of the problem; the left-hand side, containing the flexibilities, remained the same. The thought might occur that it should be possible to solve the equation set once and for all, independent of the loading conditions, rather than start each time from scratch, as was done here. Such an approach will be taken by the use of matrices in the next section.

10.2. Simple Matrix Formulation of the Force Method

It will have become apparent to the reader that the force method as discussed in Sec. 10.1 lends itself to matrix formulation, and that in this way the repetitive work contained in Ex. 10.2 might be avoided.

Equation (10.2), for instance, can be rewritten for l loading conditions as

$$\begin{bmatrix} F_{11} & F_{12} & \cdots & F_{1r} \\ F_{21} & F_{22} & & \cdot \\ \cdot & & & \cdot \\ \cdot & & & \cdot \\ F_{r1} & F_{r2} & \cdots & F_{rr} \end{bmatrix} \begin{bmatrix} X_{11} & X_{12} & \cdots & X_{1l} \\ X_{21} & X_{22} & & \cdot \\ \cdot & & & \cdot \\ \cdot & & & \cdot \\ X_{r1} & X_{r2} & \cdots & X_{rl} \end{bmatrix} + \begin{bmatrix} \Delta^\circ_{11} & \Delta^\circ_{12} & \cdots & \Delta^\circ_{1l} \\ \Delta^\circ_{21} & \Delta^\circ_{22} & & \cdot \\ \cdot & & & \\ \cdot & & & \cdot \\ \Delta^\circ_{r1} & \Delta^\circ_{r2} & \cdots & \Delta^\circ_{rl} \end{bmatrix}$$

$$= \begin{bmatrix} \Delta_{11} & \Delta_{12} & \cdots & \Delta_{1l} \\ \Delta_{21} & \Delta_{22} & & \cdot \\ \cdot & & & \cdot \\ \cdot & & & \\ \Delta_{r1} & \Delta_{r2} & \cdots & \Delta_{rl} \end{bmatrix}$$

or, in shorter form,

$$[F][X] + [\Delta^\circ] = [\Delta] \tag{10.4}$$

where the structure flexibility matrix $[F]$ is of order $(r \times r)$, and the other matrices are of order $(r \times l)$, with each column containing the values for one loading condition.

Solving for the unknown redundants, we obtain by inversion of the $[F]$ matrix

$$[X] = [F]^{-1}[[\Delta] - [\Delta^\circ]] \tag{10.5}$$

Note that once the $[F]^{-1}$ matrix has been obtained, further loading conditions require only additional matrix multiplications, which can be performed very rapidly on the computer.

The matrices of the deformations $[\Delta^\circ]$ and of the flexibilities $[F]$ can be computed by any method, but for consistency of operations the reader is reminded of the matrix procedures outlined in Secs. 9.5 and 9.6.

The following example will recast a previously worked problem in matrix form; the reader should compare the conciseness of formulation and amount of computational effort involved in the two methods.

Example 10.3: Formulate Ex. 10.2a, b, and c in terms of matrices, and solve for the redundants due to all three loading conditions. In addition, consider a fourth condition resulting from a specified built-in kink in the beam at point B, of value θ.

Solution: The structure flexibility matrix $[F]$, from Ex. 10.2, is

$$[F] = \frac{L}{6EI}\begin{bmatrix} 4 & 1 \\ 1 & 2 \end{bmatrix}$$

and the matrices $[\Delta]$ and $[\Delta^\circ]$ of Eq. (10.5) are

$$[\Delta^\circ] = \begin{bmatrix} \dfrac{1}{12}\dfrac{wL^3}{EI} & -2\dfrac{\Delta}{L} & -\dfrac{\alpha\,\Delta TL}{d} & 0 \\[2ex] \dfrac{1}{24}\dfrac{wL^3}{EI} & \dfrac{\Delta}{L} & -\dfrac{1}{2}\dfrac{\alpha\,\Delta TL}{d} & 0 \end{bmatrix}$$

and

$$[\Delta] = \begin{bmatrix} 0 & 0 & 0 & \theta \\ 0 & 0 & 0 & 0 \end{bmatrix}$$

The flexibility matrix is now inverted to yield

$$[F]^{-1} = \frac{6EI}{7L}\begin{bmatrix} 2 & -1 \\ -1 & 4 \end{bmatrix}$$

and substituting these matrices into Eq. (10.5) and performing the indicated matrix operations results in the redundants due to all four loading conditions:

$$\begin{bmatrix} X_{11} & \cdots & X_{14} \\ X_{21} & \cdots & X_{24} \end{bmatrix} = \frac{6EI}{7L}\begin{bmatrix} 2 & -1 \\ -1 & 4 \end{bmatrix}\left(\begin{bmatrix} 0 & 0 & 0 & \theta \\ 0 & 0 & 0 & 0 \end{bmatrix}\right.$$

$$\left. -\begin{bmatrix} \dfrac{1}{12}\dfrac{wL^3}{EI} & -2\dfrac{\Delta}{L} & -\dfrac{\alpha\,\Delta TL}{d} & 0 \\[2ex] \dfrac{1}{24}\dfrac{wL^3}{EI} & \dfrac{\Delta}{L} & -\dfrac{1}{2}\dfrac{\alpha\,\Delta TL}{d} & 0 \end{bmatrix}\right)$$

$$= \begin{bmatrix} -.107wL^2 & \dfrac{30}{7}\dfrac{EI}{L^2}\Delta & 1.28\dfrac{\alpha\,\Delta TL}{d} & \dfrac{12}{7}\dfrac{EI}{L}\theta \\[2ex] -.071wL^2 & -\dfrac{36}{7}\dfrac{EI}{L^2}\Delta & .85\dfrac{\alpha\,\Delta TL}{d} & -\dfrac{6}{7}\dfrac{EI}{L}\theta \end{bmatrix}$$

With these reactions, all other forces may be obtained by statics.

The matrix formulation of this section confined itself to bare-bones essentials. A more complete and formal presentation of the matrix force method suitable for computer programming will be attempted in Sec. 10.4.

10.3. Application of the Force Method to General Structures

The force method as presented in the preceding sections is applicable to all structures. The sequence outlined in Secs. 10.1 and 10.2 will in all cases lead to results. A large portion of the work will concern itself with deformation calculations of the primary structure, and fluency and good organization are necessary to achieve success. In longhand operation it is best to perform these calculations as preliminary operations to the actual indeterminate analysis, so as to break the overall problem into manageable portions. If any deformations, or flexibilities, are available from previous solutions, or from compilations, they may of course be used. Only after all numerical values of the Δ, $\Delta°$, and F deformations are available should they be inserted into the conditions of geometry.

The formulation of the problem in matrix form is primarily intended be used with a high-speed computer, and a complete computer program will contain the means of calculation of these deformations, as well as of their further processing. The matrix examples shown in this book are, in general, short enough so that they can be performed longhand. The practice of organizing the hand operations will help in the development of efficient computer routines.

It is always advisable to draw, whenever possible, the anticipated deformed shape of the indeterminate structure under load prior to proceeding to actual computations. This will convey a feel for the structural response and provide a basis for a quick check of final results by comparison of the moment diagram with anticipated curvatures. While such qualitative checks cannot be considered final, they can nevertheless give confidence in the validity of the solution.

The following examples are intended to provide further insight in the application of the force method.

Example 10.4: Analyze the fixed-ended, semicircular, uniform cross-section arch due to a uniform load of intensity w kips/ft of projected area, as shown in part (a) of the figure. Consider only flexural deformations.

Solution: To gain a feel for the structural response, we draw the anticipated shape of the arch under load; due to the assumed lack of axial extensions, a flattening of the crown must be accompanied by a spreading of the sides. The moments are associated here with the change of curvature, which shows a total of four inflection points.

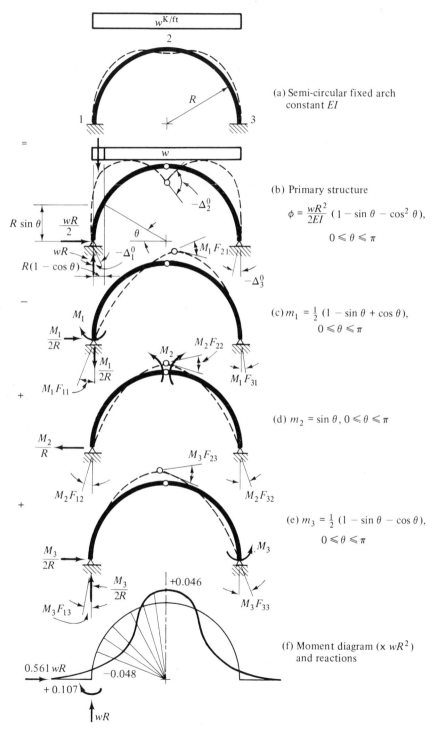

(a) Semi-circular fixed arch constant EI

(b) Primary structure

$$\phi = \frac{wR^2}{2EI} (1 - \sin \theta - \cos^2 \theta),$$

$$0 \leqslant \theta \leqslant \pi$$

(c) $m_1 = \frac{1}{2} (1 - \sin \theta + \cos \theta),$

$$0 \leqslant \theta \leqslant \pi$$

(d) $m_2 = \sin \theta, 0 \leqslant \theta \leqslant \pi$

(e) $m_3 = \frac{1}{2} (1 - \sin \theta - \cos \theta),$

$$0 \leqslant \theta \leqslant \pi$$

(f) Moment diagram ($\times wR^2$) and reactions

Fig. E10.4

237

This structure is three-times indeterminate, and the moments at the supports 1 and 3 and at the crown 2 are chosen as redundants; the primary structure is a three-hinged arch. The corresponding displacements of the primary system are the relative rotations θ_1, θ_2, and θ_3 at the three cuts, which are to be equated to zero; according to parts (b) to (e), these continuity conditions become

$$\sum \theta_1 = 0: \quad M_1 F_{11} + M_2 F_{12} + M_3 F_{13} + \Delta_1^{\circ} = 0$$

$$\sum \theta_2 = 0: \quad M_1 F_{21} + M_2 F_{22} + M_3 F_{23} + \Delta_2^{\circ} = 0$$

$$\sum \theta_3 = 0: \quad M_1 F_{31} + M_2 F_{32} + M_3 F_{33} + \Delta_3^{\circ} = 0$$

or, in matrix formulation,

$$\begin{array}{cccc} [F] & [X] & + \; [\Delta^{\circ}] & = & [0] \\ (3 \times 3) & (3 \times 1) & (3 \times 1) & (3 \times 1) \end{array}$$

In this case, the $[\Delta]$ deformations due to the loads and the $[F]$ flexibilities will be computed by the method of virtual forces in its integral formulation, Eq. (8.11), since the shape of the structure and the load are easily represented in analytical form. A matrix method such as presented in Sec. 9.5 could also be used if the element flexibility matrix $[f]$ for a curved element is known, or if the arch is broken down into a sufficient number of elements so that they can be represented by straight segments with sufficient accuracy. It is always up to the analyst's judgment to select the most appropriate method for the purpose at hand.

The moments in the primary structure due to the real load and the real unit redundants are computed in terms of the polar coordinate system shown in part (b) and the corresponding curvatures are obtained by division by the flexural beam stiffness EI. The virtual moments are identical to those of unit values of the redundants. The variations of these functions are listed in parts (b) to (e).

When substituted into the virtual work equation, the required deformations result:

$$\Delta_1^{\circ} = \int \frac{M}{EI} \cdot m_1 \, ds = \int_{\theta=0}^{\pi} \frac{wR^2}{2EI} [1 - \sin \theta - \cos^2 \theta]$$

$$\times \frac{1}{2} [1 - \sin \theta + \cos \theta] R \, d\theta$$

$$= -.048 \frac{wR^3}{EI}$$

Similarly,

$$\Delta_2^{\circ} = \int \frac{M}{EI} \cdot m_2 \, ds = -.118 \frac{wR^3}{EI}; \qquad \Delta_3^{\circ} = \int \frac{M}{EI} \cdot m_3 \, ds = -.048 \frac{wR^3}{EI}$$

Likewise,

$$F_{12} = \int_{\theta=0}^{\pi} m_1 \frac{m_2}{EI} R \, d\theta = \frac{1}{EI} \int_0^{\pi} \frac{1}{2}[1 - \sin\theta + \cos\theta][\sin\theta] R \, d\theta = .215\frac{R}{EI}$$

In general, $F_{ij} = \int m_i(m_j/EI) \, ds$, from which

$$F_{21} = F_{12} = F_{32} = F_{23} = .215\frac{R}{EI}; \qquad F_{11} = F_{33} = .571\frac{R}{EI}$$

$$F_{13} = F_{31} = -.215\frac{R}{EI}$$

$$F_{22} = 1.571\frac{R}{EI}$$

Substituting these values into the continuity conditions, we get

$$\begin{bmatrix} .571 & .215 & -.215 \\ .215 & 1.571 & .215 \\ -.215 & .215 & .571 \end{bmatrix} \begin{bmatrix} M_1 \\ M_2 \\ M_3 \end{bmatrix} = \begin{bmatrix} .048 \\ .118 \\ .048 \end{bmatrix} wR^2$$

The job of inverting the $[F]$ matrix, or otherwise solving the set of three simultaneous equations, can be simplified in this case by using the symmetry condition $M_1 = M_3$; in any case, the solution yields

$$M_1 = M_3 = .107wR^2; \qquad M_2 = .046wR^2$$

Once the redundant moments are known, the remaining reactions and moments can be calculated by statics, as discussed earlier. The moment diagram and the reactions are shown in part (f).

Example 10.5: Analyze the semicircular bow girder shown in the figure, fixed at one end, simply supported at the other, due to
a. Its dead weight, of intensity w.
b. A vertical displacement Δ of the simply supported end.

Solution: This is the same member as already considered in Ex. 8.4, of torsional stiffness $GC = .265EI$.

This structure is statically indeterminate to the first degree; we shall use the vertical reaction at the simple support as redundant, since the required deformations of the cantilever primary structure have already been calculated in Ex. 8.4.
a. The geometrical condition for zero displacement at the pin support is

$$XF + \Delta° = 0$$

in which, from Ex. 8.4 and considering upward deflections positive,

$$\Delta° = -20.56\frac{wR^4}{EI} \quad \text{(downward)}; \qquad F = 19.35\frac{R^3}{EI} \quad \text{(upward)}$$

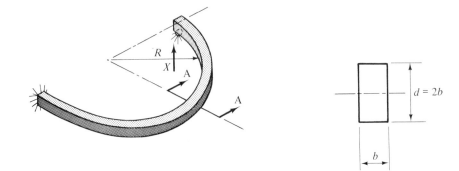

Section A-A

Fig. E10.5

so that

$$X = \frac{20.56}{19.35}wR = \underline{1.06wR} \quad \text{(upward)}$$

With the redundant known, all internal forces can be found by statics.

b. For the unloaded beam with its end displaced Δ, the condition of geometry is $XF = \Delta$, from which

$$X = \frac{\Delta \cdot EI}{19.35R^3} = \underline{.0517 \frac{EI}{R^3} \cdot \Delta}$$

Note that in the case of a single redundant, the matrices involved reduce to single numbers, or scalars, and the inversion reduces to taking the reciprocal of the flexibility.

Example 10.6: Analyze the statically-indeterminate truss shown in part (a) due to the indicated load.

Solution: With five members and five components of reaction, this truss has ten unknown forces, versus the eight force equilibrium equations furnished by four joints; the truss is therefore two times statically indeterminate. We remove two reactive components to obtain the primary structure of part (b); this happens to be the truss already treated in Ex. 8.6. We write the two equations of geometry for the cuts, referring to parts (b) to (d) of the figure:

$$\sum \Delta_1 = 0: \quad \Delta_1^\circ + X_1 F_{11} + X_2 F_{12} = 0$$
$$\sum \Delta_2 = 0: \quad \Delta_2^\circ + X_1 F_{21} + X_2 F_{22} = 0$$

or

$$\begin{bmatrix} F_{11} & F_{12} \\ F_{21} & F_{22} \end{bmatrix} \begin{Bmatrix} X_1 \\ X_2 \end{Bmatrix} = -\begin{Bmatrix} \Delta_1^\circ \\ \Delta_2^\circ \end{Bmatrix}$$

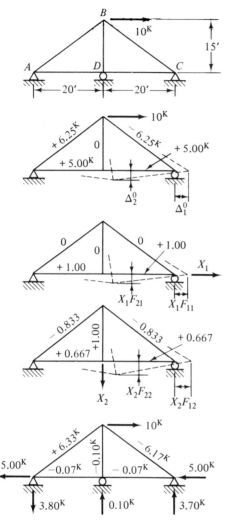

(a)   Two-times indeterminate truss
      All members AE

(b)   Primary structure under load
      Member forces S

(c)   Effect of redundant X_1
      Member forces s_1 due to $X_1 = 1$

(d)   Effect of redundant X_2
      Member forces s_2 due to $X_2 = 1$

(e)   Final forces

Fig. E10.6

The required deformations will be computed by applying the theorem of virtual forces as in Sec. 8.5. This requires three separate analyses of the systems of parts (b) to (d) of the figure; we shall call the forces due to the applied load and the unit redundants X_1 and X_2, S, s_1, and s_2, respectively. Multiplying appropriate virtual member forces by real member elongations, the deformations will be

$$\Delta_1^\circ = \sum s_1^{**} \frac{S*L}{AE}; \qquad \Delta_2^\circ = \sum s_2^{**} \frac{S*L}{AE}$$

$$F_{11} = \sum s_1^{**} \frac{s_1^* L}{AE}; \qquad F_{22} = \sum s_2^{**} \frac{s_2^* L}{AE}; \qquad F_{12} = F_{21} = \sum s_1 s_2 \frac{L}{AE}$$

These calculations are performed in the table:

MEMBER	$\left(\dfrac{L}{AE}\right) \cdot \dfrac{AE}{12}$ (IN.)	S (KIP)	s_1 (KIP/KIP)	s_2 (KIP/KIP)	
AB	25	+6.25	0	−.833	
BC	25	−6.25	0	−.833	
AD	20	+5.00	+1.00	+.667	
DC	20	+5.00	+1.00	+.667	
DB	15	0	0	+1.00	

MEMBER	$Ss_1 \dfrac{L}{AE}$	$Ss_2 \dfrac{L}{AE}$	$s_1^2 \dfrac{L}{AE}$	$s_2^2 \dfrac{L}{AE}$	$s_1 s_2 \dfrac{L}{AE}$
AB	0	−130.0	0	17.36	0
BC	0	+130.0	0	17.36	0
AD	+100.00	+ 66.67	20.00	8.92	13.33
DC	+100.00	+ 66.67	20.00	8.92	13.33
DB	0	0	0	15.00	0
	+200.00 $= \Delta_1^\circ$	+133.33 $= \Delta_2^\circ$	+40.00 $= F_{11}$	67.56 $= F_{22}$	26.67 $= F_{12} = F_{21}$

$$\left(\text{All values} \times \frac{12}{AE}\right)$$

The equations of geometry in matrix form become

$$\begin{Bmatrix} X_1 \\ X_2 \end{Bmatrix} = -\begin{bmatrix} 40.00 & 26.67 \\ 26.67 & 67.56 \end{bmatrix}^{-1} \begin{Bmatrix} 200.00 \\ -133.33 \end{Bmatrix} = -\begin{Bmatrix} 5.00 \\ 0.10 \end{Bmatrix} \text{ kips}$$

With these redundants determined, we find the final member forces by super-position:

$$S_{\text{Total}} = S - 5.00 s_1 - 0.10 s_2$$

This calculation is again best done in tabular form, leading to the member forces shown in part (e) of the figure.

This concludes the problem.

10.4. General Matrix Formulation of the Force Method

We now consider an r-times statically indeterminate structure subjected only to concentrated loads or moments, as for instance the structure of $r = 3$ shown in Fig. 10.3(a). The structure is to be analyzed and the displacements of the loaded points as well as the deflection Δ are to be found.

As before, we begin by removing r redundants to reduce the structure to

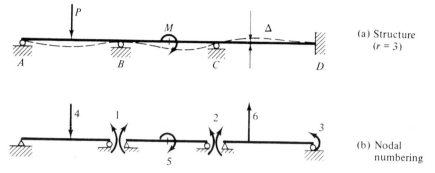

Fig. 10.3 Nodal numbering for force method.

static determinacy, resulting in the primary structure shown in Fig. 10.3(b). To this structure we attach a set of nodes, beginning the numbering with the r nodes corresponding to the unknown redundants. Following this, we introduce a node corresponding to every load component, and every displacement component to be determined, as shown by nodes 4 to 6 in Fig. 10.3(b). The number of these latter nodes will be called p, so that the total number of nodes is $(r + p)$.

We observe that the first r nodes (here, 1 to 3) correspond to known displacements Δ (here, because of the continuity of the elastic beam, $\Delta_1 = \Delta_2 = 0$, and because of the specified end fixity, $\Delta_3 = 0$), but unknown redundant forces X. The p following nodes correspond to known forces X (here, the loads $X_4 = -P$ and $X_5 = M$ are given, and $X_6 = 0$), but unknown displacements. Note that for uniformity all forces, including reactions and applied loads, are designated by the symbol X.

We recall that the force method of an r-times indeterminate structure demands satisfaction of the r conditions of geometry:

$$X_1 F_{11} + \cdots + X_r F_{1r} + \Delta_1^\circ = \Delta_1$$
$$\vdots \qquad\qquad (10.6)$$
$$X_1 F_{r1} + \cdots + X_r F_{rr} + \Delta_r^\circ = \Delta_r$$

or

$$[F][X] + [\Delta^\circ] = [\Delta]$$

where $[F]$ is the flexibility matrix, $[\Delta^\circ]$ are the deformations of the primary structure due to the applied loads, and $[\Delta]$ are the specified displacements of the primary structure at the points of redundancy.

The displacements Δ_i° at the cuts due to the applied concentrated loads X_j $[j = (r + 1), \ldots, (r + p)]$ can be written in terms of the appropriate

$$
\begin{array}{l}
\text{Effects of } r \text{ Unknown Forces (Redundants)} \qquad \text{Effects of } p \text{ Known Forces (Loads)} \\[1em]
\left.
\begin{array}{l}
X_1 F_{11} + \cdots + X_r F_{1r} + X_{(r+1)} F_{1(r+1)} + \cdots + X_{(r+p)} F_{1(r+p)} = \Delta_1 \\
\qquad\qquad \cdot \qquad\qquad\qquad\qquad\qquad \cdot \qquad\qquad \cdot \\
\qquad\qquad \cdot \qquad\qquad\qquad\qquad\qquad \cdot \qquad\qquad \cdot \\
X_1 F_{r1} + \cdots + X_r F_{rr} + X_{(r+1)} F_{r(r+1)} + \cdots + X_{(r+p)} F_{r(r+p)} = \Delta_r
\end{array}
\right\} \text{Known displacements of redundants} \\[2em]
\left.
\begin{array}{l}
X_1 F_{(r+1)1} + \cdots + X_r F_{(r+1)r} + X_{(r+1)} F_{(r+1)(r+1)} + \cdots + X_{(r+p)} F_{(r+1)(r+p)} = \Delta_{(r+1)} \\
\qquad\qquad \cdot \qquad\qquad\qquad\qquad\qquad\qquad \cdot \qquad\qquad \cdot \\
\qquad\qquad \cdot \qquad\qquad\qquad\qquad\qquad\qquad \cdot \qquad\qquad \cdot \\
X_1 F_{(r+p)1} + \cdots + X_r F_{(r+p)r} + X_{(r+1)} F_{(r+p)(r+1)} + \cdots + X_{(r+p)} F_{(r+p)(r+p)} = \Delta_{(r+p)}
\end{array}
\right\} \text{Unknown displacements of loads}
\end{array}
\tag{10.7}
$$

r Equations of geometry for unknown reactions X

p Equations for unknown deflections Δ

flexibilities F_{ij} as

$$\Delta_i^\circ = X_{(r+1)}F_{i(r+1)} + \cdots + X_{(r+p)}F_{i(r+p)}$$

Similarly, the unknown deformations Δ_i $[i = (r+1), \ldots, (r+p)]$ at the p load points are obtained by superposition of the effects of all redundant and applied forces:

$$\Delta_{(r+1)} = X_1 F_{(r+1)1} + \cdots + X_r F_{(r+1)r} + X_{(r+1)}F_{(r+1)(r+1)} + \cdots + X_{(r+p)}F_{(r+1)(r+p)}$$

$$\cdot$$
$$\cdot$$
$$\cdot$$

$$\Delta_{(r+p)} = X_1 F_{(r+p)1} + \qquad\qquad \cdots \qquad\qquad + X_{(r+p)}F_{(r+p)(r+p)}$$

We add these p equations to the r equations of geometry and get Eq. (10.7), as shown on page 244.

In matrix form, Eq. (10.7) can be written as

$$\begin{bmatrix} F_{\alpha\alpha} & F_{\alpha\beta} \\ \hline F_{\beta\alpha} & F_{\beta\beta} \end{bmatrix}\begin{bmatrix} X_\alpha \\ \hline X_\beta \end{bmatrix} = \begin{bmatrix} \Delta_\alpha \\ \hline \Delta_\beta \end{bmatrix}; \qquad \begin{matrix} (X_\alpha \text{ unknown}; \ \Delta_\alpha \text{ known}) \\ (X_\beta \text{ known}; \quad \Delta_\beta \text{ unknown}) \end{matrix} \qquad (10.8)$$

The square flexibility matrix $[F]$, of order $(r+p)$, can be determined by any method, for instance, the matrix formulation

$$[F] = [b]^T[f][b] \qquad (10.9)$$

which was discussed in Sec. 9.5. The use of this equation requires the decomposition of the structure into individual elements between load points, an appropriate nodal system corresponding to the internal element forces, and calculation of the element flexibilities $[f]$ and of the force transformation matrix $[b]$.

The flexibility matrix of Eq. 10.8 has been partitioned into $F_{\alpha\alpha}$ (order $r \times r$), $F_{\alpha\beta}$ (order $r \times p$), $F_{\beta\alpha}$ (order $p \times r$), and $F_{\beta\beta}$ (order $p \times p$) in anticipation of future operations. The $[X]$ and $[\Delta]$ matrices have been partitioned similarly so that the α portion contains the quantities at the r cuts, and the β portion contains those at the p load points.

The matrix of forces $[X]$, of order $(r+p) \times 1$ for a single loading condition, or of order $(r+p) \times l$ for l different loading conditions, contains in each column r unknown redundants, denoted by the subscript α, followed by p known loads, subscripted β.

The matrix $[\Delta]$, of order $(r+p) \times 1$ for a single and of order $(r+p) \times l$ for l loading conditions, contains in each column r specified redundant displacements Δ_α, followed by p unknown displacements Δ_β. For instance, in the particular case of the structure of Fig. 10.3, of $r = 3, p = 3$, Eq. (10.7) would look as follows:

$$r = 3 \begin{cases} \\ \\ \\ \end{cases} \quad p = 3 \begin{cases} \\ \\ \\ \end{cases} \begin{bmatrix} F_{11} & F_{12} & F_{13} & F_{14} & F_{15} & F_{16} \\ F_{21} & F_{22} & F_{23} & F_{24} & F_{25} & F_{26} \\ F_{31} & F_{32} & F_{33} & F_{34} & F_{35} & F_{36} \\ \hline F_{41} & F_{42} & F_{43} & F_{44} & F_{45} & F_{46} \\ F_{51} & F_{52} & F_{53} & F_{54} & F_{55} & F_{56} \\ F_{61} & F_{62} & F_{63} & F_{64} & F_{65} & F_{66} \end{bmatrix} \begin{bmatrix} X_1 \\ X_2 \\ X_3 \\ X_4 = -P \\ X_5 = M \\ X_6 = 0 \end{bmatrix} = \begin{bmatrix} \Delta_1 = 0 \\ \Delta_2 = 0 \\ \Delta_3 = 0 \\ \hline \Delta_4 \\ \Delta_5 \\ \Delta_6 \end{bmatrix}$$

The flexibilities F_{ij} are those corresponding to the numbered nodes of the primary system of Fig. 10.3(b).

The solution of the set of equations (10.8) is accomplished in two steps: First, we solve r equations for the unknown redundants $[X_\alpha]$:

$$[F_{\alpha\alpha}][X_\alpha] + [F_{\alpha\beta}][X_\beta] = [\Delta_\alpha]$$

or
$$[X_\alpha] = [F_{\alpha\alpha}]^{-1}[[\Delta_\alpha] - [F_{\alpha\beta}][X_\beta]] \qquad (10.10)$$

This corresponds to the earlier solution for the r redundants by classical methods. If all specified support or internal displacements of the redundants are zero, the Δ_α matrix drops out.

After the redundants $[X_\alpha]$ are known, the remaining p equations can be solved for the deformations at the load points:

$$[\Delta_\beta] = [F_{\beta\alpha}][X_\alpha] + [F_{\beta\beta}][X_\beta] \qquad (10.11)$$

It remains to determine the internal element forces $[S]$ (such as internal moments or shears). These forces had already been numbered in setting up Eq. (10.9); they are related to the forces $[X]$ by the force transformation matrix $[b]$:

$$[S] = [b][X] \qquad (10.12)$$

The following example will demonstrate the complete procedure in terms of a simple problem.

Example 10.7: It is required to find the redundant X_1, the deflections Δ_2 and Δ_3, and the moments at points 1 to 4 of the beam of part (a) of the figure due to two conditions:

a. Loading condition 1: $X_2 = 3$ kips; $X_3 = 2$ kips.
b. Loading condition 2: A support settlement $\Delta_1 = 1$.
Consider only flexural deformations of the structure.

Solution: The primary structure is as shown in part (b). The solution is accomplished in the following steps:

1. Determine the flexibility matrix $[F] = [b]^T[f][b]$. The beam is decomposed into three equal elements between load points, and the $[f]$ matrix is written by reference to Eq. (9.6). The element end moments are related to unit

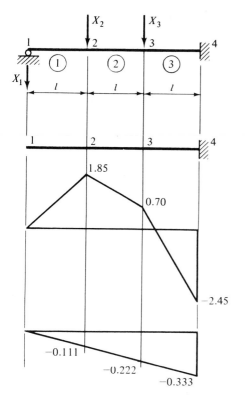

(a) Structure nodal and element numbering. EI is constant.

(b) Primary structure

(c) Moment diagram, loading condition 1 $(\times l)$

(d) Moment diagram, loading condition 2 $(\times \dfrac{EI}{l^2})$

Fig. E10.7

values of the external forces by statics and written in the force transformation matrix $[b]$:

$$[f] = \frac{l}{6EI}\begin{bmatrix} 2 & 1 & & & & \\ 1 & 2 & & & & \\ & & 2 & 1 & & \\ & & 1 & 2 & & \\ & & & & 2 & 1 \\ & & & & 1 & 2 \end{bmatrix} ; \qquad [b] = l\begin{bmatrix} 0 & 0 & 0 \\ -1 & 0 & 0 \\ -1 & 0 & 0 \\ -2 & -1 & 0 \\ -2 & -1 & 0 \\ -3 & -2 & -1 \end{bmatrix}$$

$$\begin{array}{cccc} [F] &= [b]^T & [f] & [b] \\ (3\times 3) & (3\times 6) & (6\times 6) & (6\times 3) \end{array} = \frac{l^3}{6EI}\begin{bmatrix} 54 & 28 & 8 \\ 28 & 16 & 5 \\ 8 & 5 & 2 \end{bmatrix} \quad \text{(note symmetry)}$$

2. Set up force–displacement equations for two loading conditions:

$$\begin{bmatrix} F_{\alpha\alpha} & F_{\alpha\beta} \\ F_{\beta\alpha} & F_{\beta} \end{bmatrix}\begin{bmatrix} X_\alpha \\ X_\beta \end{bmatrix} = \begin{bmatrix} \Delta_\alpha \\ \Delta_\beta \end{bmatrix}$$

(The first and second columns of the $[X]$ and $[\Delta]$ matrices denote the first and second loading conditions.)

$$\frac{l^3}{6EI}\begin{bmatrix} 54 & 28 & 8 \\ 28 & 16 & 5 \\ 8 & 5 & 2 \end{bmatrix}\begin{bmatrix} X_{11} & X_{12} \\ X_{21}=3 & X_{22}=0 \\ X_{31}=2 & X_{32}=0 \end{bmatrix} = \begin{bmatrix} \Delta_{11}=0 & \Delta_{12}=1 \\ \Delta_{21} & \Delta_{22} \\ \Delta_{31} & \Delta_{32} \end{bmatrix}$$

3. Solve for redundant X_1 due to both loading conditions according to Eq. (10.10):

$$\frac{l^3}{6EI}\left\{ 54[X_{11} \quad X_{12}] + [28 \quad 8]\begin{bmatrix} 3 & 0 \\ 2 & 0 \end{bmatrix} \right\} = [0 \quad 1]$$

Solve for $[X_{11} \quad X_{12}]$:

$$[X_{11} \quad X_{12}] = \frac{1}{54}\left\{ \frac{6EI}{l^3}[0 \quad 1] - [100 \quad 0] \right\} = \begin{bmatrix} -1.85 & .111\dfrac{EI}{l^3} \end{bmatrix}$$

Note that the redundant due to the applied loads (condition 1) is 1.85 kips upward; that due to the support settlement (condition 2) depends on the beam stiffness.

4. Solve for deflections Δ_2 and Δ_3 due to both loading conditions according to Eq. (10.11):

$$\begin{bmatrix} \Delta_{21} & \Delta_{22} \\ \Delta_{31} & \Delta_{32} \end{bmatrix} = \frac{l^3}{6EI}\left[\begin{bmatrix} 28 \\ 8 \end{bmatrix}\begin{bmatrix} -1.85 & .111\dfrac{EI}{l^3} \end{bmatrix} + \begin{bmatrix} 16 & 5 \\ 5 & 2 \end{bmatrix}\begin{bmatrix} 3 & 0 \\ 2 & 0 \end{bmatrix} \right]$$

$$= \begin{bmatrix} 1.016\dfrac{l^3}{EI} & .519 \\ .700\dfrac{l^3}{EI} & .148 \end{bmatrix}$$

Note that the deflections due to the applied loads depend on the beam stiffness; those due to the support settlement ($\Delta_1 = 1$) do not.

5. Solve for the moments at points 1 to 4 due to both loading conditions:

$$[S] = [b][X]$$

(We contract the $[b]$ matrix used in determining the structure flexibility matrix to list only *one* moment at each point.)

$$\begin{bmatrix} M_{11} & M_{12} \\ M_{21} & M_{22} \\ M_{31} & M_{32} \\ M_{41} & M_{42} \end{bmatrix} = l\begin{bmatrix} 0 & 0 & 0 \\ -1 & 0 & 0 \\ -2 & -1 & 0 \\ -3 & -2 & -1 \end{bmatrix}\begin{bmatrix} -1.85 & .111\dfrac{EI}{l^3} \\ 3.0 & 0 \\ 2.0 & 0 \end{bmatrix} = \begin{bmatrix} 0 & 0 \\ 1.85l & -.111\dfrac{EI}{l^2} \\ .70l & -.222\dfrac{EI}{l^2} \\ -2.45l & -.333\dfrac{EI}{l^2} \end{bmatrix}$$

The moment diagrams for the two loading conditions can now be plotted, as shown in parts (c) and (d).

Example 10.8: Analyze the pin-ended truss shown in part (a) of the figure for the applied load *P*. All members are of equal stiffness *AE*.

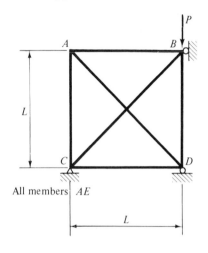

(a) Structure and load

(b) Primary structure and numbering system

Fig. E10.8

Solution: This truss is two times statically indeterminate, once internally, once externally. We can reduce it to statically determinate primary form by cutting the horizontal support at point *B* and the diagonal *AD*, as shown in part (b). Since each of these cuts corresponds to an unknown force, we number these nodes first, and follow with a vertical node at point *B* for the applied load. The numbering of nodes and truss members is also shown in part (b).

We next compute the structure flexibility matrix [*F*] by Eq. (9.13), in which the element flexibility matrix [*f*] and the force transformation matrix [*b*] are

$$
[f] = \frac{L}{AE}
\begin{bmatrix}
1 & & & & & \\
& 1 & & & & \\
& & 1 & & & \\
& & & 1 & & \\
& & & & 1.414 & \\
& & & & & 1.414
\end{bmatrix}
;\quad
[b] =
\begin{bmatrix}
0 & -.707 & 0 \\
0 & -.707 & 0 \\
-1 & -.707 & 1 \\
0 & -.707 & 0 \\
1.414 & 1 & 0 \\
0 & 1 & 0
\end{bmatrix}
$$

Therefore,

$$
[F] = [b]^T[f][b] = \frac{L}{AE}
\begin{bmatrix}
3.82 & & \text{sym} \\
2.71 & 4.82 & \\
-1.00 & -.71 & 1.00
\end{bmatrix}
$$

The displacement–force relationships are next set up:

$$\left\{\begin{array}{c} 0 \\ 0 \\ \hline \Delta_3 \end{array}\right\} = \left[\begin{array}{cc|c} 3.82 & 2.71 & -1.00 \\ 2.71 & 4.82 & -.71 \\ \hline -1.00 & -.71 & 1.00 \end{array}\right] \left\{\begin{array}{c} X_1 \\ X_2 \\ \hline -P \end{array}\right\}$$

The partitioning into the α and β parts is also shown, and Eqs. (10.10) and (10.11) are solved in turn for the unknown forces $\{X_\alpha\}$ and the unknown displacements $\{\Delta_\beta\}$:

$$\left\{\begin{array}{c} X_1 \\ X_2 \end{array}\right\} = \frac{AE}{L}\left[\begin{array}{cc} .435 & -.245 \\ -.245 & .345 \end{array}\right]\left(\left\{\begin{array}{c} 0 \\ 0 \end{array}\right\} - \frac{L}{AE}\left\{\begin{array}{c} -1.00 \\ -.71 \end{array}\right\}[-P]\right) = \left\{\begin{array}{c} -.262P \\ 0 \end{array}\right\}$$

$$\{\Delta_3\} = \left([-1.00 \quad -.71]\left\{\begin{array}{c} -.262P \\ 0 \end{array}\right\} + [1.00][-P]\right)\frac{L}{AE} = \left\{-.738\frac{PL}{AE}\right\}$$

The member forces are obtained by the equilibrium equation, Eq. (10.12):

$$\{S\} = \left\{\begin{array}{c} S_1 \\ S_2 \\ S_3 \\ S_4 \\ S_5 \\ S_6 \end{array}\right\} = [b]\{X\} = \left\{\begin{array}{c} 0 \\ 0 \\ -.738P \\ 0 \\ -.371P \\ 0 \end{array}\right\}$$

This concludes the analysis.

10.5. Extension of the Matrix Method to Distributed Loads and Self-Strains

The matrix formulation of Sec. 10.4 requires the decomposition of the structure into a number of elements between load points, even if the beam itself shows no discontinuities at these points; with many load points, this may lead to unduly large matrices. In such cases, it may be preferred to number only the redundant nodes [the α nodes of Eq. (10.8)], and go back to the simple procedure proposed in Sec. 10.2, in which the deformations $[\Delta°]$ at the points of redundancy, be they due to applied loads or any other causes, are calculated by any means. The forces and deformations at other than the points of redundancy are in this case not determined, since these nodes [the β nodes in Eq. (10.8)] were not even labeled. They can, however, be found by superposition of statically determinate and redundant effects, as was done in Ex. 10.2.

The matrix formulation of Sec. 10.4 also does not allow the analysis for distributed loads or self-straining effects, such as temperature changes or prestressing. We now seek to extend the formulation to include such loading cases.

In such cases, we can consider the total displacement $[\Delta]$ of Eq. (10.8) at any node to be split up into two components: the displacements $[\Delta_1]$ due to the distributed loads or self-strains in the primary structure, and the displacements $[\Delta_2]$ due to the forces $[X]$ (which may be redundants or applied loads). As before, these latter displacements are given by $\Delta_2 = [F][X]$.

For instance, consider the propped cantilever beam of Fig. 10.4(a) which is subjected to loads and to a temperature gradient that causes greater elongation in the top than in the bottom fibers. In the primary structure that results from removing the left-hand support, the temperature gradient causes the deformations $[\Delta_1] \equiv [\Delta^\circ]$, shown in Fig. 10.4(b), and the forces result in the deformations $[\Delta_2]$, shown in Fig. 10.4(c). The total displacements $[\Delta]$ are, then, analogous to Eq. (10.8):

$$\begin{bmatrix} \Delta_\alpha \\ \hline \Delta_\beta \end{bmatrix} = \begin{bmatrix} F_{\alpha\alpha} & \vline & F_{\alpha\beta} \\ \hline F_{\alpha\beta} & \vline & F_{\beta\beta} \end{bmatrix} \begin{bmatrix} X_\alpha \\ \hline X_\beta \end{bmatrix} + \begin{bmatrix} \Delta_\alpha^\circ \\ \hline \Delta_\beta^\circ \end{bmatrix} \qquad (10.13)$$

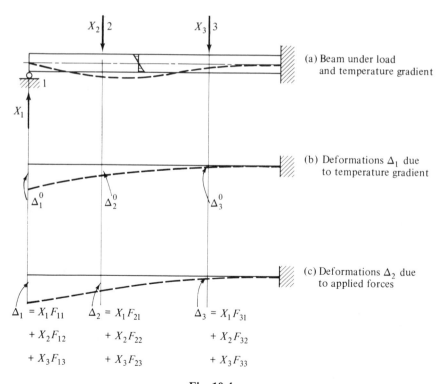

(a) Beam under load and temperature gradient

(b) Deformations Δ_1 due to temperature gradient

(c) Deformations Δ_2 due to applied forces

$\Delta_1 = X_1 F_{11}$ $\quad \Delta_2 = X_1 F_{21}$ $\quad \Delta_3 = X_1 F_{31}$

$\quad\;\; + X_2 F_{12}$ $\qquad\;\; + X_2 F_{22}$ $\qquad\;\; + X_2 F_{32}$

$\quad\;\; + X_3 F_{13}$ $\qquad\;\; + X_3 F_{23}$ $\qquad\;\; + X_3 F_{33}$

Fig. 10.4

The difference between this and Eq. (10.8) consists of the inclusion of the displacements [Δ°] due to the distributed loads or self-strains in the primary structure.

The self-strain deformations [Δ°] can be determined by the matrix procedure of Sec. 9.5. By the theorem of virtual forces,

$$[\Delta^\circ] = \int_{\text{vol}} [\sigma]^T[\epsilon]\, dV = [b]^T[\theta] \tag{10.14}$$

where [b] is the matrix of virtual element forces due to unit values of the nodal forces [X], and [θ] is the matrix of real element deformations due to the self-straining effects. In the absence of any applied loads, the [X_β] matrix will be zero.

The procedure is illustrated in the following example.

Example 10.9: The one-times statically indeterminate beam of Ex. 10.7 is prestressed as shown in part (a) of the figure. The actual prestressing of each of the three elements is simulated by the cable locations shown in part (b). The prestressing force is uniform, of value *P*. Analyze the structure due to this effect.

Solution: Since this beam, its subdivision into elements, and its nodal numbering are identical to that of Ex. 10.7, the flexibility matrix [F] of that example is applicable.

To determine the matrix [Δ°] of Eq. (10.13), we follow Eq. (10.14) and first calculate the real element end rotations [θ] by geometrical methods; the moment in each element is constant, of value $M = P \cdot e$, and the curvature is constant, of value $\Phi = (P \cdot e)/EI$. The end rotations of each element of length *l* are, by the curvature-area method, $\theta = \Phi \cdot (l/2) = (P \cdot e \cdot l/2EI)$. Substituting appropriate eccentricities from part (b), the transposed matrix of end rotations of the elements will be

$$[\theta]^T = \frac{P \cdot ld}{2EI}\left[-\frac{1}{8} \quad -\frac{1}{8} \quad -\frac{1}{4} \quad -\frac{1}{4} \quad +\frac{1}{4} \quad +\frac{1}{4}\right]$$

The sign convention of these rotations must be the same as that of the internal moments [b]; that is, compression on top is positive.

The force transformation matrix of virtual element forces is identical to that used in Ex. 10.7, so that the [Δ°] matrix will be

$$[\Delta^\circ] = [b]^T[\theta] = l\begin{bmatrix} 0 & -1 & -1 & -2 & -2 & -3 \\ 0 & 0 & 0 & -1 & -1 & -2 \\ 0 & 0 & 0 & 0 & 0 & -1 \end{bmatrix} \cdot \frac{Pld}{16EI}\begin{bmatrix} -1 \\ -1 \\ -2 \\ -2 \\ +2 \\ +2 \end{bmatrix} = \frac{Pl^2d}{16EI}\begin{bmatrix} -3 \\ -4 \\ -2 \end{bmatrix}$$

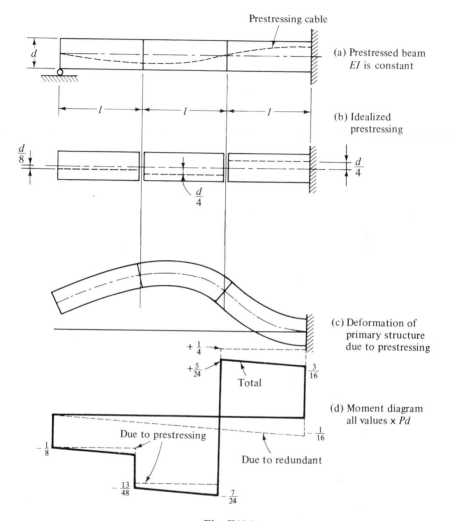

Prestressing cable

(a) Prestressed beam
EI is constant

(b) Idealized prestressing

(c) Deformation of primary structure due to prestressing

(d) Moment diagram all values × Pd

Total

Due to prestressing

Due to redundant

Fig. E10.9

It is useful to visualize these deformations as shown in part (c). Equation (10.13) will in this case appear as

$$\frac{l^3}{6EI}\begin{bmatrix} 54 & 28 & 8 \\ 28 & 16 & 5 \\ 8 & 5 & 2 \end{bmatrix}\begin{bmatrix} X_1 \\ 0 \\ 0 \end{bmatrix} + \frac{Pl^2d}{16EI}\begin{bmatrix} -3 \\ -4 \\ -2 \end{bmatrix} = \begin{bmatrix} 0 \\ \Delta_2 \\ \Delta_3 \end{bmatrix}$$

The redundant X_1 is solved first, analogously to Eq. (10.10), as

$$X_1 = \frac{1}{48} \frac{Pd}{l}$$

and the unknown deflections Δ_2 and Δ_3 are obtained analogously to Eq. (10.11) as

$$\begin{bmatrix} \Delta_2 \\ \Delta_3 \end{bmatrix} = \frac{l^3}{6EI} \begin{bmatrix} 28 \\ 8 \end{bmatrix} \cdot \frac{1}{48} \frac{Pd}{l} + \frac{Pl^2d}{16EI} \begin{bmatrix} -4 \\ -2 \end{bmatrix} = -\frac{Pl^2d}{72EI} \begin{bmatrix} 11 \\ 7 \end{bmatrix}$$

Note that these upward deflections tend to counteract those due to downward loads applied to the structure. It is the purpose of prestressing to minimize the bending deformations and moments in a loaded structure.

The internal moments due to the redundant X_1 are given by the $[b]$ matrix:

$$[S] = \begin{bmatrix} M_1 \\ M_2 \\ M_3 \\ M_4 \end{bmatrix} = l \begin{bmatrix} 0 & 0 & 0 \\ -1 & 0 & 0 \\ -2 & -1 & 0 \\ -3 & -2 & -1 \end{bmatrix} \begin{bmatrix} \dfrac{1}{48} \dfrac{Pd}{l} \\ 0 \\ 0 \\ 0 \end{bmatrix} = -\frac{Pd}{48} \begin{bmatrix} 0 \\ 1 \\ 2 \\ 3 \end{bmatrix}$$

To these moments must be added the moments $(P \cdot e)$ due to the prestressing of the individual elements of the primary structure, so that the total moments will be those shown in part (d). The moment discontinuities arise from the simplified arrangement of the prestressing cables, as shown in part (b). Note that the moments due to the prestressing tend to cancel those due to an applied downward loading, such as that of Ex. 10.7.

PROBLEMS

In all problems try to draw the deflected shape ahead of time and compare it with information obtained from a rigorous analysis. Present results in the form of shear, moment, and axial force diagrams as required.

1. The spring assembly, consisting of springs of the indicated modulus, is pulled by a force P. Determine the forces in all springs, and the resulting displacement Δ.

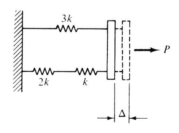

Prob. 1

2. Analyze the continuous beam, and draw shear and moment diagrams
  a. By releasing the interior supports B and C.
  b. By cutting the internal moments at supports B and C.
  c. Discuss the effect of choice of primary system on the indeterminate computations.
  d. Note the attenuation of the load effect in portions of the structure remote from the load, and draw conclusions regarding the possibility of simplified, approximate methods of analysis.

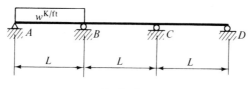

Prob. 2

3. Analyze the propped cantilever beam with elastic support at point A. Plot the variation of the moment at point A as the spring modulus k varies from zero to infinity, and discuss the results in the light of the nature of connections in real structures.

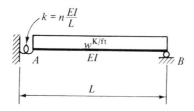

Prob. 3

4. a. Analyze the beam under uniform load, considering only flexural distortions; draw shear and moment diagrams.
  b. Analyze the beam, considering flexural and shear distortions; plot the variation of the support moment M_A with change of ratio of flexural to shear stiffnesses $EI/(AG \cdot L^2)$.
  c. For a rectangular-cross-section beam, express the ratio $EI/(AG \cdot L^2)$ in terms of the depth–span ratio d/L, and discuss the effect of shear distortions on the results of this analysis.

Prob. 4

5. Analyze the continuous beam shown due to
   a. A uniform load *w*.
   b. A concentrated load *P* at the midspan of *BC*.
   c. A support settlement Δ of point *B*.
   Draw shear and moment diagrams for each case. (*Hint:* The results of Prob. 9.8 may be used.)

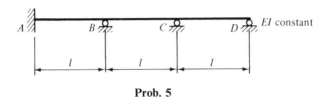

Prob. 5

6. Analyze the fixed-ended beam under uniform load, and draw shear and moment diagrams. Consider only flexural distortions. (*Hint:* The flexibility matrix of Prob. 9.9 may be used.)
   Discuss the effect of stiffening a portion of the structure on the transmission of the applied load to its supports.

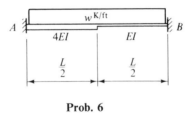

Prob. 6

7. Analyze the tapering beam shown (a "wedge beam"), fixed at both ends, due to
   a. A uniform load *w* kips/ft.
   b. A vertical support settlement Δ at end *B*.
   Consider only flexural distortions, and draw shear and moment diagrams. (*Hint:* The results of Prob. 9.4 may be used.)

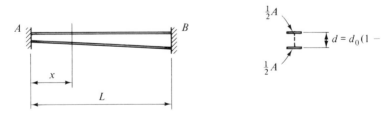

Prob. 7

8. Each half of the two-span haunched beam is described in greater detail in

Prob. 9.11, the results of which may be used here. Analyze the structure and draw shear and moment diagrams.

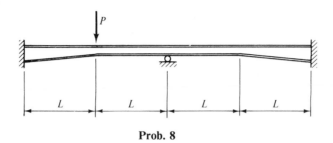

Prob. 8

9. Analyze the rigid frame shown for the effects of
   a. The uniform lateral load shown.
   b. A uniform temperature expansion of value $\alpha \, \Delta T$.
   c. A specified support rotation θ at point *D*.
   Draw shear and moment diagrams for each case.

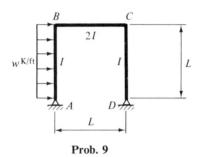

Prob. 9

10. Analyze the two-bay rigid frame under the load shown. Draw shear and moment diagrams.

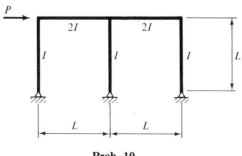

Prob. 10

11. Analyze the structure under the load shown, considering only flexural dis-

tortions. Draw shear, moment, and axial force diagrams. (*Hint:* The results of Prob. 9.10 may be used.)

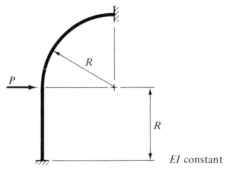

Prob. 11

12. a. Analyze the two-hinged semicircular arch under uniform load, considering only flexural distortions.

   b. Analyze the arch, considering both axial and flexural distortions. Plot the variation of the moment M_B at the crown with respect to the ratio I/AR^2 of the flexural to axial stiffness, and draw conclusions regarding the effect of axial shortening.

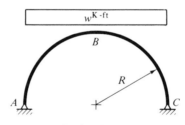

Prob. 12

13. Analyze the two-hinged semicircular arch, and compute the rotation at point B due to
   a. The applied moment M_B.
   b. The load P applied at point α.

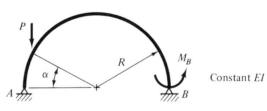

Prob. 13

14. Using the results of Prob. 13, analyze the continuous arch structure. Plot the moments in the structure.

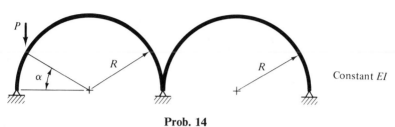

Constant *EI*

Prob. 14

15. Analyze the structure shown by the matrix force method. Reduce the structure to an appropriate primary form, number the nodes in an appropriate fashion (listing the redundant nodes first), compute the structure flexibility matrix, set up a set of equations in matrix form, and solve for the following quantities:

a. The redundant forces,

b. The vertical deflection at point *B*,

c. The internal moments,

due to two loading conditions:

1. A vertical force *P* applied at point *B*.

2. A vertical support displacement at point *C*.

Neglect axial deformations of the structure.

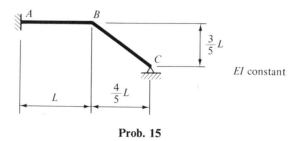

EI constant

Prob. 15

16. Analyze the circular ring due to the diametrically opposed loads *P*. Consider only flexural distortions. Draw a moment diagram in polar coordinates. (*Hint:* Cut the structure so you can use symmetry.)

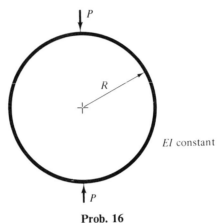

Prob. 16

17. Analyze the rigid gable frame due to
    a. Uniform roof load, of intensity w lb/ft of horizontal projection.
    b. Lateral load P applied horizontally at point B.
    c. A temperature gradient $\alpha\,\Delta T/d$, acting on the gable members BC and CD only.
    Draw shear and moment diagrams for each case.

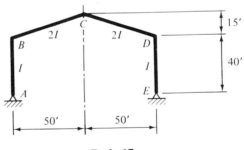

Prob. 17

18. For the ideal truss subject to the concentrated load shown, calculate
    a. All reactions and member forces, indicating them on an appropriate sketch of the structure.
    b. The vertical deflection of point B of the truss.
    All members are of equal cross-sectional area A, and of modulus E.

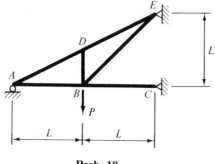

Prob. 18

19. Analyze the truss due to
  a. The indicated load *P*.
  b. A uniform temperature elongation $\alpha \, \Delta T$ of the top chord only.

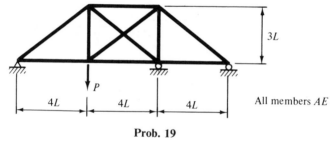

Prob. 19

20. Joint *C* of the pin-connected truss is held against displacement in the *Y* direc-

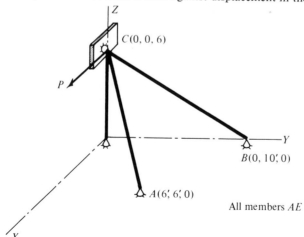

Prob. 20

tion. Calculate all member forces and the Z component of displacement of joint C, due to the applied load P, parallel to the X axis. (*Note:* The results of Ex. 9.7 may be used.)

21. The bent pipe structure, fully fixed at ends A and C, lies in the horizontal plane. $E = 2G$. Analyze the structure, and plot bending and torsional moments due to
    a. A vertical load P applied at point B.
    b. The dead weight of the structure, of intensity w.
    c. An imposed vertical displacement Δ at point C.
    d. A uniform vertical temperature gradient $\alpha \, \Delta T/d$ over the entire structure.
    (*Hint:* Use results of Ex. 9.8.)

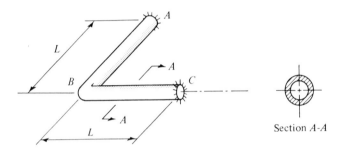

Section A-A

Prob. 21

22. The tapering beam, which has been analyzed exactly in Prob. 7, is to be analyzed approximately by representing it by three prismatic elements of appropriate stiffness, as shown. Apply the matrix force method to this substitute structure, subject to
    a. A uniform load w kips/ft.
    b. A vertical support settlement Δ at end B.

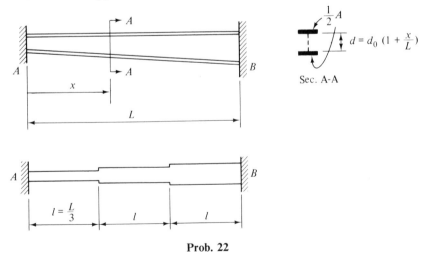

$d = d_0 \left(1 + \dfrac{x}{L}\right)$

Sec. A-A

$l = \dfrac{L}{3}$

Prob. 22

Compare approximate results with the exact ones of Prob. 7, and draw conclusions regarding the feasibility of discretizing variable stiffness members.

23. a. The helicoidal girder is fully fixed at both ends. Analyze it due to its dead
weight, of intensity w lb/ft. Let $H = R$ and $G = .385E$. (*Hint:* Refer to Ex.
8.4, and Prob. 8.15.)

    b. Approximate the helicoidal girder by six straight elements of equal length,
analyze by the matrix force method, and compare results.

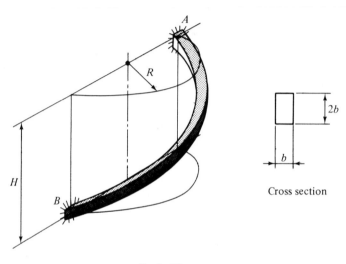

Cross section

Prob. 23

CHAPTER 11

Fixed-End Forces and Member Stiffness

In this chapter we shall define two concepts, fixed-end forces and stiffnesses, which will be used as building blocks in the displacement method to be discussed in Part V. Their calculation will be effected by use of the force method, whereby the decision whether or not to cast the work in matrix formulation is in all cases up to the analyst.

The contents of this chapter should be considered both as further practice in the use of the force method and as a foundation for analysis by the displacement method.

11.1. Fixed-End Forces

Fixed-end forces are the end reactions of a member resulting from a load when the member ends are completely restrained against motion. Such fixed-end forces are one of the building blocks of the displacement method; their nature and computation should be fully understood at this time so they will be available when needed.

Figure 11.1 shows three different types of loaded members whose ends are fully fixed. The indicated reactions are, according to the previous definition, the fixed-end forces (including both actual forces and moments). They could be caused either by applied loads or by self-straining effects, such as temperature changes or prestressing. As will be recognized from Fig. 11.1, the

(a) Fixed-ended beam

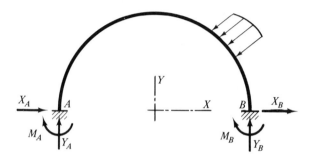

(b) Fixed-ended arch

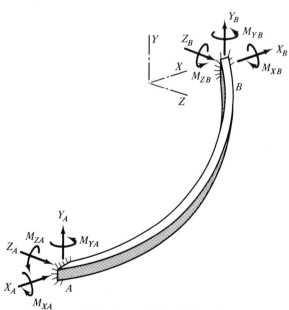

(c) Fixed-ended helicoidal girder

Fig. 11.1 Fixed-end forces.

determination of fixed-end forces constitutes a statically indeterminate problem, which we shall here attack by the force method.

Note in particular that Ex. 10.4 is typical of the work to be performed; the reactions shown in part (f) of the figure of that example constitute the fixed-end forces due to the specified load. Note that the conditions of geometry yield only a number of fixed-end forces equal to the degree of static indeterminacy; the remaining forces, if required, will have to be computed from statics.

Sometimes not all fixed-end forces are needed; for instance, in the special versions of the displacement method discussed in Secs. 14.1 and 14.2, only the fixed-end moments (not forces) are required. In such cases, it seems appropriate to cut the statically indeterminate structure by deleting these moments so that they will appear as redundants, and no further statical calculations are necessary.

The condition of the structure in which all its support displacements take the value zero is called *geometrically determinate*, or *kinematically determinate*. This is in analogy to the condition in which all redundant support forces take the value zero, in which case the structure is called *statically determinate*. A clear understanding of this analogy will help us to grasp some important relations between the force and the displacement methods.

The following example illustrates the method further.

Example 11.1: Calculate the fixed-end forces at ends A and B of the prismatic beam of part (a) of the figure due to a load P at x from the left end.

Solution: The structure is twice indeterminate; we consider the end moments as redundants, and accordingly number these nodes as 1 and 2. The end reactions will be numbered 3 and 4, as shown in part (a). The deformations due to loads and the flexibilities of the primary structure are computed by any method and inserted into the $\{\Delta^\circ\}$ and $[F]$ matrices:

$$\{\Delta^\circ\} = \frac{P \cdot x \cdot (L - x)}{6EI \cdot L} \begin{bmatrix} 2L - x \\ -(L + x) \end{bmatrix}; \qquad [F] = \frac{L}{6EI} \begin{bmatrix} 2 & -1 \\ -1 & 2 \end{bmatrix}$$

The equations of geometry are set up according to Eq. (10.4), and solved for the fixed-end moments according to Eq. (10.5):

$$\begin{Bmatrix} X_1 \\ X_2 \end{Bmatrix} \equiv \begin{Bmatrix} M_A^F \\ M_B^F \end{Bmatrix} = [F]^{-1}[-\{\Delta^\circ\}] = \frac{P}{L^2} \begin{Bmatrix} -x(L - x)^2 \\ x^2(L - x) \end{Bmatrix} \qquad (11.1)$$

The fixed-end reactive forces X_3 and X_4 are found by the statics of part (b):

$$\sum \overset{\curvearrowright}{M_B} = 0: \qquad -P(L - x) - \frac{P}{L^2}x(L - x)^2 + \frac{P}{L^2}x^2(L - x) + X_3 L = 0$$

$$X_3 = R_A^F = P\left[1 - 3\left(\frac{x}{L}\right)^2 + 2\left(\frac{x}{L}\right)^3\right] \qquad (11.2)$$

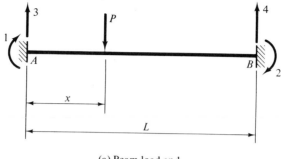

(a) Beam load and
nodal numbering

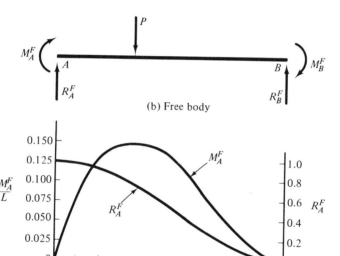

(b) Free body

(c) Influence lines for M_A^F and R_A^F

Fig. E11.1

Similarly,

$$X_4 = R_B^F = P\left[3\left(\frac{x}{L}\right)^2 - 2\left(\frac{x}{L}\right)^3\right]$$

Equations (11.1) and (11.2), with the value of P equal to unity, are influence functions for the fixed-end forces; those for M_A^F and R_A^F are shown plotted as influence lines in part (c).

Once we have found the effects due to an arbitrarily located load on a structure, we can find the corresponding effects due to a distributed load by integration, as already discussed in Sec. 3.1. For instance, in Ex. 11.1 the fixed-end moment M_B^F due to a concentrated load P located at a distance x of a beam has been found to be

$$M_B^F = f(P, x) = \frac{P}{L^2}x^2(L - x)$$

We now consider the distributed load $w(x)$ acting on the same beam, as shown in Fig. 11.2; the small portion dP of the load acting on an infinitesimal length dx is $dP = w(x)\,dx$. The moment due to this infinitesimal load is then that due to $P = dP = w(x)\,dx$:

$$dM_B^F = f[w(x)\,dx, x] = \frac{w(x)\,dx}{L^2}(Lx^2 - x^3)$$

and the moment due to the entire load on the beam is obtained by integration:

$$M_B^F = \int_{x=0}^{L} dM_B^F = \int_0^L f[w(x)\,dx, x] = \int_0^L \frac{w(x)\,dx}{L^2}(Lx^2 - x^3)$$

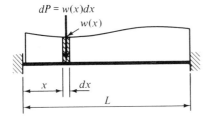

Fig. 11.2 Integration of distributed load.

In the particular case of a *uniform* load w on the beam, for instance, the constant w can be removed from under the integral, resulting in

$$M_B^F = \frac{w}{L^2}\int_0^L (Lx^2 - x^3)\,dx = \frac{wL^2}{12} \tag{11.3}$$

The same solution could of course have been obtained by starting from scratch.

Self-straining effects, such as temperature gradients, prestressing, or the like, would also cause fixed-end forces in structures. In all such cases, an appropriate indeterminate analysis is required, in accordance with the approach of this chapter.

11.2. Member or Element Stiffness

To define the concept of stiffness of a member in a clear-cut manner, we resort to numbering of nodes. For each reactive component on the member, a node is attached to the structure and suitably numbered. Figure 11.3(a), for

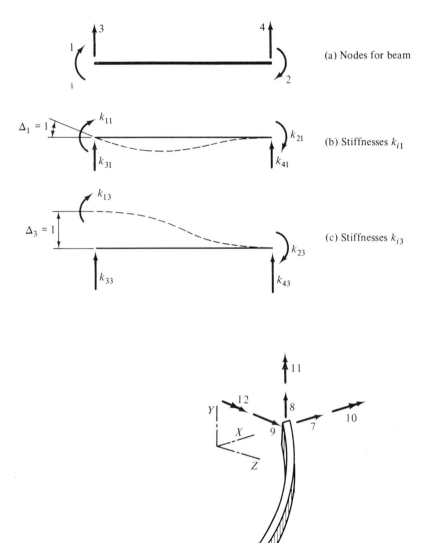

(a) Nodes for beam

(b) Stiffnesses k_{i1}

(c) Stiffnesses k_{i3}

(d) Nodal numbering for helicoidal girder

Fig. 11.3 Nodal numbering for stiffness.

instance, shows the nodal numbering for a plane beam, and Fig. 11.3(d) that for a general curvilinear member in space. It was discussed earlier that each such node represents a force and a corresponding displacement.

We can now define the member, or element, stiffnesses in the following manner:

The stiffness k_{ij} is the force arising at node i due to a unit displacement of node j, all other nodal displacements equal to zero.

Figure 11.3b, for instance, shows the deformed shape of the beam due to a unit rotation $\Delta_1 = 1$ of node 1 (all other nodal displacements zero), and the corresponding stiffnesses k_{i1}; similarly, Fig. 11.3c shows the stiffnesses k_{i3} due to a unit displacement of node 3, all other nodal displacements zero.

The stiffness k_{ij} ($i = j$), that is, k_{11} in Fig. 11.3(b) and k_{33} in Fig. 11.3(c), may be visualized as the loads causing the unit displacement, and k_{ij} ($i \neq j$), that is, all other stiffnesses, can be looked at as the reactions of the suitably restrained member. The determination of member stiffnesses can be recognized as a statically indeterminate problem, in which all reactions due to specified support displacements of value unity, applied one at a time, are to be found. We shall tackle this problem by the force method, as demonstrated in the following examples.

Element stiffnesses constitute one of the building blocks of the displacement (or stiffness) method of analysis. Their meaning and calculation should be learned at this time so that they will be well understood when they are needed in Chapter 13.

Example 11.2: Given is the straight prismatic beam of length L, stiffness EI, and nodal numbering shown in part (a) of the Figure. Calculate the end stiffnesses k_{i1}.

Solution: The stiffnesses k_{i1} are caused by a unit rotation of node 1, all other nodal displacements zero, as shown in part (b). The stiffnesses k_{11} and k_{21} are the end moments, and k_{31} and k_{41} the end reactive forces. We shall consider the end moments as the redundants of this two-times indeterminate structure; that is, the primary structure consists of the simple beam of part (c). There is no load on the structure, and satisfaction of the specified end rotations is assured by the conditions of geometry.

$$\Sigma\,\theta_A = 1: \qquad k_{11}\frac{L}{3EI} - k_{21}\frac{L}{6EI} = 1$$

$$\Sigma\,\theta_B = 0: \qquad -k_{11}\frac{L}{6EI} + k_{21}\frac{L}{6EI} = 0$$

The flexibilities are taken from previous work [see Eq. (9.6)]. Solving simul-

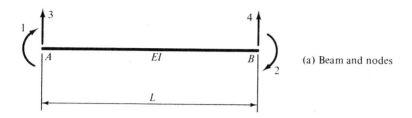

(a) Beam and nodes

(b) Nodal displacements

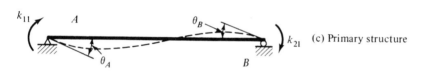

(c) Primary structure

Fig. E11.2

taneously for the stiffnesses, we obtain

$$k_{11} = \frac{4EI}{L}; \qquad k_{21} = \frac{2EI}{L}$$

With the redundants known, we can solve for the remaining reactions by statics:

$$k_{31} = -\frac{6EI}{L^2}; \qquad k_{41} = \frac{6EI}{L^2}$$

We now have the four stiffness factors k_{i1} due to a unit rotation $\Delta_1 = 1$; we recognize that, in general, the work must now be repeated for all the remaining nodal displacements $\Delta_j = 1$ to determine all stiffnesses. It is apparent that this highly repetitive work is suitable for matrix formulation. The following example will continue the work of Ex. 11.2 in this particular fashion. It should be noted that the reasoning is identical in the two examples; only the bookkeeping scheme differs.

Example 11.3: Find the complete stiffness matrix for the four numbered nodes of the straight prismatic beam of part (a) of the figure.

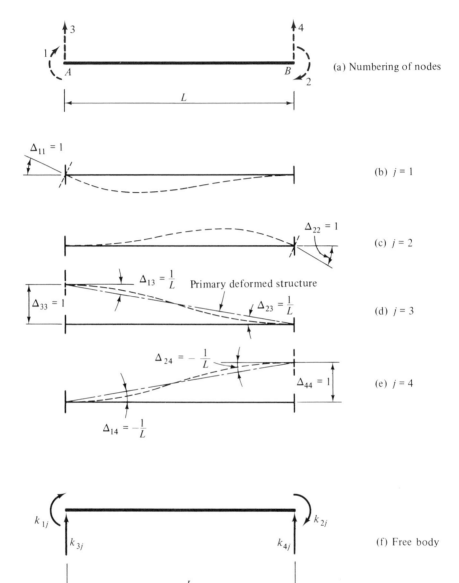

(a) Numbering of nodes

(b) $j = 1$

(c) $j = 2$

(d) $j = 3$

(e) $j = 4$

(f) Free body

Fig. E11.3

Solution: The equations of geometry of Ex. 11.2 are now cast in matrix form, where we now consider four different loading conditions (the loading here consists of specified unit support displacements). In terms of the previously used redundant end rotations Δ_1 and Δ_2 of the primary structure, the specified displacements are, referring to parts (b) through (e),

$$
\begin{array}{llll}
\text{loading 1} & j = 1: & \Delta_{11} = 1; & \Delta_{21} = 0 \\[6pt]
\text{loading 2} & j = 2: & \Delta_{12} = 0; & \Delta_{22} = 1 \\[6pt]
\text{loading 3} & j = 3: & \Delta_{13} = \dfrac{1}{L}; & \Delta_{23} = \dfrac{1}{L} \\[10pt]
\text{loading 4} & j = 4: & \Delta_{14} = -\dfrac{1}{L}; & \Delta_{24} = -\dfrac{1}{L}
\end{array}
$$

For each of these four loading conditions, a separate column is provided for the redundant end moments k_{1j} and k_{2j}, and for the specified displacements Δ_{1j} and Δ_{2j}. The flexibility matrix consists of the same flexibilities used in Ex. 11.2. The conditions of geometry, Eq. (10.4), will now become

$$
\frac{L}{6EI}
\begin{bmatrix} 2 & -1 \\ -1 & 2 \end{bmatrix}
\begin{bmatrix} k_{11} & k_{12} & k_{13} & k_{14} \\ k_{21} & k_{22} & k_{23} & k_{24} \end{bmatrix}
=
\begin{bmatrix} 1 & 0 & \dfrac{1}{L} & -\dfrac{1}{L} \\ 0 & 1 & \dfrac{1}{L} & -\dfrac{1}{L} \end{bmatrix}
$$

Inverting the $[f]$ matrix and multiplying as per Eq. (10.5), we obtain the rotational end stiffnesses:

$$
\begin{bmatrix} k_{11} & k_{12} & k_{13} & k_{14} \\ k_{21} & k_{22} & k_{23} & k_{24} \end{bmatrix}
= \frac{EI}{L^3}
\begin{bmatrix} 4L^2 & 2L^2 & -6L & 6L \\ 2L^2 & 4L^2 & -6L & 6L \end{bmatrix}
$$

From these redundant end moments, the stiffnesses k_{3j} and k_{4j}, representing the reactive end forces, may be obtained by statics, using for instance the force transformation matrix $[b]$ relating end reactions to end moments; from the equilibrium of the free body of part (f),

$$
\begin{bmatrix} k_{3j} \\ k_{4j} \end{bmatrix}
=
\begin{bmatrix} -\dfrac{1}{L} & -\dfrac{1}{L} \\ \dfrac{1}{L} & \dfrac{1}{L} \end{bmatrix}
\begin{bmatrix} k_{1j} \\ k_{2j} \end{bmatrix}
\equiv [b]
\begin{bmatrix} k_{1j} \\ k_{2j} \end{bmatrix}
$$

Substituting the previously obtained matrix $\begin{bmatrix} k_{1j} \\ k_{2j} \end{bmatrix}$, we obtain the remaining stiffnesses:

$$
\begin{aligned}
\begin{bmatrix} k_{3j} \\ k_{4j} \end{bmatrix}
&= \frac{1}{L}
\begin{bmatrix} -1 & -1 \\ 1 & 1 \end{bmatrix}
\frac{EI}{L^3}
\begin{bmatrix} 4L^2 & 2L^2 & -6L & 6L \\ 2L^2 & 4L^2 & -6L & 6L \end{bmatrix} \\[6pt]
&= \frac{EI}{L^3}
\begin{bmatrix} -6L & -6L & 12 & -12 \\ 6L & 6L & -12 & 12 \end{bmatrix}
\end{aligned}
$$

Writing now the rotational end stiffnesses k_{1j} and k_{2j} and the translational end stiffnesses k_{3j} and k_{4j} into one stiffness matrix $[k]$, we obtain

$$[k] = \frac{EI}{L^3} \begin{bmatrix} 4L^2 & 2L^2 & -6L & 6L \\ 2L^2 & 4L^2 & -6L & 6L \\ -6L & -6L & 12 & -12 \\ 6L & 6L & -12 & 12 \end{bmatrix} \tag{11.4}$$

This square symmetric stiffness matrix of order equal to the number of nodes considered contains the end stiffnesses of a straight prismatic beam. The analyst should be able to explain the meaning of each of the elements of this matrix in terms of a sketch similar to Figs. 11.3(b) to (e).

The stiffness matrix $[k]$ relates the nodal forces $[X]$ to the nodal displacements $[\Delta]$:

$$[X] = [k][\Delta] \tag{11.5}$$

For a straight prismatic beam with ends A and B, for instance, Eq. (11.4) yields the end actions resulting from given end rotations and translations:

$$\begin{Bmatrix} M_A \\ M_B \\ Y_A \\ Y_B \end{Bmatrix} = \frac{EI}{L^3} \begin{bmatrix} 4L^2 & 2L^2 & -6L & 6L \\ 2L^2 & 4L^2 & -6L & 6L \\ -6L & -6L & 12 & -12 \\ 6L & 6L & -12 & 12 \end{bmatrix} \begin{Bmatrix} \theta_A \\ \theta_B \\ \Delta_A \\ \Delta_B \end{Bmatrix}$$

The stiffness matrix thus represents a *force–deformation relationship*.

Before proceeding to a more formal development of the stiffness matrix, we shall consider another problem to strengthen our insight into the concept.

Example 11.4: Calculate the rotational stiffness of the right end of the semicircular arch shown in the figure.

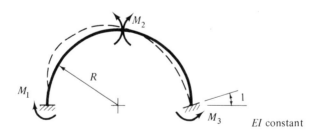

Fig. E11.4

Solution: This three-times indeterminate structure has already been analyzed in Ex. 10.4; here, we remove the terms due to the load in the three conditions of geometry of that problem, and set the sum of the rotations at point 3 equal to 1, all others equal to zero (note that here the stiffness EI cannot be canceled):

$$.571M_1 + .215M_2 - .215M_3 = 0$$

$$.215M_1 + 1.571M_2 + .215M_3 = 0$$

$$-.215M_1 + .215M_2 + .571M_3 = 1 \cdot \frac{EI}{R}$$

$$M_3 = k_{33} = 2.32\frac{EI}{R}$$

In terms of the numbering of the nodes shown in the figure, we can also identify the moment M_1 as the stiffness k_{13}:

$$M_1 = k_{13} = 1.05\frac{EI}{R}$$

To establish a relation between the stiffness matrix $[k]$ and the flexibility matrix $[f]$ discussed in Chapter 9, we recall that the stiffnesses k_{ij}, contained in column j of the stiffness matrix, are forces $[X]$ due to an imposed displacement $\Delta_j = 1$, all other Δ's zero. We visualize displacements of this type imposed at all nodes, one at a time, for a total of n loading conditions; the external loads are zero, so that the redundant displacements $[\Delta°]$ due to loads in Eq. (10.4) vanish. Equation (10.4) then becomes

$$
\begin{bmatrix} f_{11} & f_{12} & \cdots & f_{1n} \\ & & \vdots & \\ f_{n1} & \cdots & & f_{nn} \end{bmatrix}
\begin{bmatrix} k_{11} & k_{12} & \cdots & k_{1n} \\ & & \vdots & \\ k_{n1} & \cdots & & k_{nn} \end{bmatrix}
=
\begin{bmatrix} 1 & 0 & \cdots & 0 \\ 0 & 1 & & \\ & & \ddots & \\ 0 & & & 1 \end{bmatrix}
$$

or

$$[f][k] = [I]$$

from which, by the definition of matrix inversion,

$$[k] = [f]^{-1} \tag{11.6}$$

Equation (11.6) could also be derived simply by comparing Eqs. (9.9) and (11.5).

Equation (11.6) represents the force-method solution by means of a specific statically determinate primary system obtained by introducing r redundant cuts. For this reason, the maximum order of the $[f]$ and $[k]$ matrices that can be related by Eq. (11.6) is equal to the degree of static indeterminacy r. Such matrices of order greater than r are singular, because they represent the properties of an unstable structure resulting from an excessive number of cuts. Singular matrices cannot be inverted. However, the remaining stiffnesses can always be determined from statics, as was done in Ex. 11.3.

According to Eq. (9.10), the flexibility matrix $[f]$ was shown to be symmetric; since the inverse of a symmetric matrix is also symmetric, it follows that the stiffness matrix $[k]$ is also symmetric; that is

$$k_{ij} = k_{ji} \qquad (11.7)$$

In words, this means that the force at node i due to a unit displacement at node j is equal to the force at node j due to a unit displacement at node i.

Example 11.5: From the flexibility matrix $[f]$ for nodes 1, 2, and 3 of the semicircular fixed prismatic arch shown in the figure, compute the stiffness matrix $[k]$.

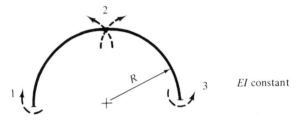

Fig. E11.5

Solution: In Exs. 10.4 and 11.4, the flexibility matrix is

$$[f] = \frac{R}{EI} \begin{bmatrix} .571 & .215 & -.215 \\ .215 & 1.571 & .215 \\ -.215 & .215 & .571 \end{bmatrix}$$

Inverting this matrix, we compute the stiffness matrix as

$$[k] = [f]^{-1} = \frac{EI}{R} \begin{bmatrix} 2.32 & -.466 & 1.05 \\ -.466 & .763 & -.466 \\ 1.05 & -.466 & 2.32 \end{bmatrix}$$

PROBLEMS

1. Calculate the fixed-end forces for the numbered nodes for the linearly varying load shown.

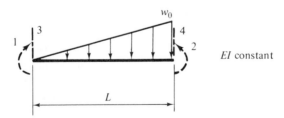

Prob. 1

2. Calculate the fixed-end forces for the numbered nodes due to the self-equilibrating force system shown.

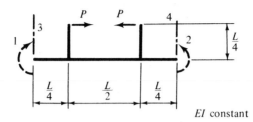

EI constant

Prob. 2

3. Calculate the fixed-end forces along nodes 1 to 4 due to a temperature gradient ΔT between top and bottom of the beam. The coefficient of thermal expansion is α. The flexural stiffness is EI.

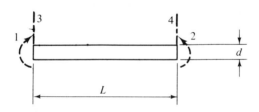

Prob. 3

4. The prismatic beam is prestressed by means of a parabolically draped cable of eccentricity $e = e_0[1 - (2x/L)^2]$ under constant prestressing force P. Calculate the fixed-end forces due to this effect. (*Hint:* Prestressing causes moment $M = P \cdot e$.)

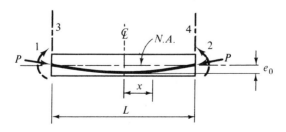

Prob. 4

5. Calculate the fixed-end forces on the tapering beam shown due to a uniform load of intensity w. (*Hint:* Use the results of Prob. 9.4.)

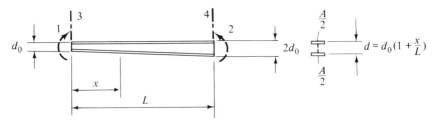

Prob. 5

6. Calculate the fixed-end forces along the numbered nodes of the prismatic circular member subtending an angle α due to a single concentrated radial load as shown. Consider only flexural distortions.

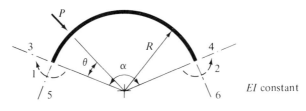

Prob. 6

7. Determine the fixed-end forces of the bent pipe structure in the horizontal plane due to the vertical load shown. (*Note:* This is the same structure as in Ex. 9.8.)

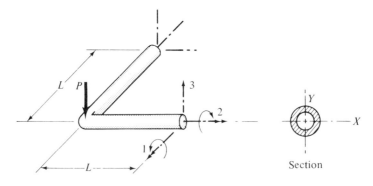

Prob. 7

8. Compute the element stiffness matrix $[k]$, of order (2×2), for the prismatic beam of flexural stiffness EI. Neglect axial distortions. Proceed in two ways:
a. Solve as a statically indeterminate problem.
b. Solve by inverting the flexibility matrix of Prob. 9.2.

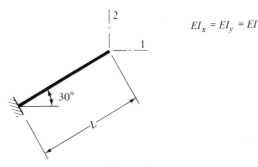

$EI_x = EI_y = EI$

Prob. 8

9. Compute the stiffness matrix, of order (2×2), for the numbered nodes of the single beam by
   a. A formal force-method solution.
   b. By inversion of the flexibility matrix.
   Show that the two methods are identical.

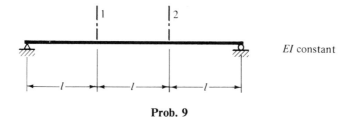

EI constant

Prob. 9

10. Determine the (2×2) stiffness matrix for the prismatic, pin-ended two-force member shown:
    a. For the nodes in part (a) of the figure.
    b. For the nodes in part (b).
    (*Hint*: Use geometric relations to determine axial strains and stresses resulting from a unit nodal displacement.)

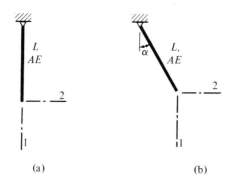

(a)   (b)

Prob. 10

11. Compute the (3×3) stiffness matrix, considering only flexural distortions. (*Hint:* Consider the results of Prob. 9.5.)

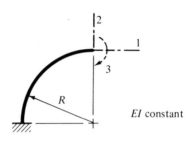

EI constant

Prob. 11

CHAPTER 12

Additional Topics in the Force Method

In this chapter we shall consider some supplementary topics in the force method of indeterminate analysis. Sections 12.1 to 12.3 concern themselves with the process of orthogonalization of the set of equations of geometry, leading to the very convenient method of column analogy for the solution of an important class of indeterminate structures, which avoids the solution of simultaneous equations. This method is particularly convenient for the determination of fixed-end forces and stiffnesses.

In Sec. 12.4 the force method is applied to the determination of influence lines for statically indeterminate structures, which are essential in the design of continuous bridge structures.

12.1. Orthogonalization and the Elastic Center

The force-method approach consists basically of the solution of a set of equations

$$
\begin{bmatrix}
F_{11} & F_{12} & \cdots & F_{1r} \\
F_{21} & F_{22} & & \cdot \\
\cdot & & & \cdot \\
\cdot & & & \cdot \\
\cdot & & & \cdot \\
F_{r1} & F_{r2} & \cdots & F_{rr}
\end{bmatrix}
\begin{Bmatrix}
X_1 \\
X_2 \\
\cdot \\
\cdot \\
\cdot \\
X_r
\end{Bmatrix}
+
\begin{Bmatrix}
\Delta_1^\circ \\
\Delta_2^\circ \\
\cdot \\
\cdot \\
\cdot \\
\Delta_r^\circ
\end{Bmatrix}
=
\begin{Bmatrix}
\Delta_1 \\
\Delta_2 \\
\cdot \\
\cdot \\
\cdot \\
\Delta_r
\end{Bmatrix}
\qquad (12.1)
$$

By redefining the redundants $\{X\}$ in a suitable manner (let the redefined set of equations be denoted by barred quantities: $\{\bar{X}\}$, $[\bar{F}]$, etc.), it is always possible to obtain a set of equations of the form

$$
\begin{bmatrix}
\bar{F}_{11} & 0 & \cdots & 0 \\
0 & \bar{F}_{22} & & \\
\vdots & & \ddots & \\
0 & & & \bar{F}_{rr}
\end{bmatrix}
\begin{Bmatrix}
\bar{X}_1 \\
\bar{X}_2 \\
\vdots \\
\bar{X}_r
\end{Bmatrix}
+
\begin{Bmatrix}
\bar{\Delta}_1^\circ \\
\bar{\Delta}_2^\circ \\
\vdots \\
\bar{\Delta}_r^\circ
\end{Bmatrix}
=
\begin{Bmatrix}
\bar{\Delta}_1 \\
\bar{\Delta}_2 \\
\vdots \\
\bar{\Delta}_r
\end{Bmatrix}
\qquad (12.2)
$$

or
$$
[\bar{F}]\{\bar{X}\} + \{\bar{\Delta}^\circ\} = \{\bar{\Delta}\}
$$

in which all off-diagonal terms $\bar{F}_{ij}$ $(i \neq j)$ of the $[\bar{F}]$ matrix are zero. Such a matrix is called a *diagonal* or *orthogonal* matrix, and the process of achieving this is called *orthogonalization*. In such a diagonalized set of equations, the unknowns $\{\bar{X}\}$ can be obtained directly without simultaneous solution.

Physically, this idea could be explained in terms of the two-times indeterminate beam of Fig. 12.1(a), with the redundant reactions X_1 and X_2. We can now visualize a certain ratio of these redundants $k_1 = X_2/X_1$, so that when these two redundants are applied *jointly* the deformed shape shown in Fig. 12.1(b) results, for which $\bar{F}_{11} \neq 0$, but $\bar{F}_{21} = 0$. The *group redundant* $\bar{X}_1$, which causes these deformations, is composed of the force X_1 at cut 1 and the force $k_1 X_1$ at cut 2. Similarly, a combination of redundants $\bar{X}_2$, as shown in Fig. 12.1(c), could be devised that causes $\bar{F}_{22} \neq 0$, $\bar{F}_{12} = 0$. By redefining the redundants in this manner, all off-diagonal flexibilities of the $[\bar{F}]$ matrix have been eliminated, so that the equations of geometry take the form of Eq. (12.2).

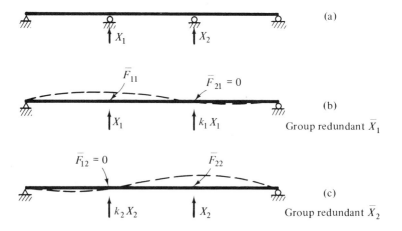

Fig. 12.1 Diagonalized redundants.

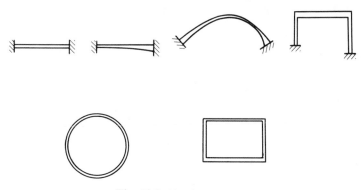

Fig. 12.2 Circuit structures.

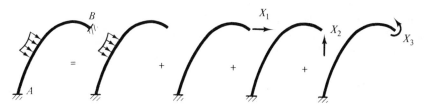

Fig. 12.3 General redundants.

Mathematical procedures for such orthogonalizations are available for matrices of any order, but we shall restrict our attention here to planar, *circuit structures*, which have at most two supports and three degrees of static indeterminacy. This class of structures includes all straight and curved beams, single arches and rigid frames, and closed rings, as shown in Fig. 12.2.

We consider a general case of such a three-times indeterminate structure, shown in Fig. 12.3. We consider only effects of flexural distortions in the analysis. The classical approach would now be to release one end B, removing the redundants X_1, X_2, and X_3, to create the primary structure, as in Fig. 12.3, and to write the three conditions of geometry as per Eq. (12.1), with $r = 3$.

With a view to the possibility of reducing Eqs. (12.1) to the diagonal form of Eq. (12.2), we now choose a new point of redundancy, labeled C in Fig. 12.4, and a new set of axes U, V through this point. To connect this point to the structure, we envision a rigid arm connecting points C and B, so that the specified displacements at end B of the actual structure can be related to the movement of point C.

At the new point C of redundancy, we apply redundants $\bar{X}_1$, $\bar{X}_2$, and $\bar{X}_3$, as shown in Fig. 12.4. If these redundants are known, the forces in the structure can be found by statics; for instance, the moments at any point of the structure defined by the coordinates u, v are given by

$$M = M_0 + \bar{X}_1 \cdot v - \bar{X}_2 \cdot u + \bar{X}_3 \qquad (12.3)$$

where M_0 denotes the moments in the primary structure due to the applied load.

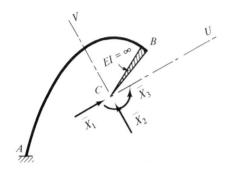

Fig. 12.4 Diagonalized redundants.

To determine the displacements at the point of redundancy C of the primary structure, we recall the general arch equations, Eqs. (7.8), (7.9), and (7.10) from Sec. 7.4. With respect to the U, V axes, they become

$$\bar{\Delta}_u = \int_L \Phi v \, ds; \qquad \bar{\Delta}_v = -\int_L \Phi u \, ds; \qquad \bar{\theta} = \int_L \Phi \, ds \qquad (12.4)$$

These deformations of the primary structure due to unit values of the redundants, applied one at a time, are the flexibilities $[\bar{F}]$; for their determination, we insert the curvatures $\Phi = M/EI$ in Eqs. (12.4), where the moment M is obtained from Eq. (12.3):

due to $\bar{X}_1 = 1$: $\Phi = \dfrac{M}{EI} = \dfrac{1 \cdot v}{EI}$;   therefore

$$\bar{\Delta}_u = \bar{F}_{11} = \int_L \frac{v^2 \, ds}{EI}; \qquad \bar{\Delta}_v = \bar{F}_{21} = -\int_L \frac{uv \, ds}{EI}$$

$$\bar{\theta} = \bar{F}_{31} = \int_L \frac{v \, ds}{EI}$$

due to $\bar{X}_2 = 1$: $\Phi = -\dfrac{1 \cdot u}{EI}$;   therefore

$$\bar{\Delta}_u = \bar{F}_{12} = -\int_L \frac{uv \, ds}{EI}; \qquad \bar{\Delta}_v = \bar{F}_{22} = \int_L \frac{u^2 \, ds}{EI} \qquad (12.5)$$

$$\bar{\theta} = \bar{F}_{32} = -\int_L \frac{u \, ds}{EI}$$

due to $\bar{X}_3 = 1$: $\Phi = \dfrac{1}{EI}$;   therefore

$$\bar{\Delta}_u = \bar{F}_{13} = \int_L \frac{v \, ds}{EI}; \qquad \bar{\Delta}_v = \bar{F}_{23} = -\int_L \frac{u \, ds}{EI}$$

$$\bar{\theta} = \bar{F}_{33} = \int_L \frac{ds}{EI}$$

Similarly, the deformations $\{\bar{\Delta}^\circ\}$ at point C of the primary structure due to the applied load are caused by curvatures $\Phi = M_0/EI$, so Eqs. (12.4) yield

$$\{\bar{\Delta}^\circ\} = \begin{Bmatrix} \bar{\Delta}_1^\circ \\ \bar{\Delta}_2^\circ \\ \bar{\Delta}_3^\circ \end{Bmatrix} = \begin{Bmatrix} \displaystyle\int_L \frac{M_0 v \, ds}{EI} \\ -\displaystyle\int_L \frac{M_0 u \, ds}{EI} \\ \displaystyle\int_L \frac{M_0 \, ds}{EI} \end{Bmatrix} \tag{12.6}$$

To keep matters concise, we shall not for the time being consider any support displacements $\{\bar{\Delta}\}$.

We now recall that the purpose of the exercise was to make all off-diagonal $\bar{F}_{ij}$'s $(i \neq j)$ equal to zero in order to orthogonalize the set of equations. Referring back to Eqs. (12.5), we now seek to locate the point C and the axes U, V such that

$$\bar{F}_{23} = \bar{F}_{32} = -\int_L \frac{u \, ds}{EI} = 0; \qquad \bar{F}_{13} = \bar{F}_{31} = \int_L \frac{v \, ds}{EI} = 0$$

$$\bar{F}_{21} = \bar{F}_{12} = -\int_L \frac{uv \, ds}{EI} = 0 \tag{12.7}$$

If these conditions are fulfilled, we call point C the *elastic center* and axes U, V the *conjugate axes* of the structure. The nature of the flexibilities resulting from the corresponding set of redundants is shown in Fig. 12.5.

To satisfy the three Eqs. (12.7), we have three unknowns, the two coordinates of the elastic center C and the inclination of the axes; therefore, it will always be possible to achieve this orthogonalization. The conjugate axis system can be found by the purely mathematical procedures of solving Eqs.

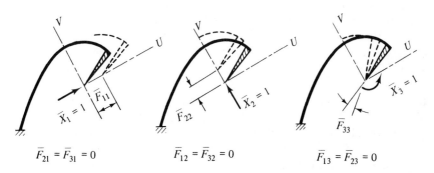

Fig. 12.5 Diagonalized flexibilities.

(12.7), but this would not help matters much, because it still involves the solution of three simultaneous equations. Rather, the usefulness of the method depends on simpler means of locating the elastic center and the conjugate axes; a discussion of this will be postponed a little so as not to disrupt the train of thought for the present.

With respect to the conjugate axes, and assuming no support displacements $\{\bar{\Delta}\}$, Eqs. (12.2) now become, under inclusion of Eqs. (12.5), (12.6), and (12.7),

$$
\begin{bmatrix}
\int_L \dfrac{v^2\,ds}{EI} & 0 & 0 \\[2mm]
0 & \int_L \dfrac{u^2\,ds}{EI} & 0 \\[2mm]
0 & 0 & \int_L \dfrac{ds}{EI}
\end{bmatrix}
\begin{Bmatrix}
\bar{X}_1 \\[2mm] \bar{X}_2 \\[2mm] \bar{X}_3
\end{Bmatrix}
= -
\begin{Bmatrix}
\int_L \dfrac{M_0 v\,ds}{EI} \\[2mm]
-\int_L \dfrac{M_0 u\,ds}{EI} \\[2mm]
\int_L \dfrac{M_0\,ds}{EI}
\end{Bmatrix}
\tag{12.8}
$$

from which

$$
\bar{X}_1 = -\frac{\displaystyle\int_L \frac{M_0 v\,ds}{EI}}{\displaystyle\int_L \frac{v^2\,ds}{EI}}; \qquad
\bar{X}_2 = \frac{\displaystyle\int_L \frac{M_0 u\,ds}{EI}}{\displaystyle\int_L \frac{u^2\,ds}{EI}}; \qquad
\bar{X}_3 = -\frac{\displaystyle\int_L \frac{M_0\,ds}{EI}}{\displaystyle\int_L \frac{ds}{EI}}
\tag{12.9}
$$

Once the redundants are thus determined, the remaining internal forces are found by statics; for the moments, for instance, Eq. (12.3) yields

$$
M = M_0 - \left(
\frac{\displaystyle\int_L \frac{M_0 v\,ds}{EI}}{\displaystyle\int_L \frac{v^2\,ds}{EI}} \cdot v
+ \frac{\displaystyle\int_L \frac{M_0 u\,ds}{EI}}{\displaystyle\int_L \frac{u^2\,ds}{EI}} \cdot u
+ \frac{\displaystyle\int_L \frac{M_0\,ds}{EI}}{\displaystyle\int_L \frac{ds}{EI}}
\right)
\tag{12.10}
$$

M_0 is the statically determinate portion of the moment, and the bracketed quantity is the portion due to the redundants on the statically indeterminate structure. From the moments, all other forces can be obtained by statics.

As discussed above, it remains to discuss convenient ways of locating the elastic center and the conjugate axes so that the diagonalizing conditions, Eq. (12.7), are satisfied. To do this, we visualize a *conjugate structure* that follows the center line of the real structure, and has at any point a width equal to the bending flexibility $1/EI$ of the real structure, as shown in Fig. 12.6(b). The "width" $1/EI$ does not represent an actual geometrical dimension; rather, we can consider it as a weighting or density quantity of the line element, equal to its bending flexibility. Nevertheless, we shall talk of the conjugate structure as possessing area properties.

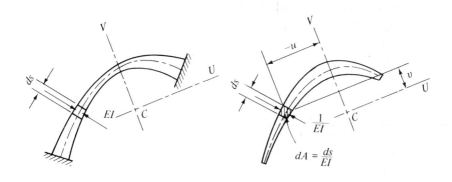

(a) Real structure   (b) Conjugate structure

Fig. 12.6 Real and conjugate structure.

A single element of length ds in Fig. 12.6(b) has an area $dA = ds/EI$. The entire area of the conjugate structure is then $\int_L (ds)/EI$, its first moments with respect to the U and V axes are, respectively, $\int_L (v\,ds)/EI$ and $\int_L (u\,ds)/EI$, and its product of inertia is $\int_L (uv\,ds)/EI$. But, according to Eq. (12.7), it is exactly these quantities that are to be set equal to zero. Therefore, the elastic center C is identified as the centroid of the area of the conjugate structure, which is defined by $\int_L (u\,ds)/EI = \int_L (v\,ds)/EI = 0$. Similarly, the conjugate axes are identified as the principal axes of this area, defined by the vanishing of the product of inertia:

$$\int_L \frac{uv\,ds}{EI} = 0$$

It is concluded that the conjugate axes U, V are the *centroidal principal axes* of the area of the conjugate structure; the problem is thus reduced to the finding of the centroidal principal axes of the area of Fig. 12.6(b). Methods for doing this are covered in books on rigid-body mechanics.

In the particularly frequent case of a structure with at least one axis of symmetry, one of the principal axes will always coincide with this axis of symmetry. Then it only remains to locate one coordinate of the centroid by analytical, numerical, or graphical means.

To summarize the procedure, we recapitulate the calculation steps:

1. Reduce the actual structure to static determinacy by releasing one

support, and calculate the statically determinate moments M_0 in the primary structure due to the applied load.

2. Draw the conjugate structure, locate its centroidal principal axes, and compute the flexibilities $\bar{F}_{ij}$ ($i = j$) according to Eqs. (12.5). These are the denominators of the terms of Eq. (12.10).

3. Calculate the deformations of the elastic center due to the applied loading according to Eqs. (12.6). These are the numerators of the terms of Eq. (12.10).

4. Insert these quantities into Eq. (12.10) to compute the internal moments.

5. Compute all other internal or reactive forces by statics.

We note that for different loading conditions, only the terms of Eq. (12.10) involving M_0 must be changed; the method is thus particularly efficient for multiple loading conditions, such as influence lines.

The elastic-center method will be demonstrated by the following example of a fixed-ended arch. A conventional analysis by the force method, as was done for this case in Ex. 10.4, would require the solution of three simultaneous equations, which is avoided by the orthogonalization of this section.

Example 12.1: Analyze the fixed-ended, prismatic, semicircular arch under uniform load shown in part (a) of the figure by use of the elastic-center concept. Consider only flexural distortions.

Solution: Following the preceding outline, we reduce the structure to statically determinate primary form, as shown in part (b) (note that this differs from that of Ex. 10.4), and write the statically determinate moments in polar coordinates as

$$M_0 = \frac{wR^2}{2} \sin^2 \theta$$

Next, we draw the conjugate structure as shown in part (c); because of the constant member stiffness EI, the conjugate structure is also prismatic, with width $1/EI$. Because of symmetry, this structure has principal axes U and V, as shown, passing through the centroid C whose location is found by computation, or by handbook, as

$$\bar{x} = 0; \quad \bar{y} = \frac{2}{\pi} R$$

This is the *elastic center*.

The diagonalized flexibilities $\bar{F}_{ij}$ ($i = j$) are found by means of the integrals of Eqs. (12.5):

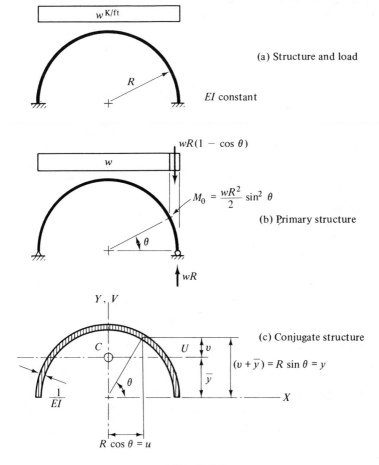

(a) Structure and load

EI constant

(b) Primary structure

(c) Conjugate structure

$(v + \bar{y}) = R \sin \theta = y$

Fig. E12.1

$$\bar{F}_{11} = \int_L \frac{v^2 \, ds}{EI} = \frac{1}{EI} \int_{\theta=0}^{\pi} (y - \bar{y})^2 R \, d\theta = \frac{1}{EI} \int_{\theta=0}^{\pi} \left(R \sin \theta - \frac{2}{\pi} R \right)^2 R \, d\theta$$

$$= \frac{R^3}{EI} \left[\frac{\pi}{2} - \frac{4}{\pi} \right] = .297 \frac{R^3}{EI}$$

$$\bar{F}_{22} = \int_L \frac{u^2 \, ds}{EI} = \frac{1}{EI} \int_{\theta=0}^{\pi} (R \cos \theta)^2 R \, d\theta = \frac{\pi}{2} \frac{R^3}{EI}$$

$$\bar{F}_{33} = \int_L \frac{ds}{EI} = \frac{1}{EI} \int_{\theta=0}^{\pi} R \, ds = \pi \frac{R}{EI}$$

These values will be inserted in the denominators of Eq. (12.10).

Next we compute the numerators of Eq. (12.10), which represent the displacements of the elastic center due to the applied loads; according to Eq. (12.6),

$$\bar{\Delta}_1^\circ = \int_L \frac{M_0 v \, ds}{EI} = \int_L \frac{M_0}{EI}(y - \bar{y}) \, ds$$

$$= \frac{1}{EI} \int_{\theta=0}^{\pi} \left(\frac{wR^2}{2} \sin^2 \theta\right)\left(R \sin \theta - \frac{2R}{\pi}\right) R \, d\theta = .167 \frac{wR^4}{EI}$$

$$\bar{\Delta}_2^\circ = -\int_L \frac{M_0 u \, ds}{EI} = \frac{1}{EI} \int_{\theta=0}^{\pi} \left(\frac{wR^2}{2} \sin^2 \theta\right)(R \cos \theta) R \, d\theta = 0$$

$$\bar{\Delta}_3^\circ = \int_L \frac{M_0 \, ds}{EI} = \frac{1}{EI} \int_{\theta=0}^{\pi} \left(\frac{wR^2}{2} \sin^2 \theta\right) R \, d\theta = \frac{\pi}{4} \frac{wR^3}{EI}$$

Inserting these into Eq. (12.10), we obtain

$$M = wR^2 \left\{\frac{1}{2} \sin^2 \theta - \left[\frac{.167}{.297R}v + 0 + \frac{\pi/4}{\pi}\right]\right\}$$

$$= wR^2 \left\{\frac{1}{2} \sin^2 \theta - \left[.562\left(\sin \theta - \frac{2}{\pi}\right) + \frac{1}{4}\right]\right\}$$

$$= \underline{wR^2\left(\frac{1}{2} \sin^2 \theta - .562 \sin \theta + .107\right)}$$

The moment diagram has already been plotted in Ex. 10.4.

We have here discussed and demonstrated the elastic-center method in terms of a general planar circuit structure. The application of the method to straight beams follows in a straightforward manner, but rather than pursuing this here we shall postpone this until we have cast the operations in terms of the convenient method of column analogy, which will be discussed in Sec. 12.3.

12.2. Elastic-Center Method—Effect of Support Displacements and Calculation of Stiffnesses

We shall now consider the case of circuit structures without applied load; that is, $\{\Delta^\circ\} = 0$, but with imposed support displacements $\{\bar{\Delta}\}$. The barred matrix $\{\bar{\Delta}\}$ contains the specified displacements of the elastic center. Since, in general, any support displacements are specified for the end of the actual structure, point B, not the elastic center, we study the geometrical relations shown in Fig. 12.7 between the specified displacements $\{\Delta\}$ at point B and those at point C, $\{\bar{\Delta}\}$. The coordinates of point B within the U, V axis system are designated by $\bar{u}$, $\bar{v}$.

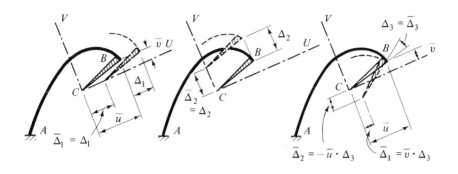

(a) Due to Δ_1 (b) Due to Δ_2 (c) Due to Δ_3

Fig. 12.7 Relations between displacements Δ_i at point B and $\overline{\Delta}_i$ at point C.

The displacements of points B and C are related by the rigid-body movement of the rigid arm BC. Figures 12.7(a), (b), and (c) show the deformations $\{\overline{\Delta}\}$ of the elastic center due to one of the specified displacement components $\{\Delta\}$ of point B. Combining these relations in a matrix equation, we obtain

$$\begin{Bmatrix} \overline{\Delta}_1 \\ \overline{\Delta}_2 \\ \overline{\Delta}_3 \end{Bmatrix} = \begin{bmatrix} 1 & 0 & \overline{v} \\ 0 & 1 & -\overline{u} \\ 0 & 0 & 1 \end{bmatrix} \begin{Bmatrix} \Delta_1 \\ \Delta_2 \\ \Delta_3 \end{Bmatrix} \tag{12.11}$$

or
$$\{\overline{\Delta}\} = [T]\{\Delta\}$$

The $[T]$ matrix is a geometrical transformation matrix that relates the displacements at points C and B. If the specified displacements at support B are zero, the $\{\Delta\}$ and $\{\overline{\Delta}\}$ matrices will of course vanish.

Under substitution of Eqs. (12.8) and (12.11) into Eq. (12.2), we can now solve for the redundants $\{\overline{X}\}$ due to displacements at point B:

$$[\overline{X}] = [\overline{F}]^{-1}[T][\Delta]$$

or
$$\begin{Bmatrix} \overline{X}_1 \\ \overline{X}_2 \\ \overline{X}_3 \end{Bmatrix} = \begin{Bmatrix} \dfrac{1}{\displaystyle\int_L (v^2\, ds/EI)}(\Delta_1 + \overline{v} \cdot \Delta_3) \\[2em] \dfrac{1}{\displaystyle\int_L (u^2\, ds/EI)}(\Delta_2 - \overline{u} \cdot \Delta_3) \\[2em] \dfrac{1}{\displaystyle\int_L (ds/EI)}\Delta_3 \end{Bmatrix} \tag{12.12}$$

Substituting these values into the equilibrium equation (12.3) to find the internal moments (note that M_0 is now zero in the absence of applied loads), we obtain

$$M = \frac{\Delta_1 + \bar{v} \cdot \Delta_3}{\int_L (v^2 \, ds/EI)} \cdot v - \frac{\Delta_2 - \bar{u} \cdot \Delta_3}{\int_L (u^2 \, ds/EI)} \cdot u + \frac{\Delta_3}{\int_L (ds/EI)} \qquad (12.13)$$

The calculation of $\bar{u}$ and $\bar{v}$ and of the integrals has been discussed earlier. The moments due to applied loads, Eq. (12.10), and due to support displacements, Eq. (12.13), can of course be superposed.

The determination of stiffnesses represents a special case of the preceding, in which we desire the reactions $[X] \equiv [k]$ at B associated with unit displacements at B, applied one at a time:

$$[\Delta] = \begin{bmatrix} 1 & 0 & 0 \\ 0 & 1 & 0 \\ 0 & 0 & 1 \end{bmatrix} \qquad (12.14)$$

Equation (12.12) enables us to find the redundants at the elastic center C; it remains to relate these redundants to those at point B by the statics of the rigid arm BC. Figure 12.8 shows the forces on this arm. Summing forces along the U and V axes and moments, we obtain the equations of statics:

$$\begin{Bmatrix} X_1 \\ X_2 \\ X_3 \end{Bmatrix} = \begin{bmatrix} 1 & 0 & 0 \\ 0 & 1 & 0 \\ \bar{v} & -\bar{u} & 1 \end{bmatrix} \begin{Bmatrix} \bar{X}_1 \\ \bar{X}_2 \\ \bar{X}_3 \end{Bmatrix} \qquad (12.15)$$

or, because of the orthogonality of the transformation,

$$\{X\} = [T]^T \{\bar{X}\}$$

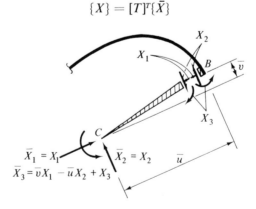

Fig. 12.8 Relations between forces X_i at point B and $\bar{X}_i$ at point C.

The relation between the geometrical transformation matrix $[T]$ of Eq. (12.11) and the statical transformation matrix, Eq. (12.15), was discussed earlier as a consequence of the theorem of virtual work.

In any case, by substituting Eq. (12.12) and (12.14) into Eq. (12.15), we obtain for the stiffnesses $[k] \equiv [X]$ of point B

$$[k] = [T]^T[\bar{F}]^{-1}[T] \tag{12.16}$$

Substituting the appropriate matrices defined by Eqs. (12.8) and (12.11), and carrying out the indicated matrix operations, we obtain the stiffness matrix

$$[k] = \begin{bmatrix} \dfrac{1}{\displaystyle\int_L \dfrac{v^2\,ds}{EI}} & 0 & \dfrac{\bar{v}}{\displaystyle\int_L \dfrac{v^2\,ds}{EI}} \\[4ex] 0 & \dfrac{1}{\displaystyle\int_L \dfrac{u^2\,ds}{EI}} & -\dfrac{\bar{u}}{\displaystyle\int_L \dfrac{u^2\,ds}{EI}} \\[4ex] \dfrac{\bar{v}}{\displaystyle\int_L \dfrac{v^2\,ds}{EI}} & -\dfrac{\bar{u}}{\displaystyle\int_L \dfrac{u^2\,ds}{EI}} & \left(\dfrac{\bar{v}^2}{\displaystyle\int_L \dfrac{v^2\,ds}{EI}} + \dfrac{\bar{u}^2}{\displaystyle\int_L \dfrac{u^2\,ds}{EI}} + \dfrac{1}{\displaystyle\int_L \dfrac{ds}{EI}} \right) \end{bmatrix} \tag{12.17}$$

Note that here the stiffnesses at point B are defined along the directions of the principal axes. If the stiffnesses in any other direction are desired, an additional transformation is required, but this will not be discussed at this time. Section 13.5 will cover ways of performing such transformations.

Example 12.2: Compute the (3×3) stiffness matrix $[k]$ for the numbered nodes at point B of the semicircular prismatic arch shown.

Solution: We apply Eq. (12.17) to this structure, which has already been discussed in Ex. 12.1, where all the necessary denominator terms of this matrix

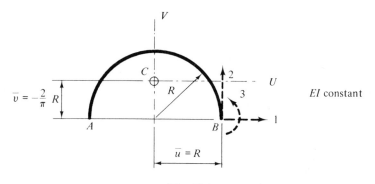

Fig. E12.2

had already been computed. In addition, we need the coordinates $\bar{u}, \bar{v}$ of point B:

$$\bar{u} = R; \qquad \bar{v} = -\frac{2}{\pi} R$$

With these, the stiffness matrix $[k]$ is written according to Eq. (12.17):

$$[k] = \begin{bmatrix} \dfrac{1}{.297 \dfrac{R^3}{EI}} & & \text{sym} \\[2em] 0 & \dfrac{1}{\dfrac{\pi}{2} \dfrac{R^3}{EI}} & \\[2em] \dfrac{-\dfrac{2}{\pi} R}{.297 \dfrac{R^3}{EI}} & -\dfrac{R}{\dfrac{\pi}{2} \dfrac{R^3}{EI}} & \left(\dfrac{\left(-\dfrac{2}{\pi} R\right)^2}{.297 \dfrac{R^3}{EI}} + \dfrac{R^2}{\dfrac{\pi}{2} \dfrac{R^3}{EI}} + \dfrac{1}{\pi \dfrac{R}{EI}} \right) \end{bmatrix}$$

$$= \frac{EI}{R^3} \begin{bmatrix} 3.36 & & \text{sym} \\ 0 & .637 & \\ -2.15R & -.637R & 2.323R^2 \end{bmatrix}$$

A portion of this stiffness matrix was already solved in Ex. 11.4.

12.3. Column Analogy

An interesting and useful analogy exists between the moments in a statically indeterminate circuit structure and the stress in a short compression block (erroneously called a "column"), or the pressures underneath a rigid foundation under load. This analogy is useful for persons who are more familiar with such stress computations than with indeterminate analysis. But even for experienced analysts it furnishes a convenient and familiar symbolism for the integrals occurring in the expressions of Secs. 12.1 and 12.2, and is thus helpful in computations. The method is particularly useful in the calculation of stiffnesses and fixed-end forces of members, which can then be used in the displacement method presented in Part V of this text.

We refer back to the conjugate structure of Fig. 12.6, and redraw it in Fig. 12.9(a) as a rigid footing under a general loading f. The resulting foundation pressures p are then given by the general equation of elementary bending theory:

$$p = \frac{P}{A} + \frac{M_u}{I_u} \cdot v + \frac{M_v}{I_v} \cdot u \qquad (12.18)$$

where P, M_u, and M_v are, respectively, the resultant force $\displaystyle\int_L f \, ds$ and the

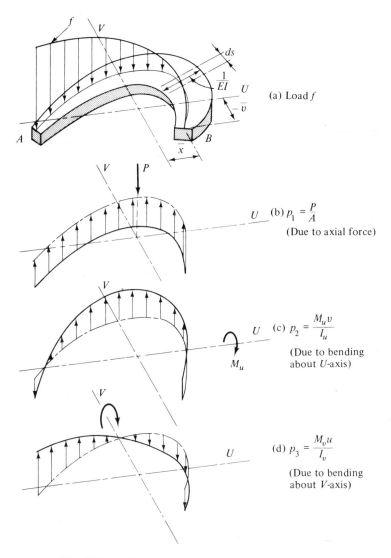

(a) Load f

(b) $p_1 = \dfrac{P}{A}$

(Due to axial force)

(c) $p_2 = \dfrac{M_u v}{I_u}$

(Due to bending about U-axis)

(d) $p_3 = \dfrac{M_v u}{I_v}$

(Due to bending about V-axis)

Fig. 12.9 Load and pressures on "analogous column."

moments $\displaystyle\int_L f \cdot v \, ds$ and $\displaystyle\int_L f \cdot u \, ds$ about the centroidal principal axes U and V of the area of the footing, and A, I_u, and I_v are the area properties with respect to the principal axes.

We now consider the footing subject to a distributed line load equal to the curvature due to the applied load on the primary structure: $f = \Phi_0 = $

M_0/EI. Then, the applied resultant force and moments will be

$$P = \int_L \frac{M_0 \, ds}{EI}; \qquad M_u = \int_L \frac{M_0 v \, ds}{EI}; \qquad M_v = \int_L \frac{M_0 u \, ds}{EI}$$

Note that M_u and M_v are only quantities in the analogous column and are not to be confused with real moments in the structure to be analyzed.

Recalling that the analogous foundation had a width of $1/EI$, an element of its area is equal to ds/EI, and the area properties are, by their definition,

$$A = \int_L \frac{ds}{EI}; \qquad I_u = \int_L \frac{v^2 \, ds}{EI}; \qquad I_v = \int_L \frac{u^2 \, ds}{EI}$$

When we substitute these integrals into Eq. (12.18), we obtain

$$p = \frac{\displaystyle\int_L \frac{M_0 \, ds}{EI}}{\displaystyle\int_L \frac{ds}{EI}} + \frac{\displaystyle\int_L \frac{M_0 v \, ds}{EI}}{\displaystyle\int_L \frac{v^2 \, ds}{EI}} \cdot v + \frac{\displaystyle\int_L \frac{M_0 u \, ds}{EI}}{\displaystyle\int_L \frac{u^2 \, ds}{EI}} \cdot u$$

which is identical with the statically indeterminate part in the brackets of Eq. (12.10). It follows that the total moments in the statically indeterminate structure are given by

$$M = M_0 - p \tag{12.19}$$

where p is the foundation pressure per unit of length due to a line loading f equal to the curvature of the primary structure, M_0/EI.

The method of column analogy is summarized by the following steps:

1. Reduce the structure to a primary system and compute the moments M_0 and the curvature $\Phi_0 = M_0/EI \equiv f$ due to the applied load.

2. Draw the conjugate structure (or "analogous column") and determine by appropriate means its centroidal principal axes and its principal area properties A, I_u, and I_v.

3. Calculate the resultant force and moments with respect to the principal centroidal axes due to a line loading $f = M_0/EI$.

4. Calculate the foundation pressures p by elementary bending theory according to Eq. (12.18).

5. Compute the total moments in the statically indeterminate structure according to Eq. (12.19).

6. Calculate any other desired forces by statics.

The main attraction of this method lies in the fact that the symbols of Eq. (12.18) are often more familiar to the engineer than the integrals of Eq. (12.10), and, accordingly, will be computed with greater competence and versatility of methods. By the same token, however, it is more remote from, and less revealing about, actual structural behavior, and should really only be considered a "crutch."

The following examples will illustrate the method. The first example will rework a previous problem for the sake of comparison, following which we apply the method to a straight beam; the conciseness of the method should be observed here.

The spirit of the analogy demands that attention be focused on the calculation of pressures acting on a loaded foundation; those who object to the consideration of such extraneous matters should solve these examples by the orthogonalization procedure of Secs. 12.1 and 12.2. Their computations will be identical to those demonstrated here.

Example 12.3: Analyze the fixed-ended prismatic semicircular arch shown in part (a) of the figure due to the uniform load w.

Solution: This problem was already solved in Ex. 12.1; it might be instructive to compare the two solutions to verify that the only difference lies in the symbols used.

We follow the preceding outline step by step:

1. The primary structure is shown in part (b); the curvature due to the load is

$$\Phi_0 = \frac{M_0}{EI} = \frac{wR^2}{2EI} \sin^2 \theta \equiv f$$

This will be the load on the analogous column.

2. Part (c) shows the analogous footing; its properties are calculated by appropriate means; for instance, use of the transfer theorem for area properties and the fact that $I_x + I_y = J$ are of use here.

Location of centroid:

$$\bar{x} = 0; \qquad \bar{y} = \frac{2}{\pi} R = .637R$$

Area of analogous foundation:

$$A = \frac{\pi R}{EI} = 3.14 \frac{R}{EI}$$

Moments of inertia:

$$I_u = I_x - A \cdot \bar{y}^2 = \frac{1}{4}\left(\frac{2\pi R}{EI} \cdot R^2\right) - \frac{\pi R}{EI}\left(\frac{2}{\pi} R\right)^2 = .297 \frac{R^3}{EI}$$

$$I_v = \frac{1}{4}\left(\frac{2\pi R}{EI} \cdot R^2\right) = 1.57 \frac{R^3}{EI}$$

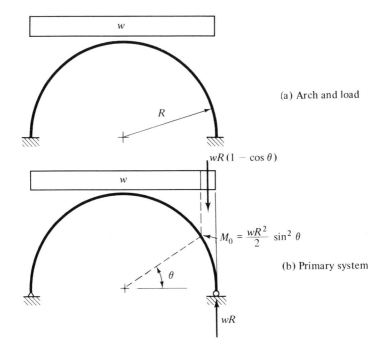

(a) Arch and load

(b) Primary system

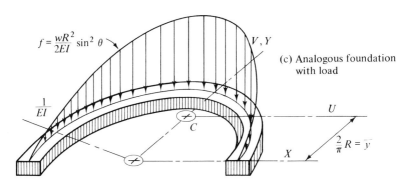

(c) Analogous foundation with load

Fig. E12.3

3. Loads on analogous foundation:

$$P = \int_L f \, ds = \int_{\theta=0}^{\pi} \left(\frac{wR^2}{2EI} \sin^2 \theta\right) R \, d\theta = .785 \frac{wR^3}{EI}$$

$$M_u = \int_L f \cdot v \cdot ds = \int_{\theta=0}^{\pi} \left(\frac{wR^2}{2EI} \sin^2 \theta\right) \left(R \sin \theta - \frac{2}{\pi} R\right) R \, d\theta = .167 \frac{wR^4}{EI}$$

$$M_v = 0 \qquad \text{by symmetry}$$

4. Pressure under analogous footing:

$$p = \frac{P}{A} + \frac{M_u}{I_u}v + 0 = \left[\frac{.785}{3.14} + \frac{.167}{.297}\frac{v}{R}\right]wR^2$$
$$= [.250 + .562(\sin\theta - .637)]wR^2 = (.562\sin\theta - .107)wR^2$$

5. Moments in real structure:

$$M = M_0 - p = \{\tfrac{1}{2}\sin^2\theta - [.562\sin\theta - .107]\}wR^2$$
$$= (.500\sin^2\theta - .562\sin\theta + .107)wR^2$$

All other statical quantities can now be determined by equilibrium calculations.

The following example specializes the method to the simpler case of a straight beam; in this case the analogous foundation is a strip footing of variable width, and bending about only one axis takes place due to a line load. This example also shows that the computational methods should be matched to the problem; here, formal integrations are unnecessary and can be avoided.

Example 12.4: Calculate the fixed-end moments at *A* and *B* of the nonprismatic beam under the load shown in part (a) of the figure.

Solution: Part (b) shows the primary system, and part (c) the curvature plot M_0/EI; this will be the loading on the analogous foundation of part (d).

Next, we determine the properties of the analogous foundation; a tabular form is used to add properties of the individual portions of part (d):

PART	A	x	Ax	u	u^2	$A\cdot u^2$	$\bar{I}$
1	$.250\dfrac{L}{EI}$	$.25L$	$.0625\dfrac{L^2}{EI}$	$-.333L$	$.111L^2$	$.0278\dfrac{L^3}{EI}$	$.0052\dfrac{L^3}{EI}$
2	$.500\dfrac{L}{EI}$	$.75L$	$.3750\dfrac{L^2}{EI}$	$+.167L$	$.028L^2$	$.0139\dfrac{L^3}{EI}$	$.0104\dfrac{L^3}{EI}$

$$\Sigma A = .750\frac{L}{EI}; \quad \Sigma Ax = .4375\frac{L^2}{EI};$$
$$\bar{x} = \frac{\Sigma Ax}{\Sigma A} = .583L$$
$$u = x - .583L$$

$$I_v = .0417\frac{L^3}{EI} + .0156\frac{L^3}{EI}$$
$$= .0573\frac{L^3}{EI}$$

The loads on the analogous foundation, using the resultants of the load shown in part (d), are

$$P = \frac{1}{32}\frac{PL^2}{EI} + \frac{1}{16}\frac{PL^2}{EI} = .0938\frac{PL^2}{EI}$$

$$M_u = \left[\left(\tfrac{1}{32}\right)\left(-\tfrac{3}{12}\right) + \left(\tfrac{1}{16}\right)\left(\tfrac{1}{12}\right)\right]\frac{PL^3}{EI} = -.0026\frac{PL^3}{EI}$$

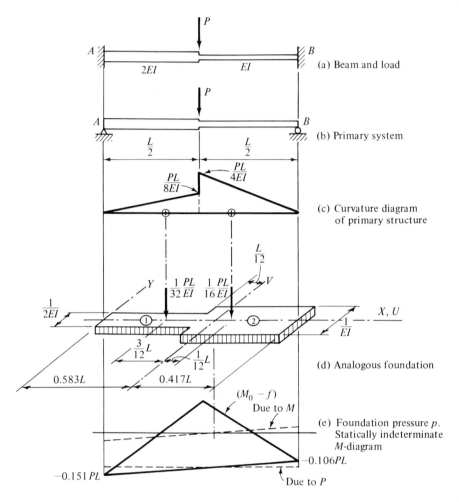

Fig. E12.4

The pressure under the analogous foundation is calculated by elementary beam theory:

$$p = \frac{P}{A} + \frac{M_v}{I_v}u = \frac{.0938}{.75}PL - \frac{.0026}{.0573}P\cdot u = .125PL - .0453P\cdot u$$

The fixed-end moments at points A and B are computed by Eq. (12.19), wherein the statically determinate moment $M_0 = 0$ at the beam ends:

point A: $u = -.583L$; $M_A^F = M_0 - p = 0 - (.125 + .0453\cdot.583)PL$

$$= -.151PL$$

point B: $u = +.417L$; $M_B^F = M_0 - p = 0 - (.125 - .0453 \cdot .417)PL$

$$= -.106PL$$

The final moment diagram is shown in part (e). This concludes the problem.

The method of column analogy leads to an interesting and concise way of calculating stiffness factors. We consider Eq. (12.17) and substitute the appropriate area properties of the analogous column (or rigid footing) to obtain the stiffness matrix for point B:

$$
[k] =
\begin{array}{ccc}
M_u = 1 & M_v = 1 & P = 1 \\
\end{array}
\begin{bmatrix}
\dfrac{1}{I_u} & 0 & \dfrac{\bar{v}}{I_u} \\[2mm]
0 & \dfrac{1}{I_v} & -\dfrac{\bar{u}}{I_v} \\[2mm]
\dfrac{\bar{v}}{I_u} & -\dfrac{\bar{u}}{I_v} & \left(\dfrac{1}{A} + \dfrac{\bar{v}^2}{I_u} + \dfrac{\bar{u}^2}{I_v}\right)
\end{bmatrix}
\begin{array}{c}
\text{cause}/\text{effect} \\
\theta_u \\
\theta_v \\
f
\end{array}
\qquad (12.20)
$$

This matrix contains only properties of the analogous column and can be used directly. The comments above and to the right of the matrix will, hopefully, become clear a little later.

We now assign a unit value to the subsoil modulus $E = 1$ under the rigid footing, and visualize the following imaginary loading conditions on the analogous column:

1. An applied unit moment about the U axis, in a sense following the right-hand rule [Fig. 12.9(c)]. The resulting rotation about axis U of the analogous foundation is then $\theta_u = M_u/EI_u = 1/(1 \cdot I_u)$. The rotation about axis V is zero. The resulting foundation displacement at point B of coordinate $\bar{v}$ is then $\Delta_B = p/E = (1/E) \cdot (M_u \bar{v}/I_u) = \bar{v}/I_u$. But these values correspond exactly to the terms in the first column ($j = 1$) of Eq. (12.20). It follows that the stiffnesses associated with a unit displacement of point B along the U axis can be represented by the rotations and displacement of the analogous foundation due to a unit moment about this axis.

2. An applied unit moment about the V axis [Fig. 12.9(d)]. The resulting rotation about the V axis is given by $\theta_v = 1/(1 \cdot I_v)$; that about the U axis is zero. The resulting foundation displacement at point B of coordinate $\bar{u}$ is $(1 \cdot \bar{u})/(1 \cdot I_v)$. But these values correspond exactly to the terms in the second column ($j = 2$) of Eq. (12.20). It follows that the stiffnesses associated with a unit displacement of point B along the V axis can be represented by the rotations and displacement of the analogous footing due to a unit moment about this axis.

3. An applied unit load normal to the analogous foundation at point B

of coordinates $\bar{u}$, $\bar{v}$ [Fig. 12.9(b)]. The resulting principal moments will then be $M_u = 1 \cdot \bar{v}$ and $M_v = -1 \cdot \bar{u}$. The rotations about the U and V axes will be, respectively, $\bar{v}/(1 \cdot I_u)$ and $-\bar{u}/(1 \cdot I_v)$, and the foundation displacement at point B will be given by Eq. (12.18) as

$$\Delta_B = \frac{p}{E} = \left(\frac{1}{A} + \frac{\bar{v}^2}{I_u} + \frac{\bar{u}^2}{I_v} \right)$$

But these values correspond exactly to the terms of the third column ($j = 3$) of Eq. (12.20). It follows that the stiffnesses associated with a unit rotation of point B can be represented by the rotations and displacement of the analogous footing due to a unit force in the sense of the rotation vector.

To summarize our conclusions, we can state:

Unit displacements parallel to the conjugate axes of a real structure are simulated by unit moments applied to the analogous footing about these axes. A unit rotation of a point of the real structure is simulated by a unit load applied to the analogous footing along the axis of rotation.

The resulting rotations of the analogous structure represent the force stiffnesses of the real structure. The resulting displacement in the analogous structure represents the moment stiffness of the real structure at that point.

Again, the calculation process for stiffness determination by column analogy will be summarized in the following steps:

1. Determine the area properties of the analogous structure, as before.

2. For each desired stiffness, apply the appropriate unit load or moment to the analogous structure in accordance will the above rules.

3. For each desired stiffness, compute the resulting rotations and displacement of the analogous structure by elementary bending theory.

Example 12.5

a. Compute the stiffness matrix $[k]$ for the numbered nodes of the semicircular prismatic arch shown in part (a) of the figure.

b. Compute the stiffness k_{21} for the numbered nodes of part (b). Consider only flexural deformations.

Solution

a. This structure was already analyzed in Ex. 12.3 and all properties of the analogous foundation can be taken from there and inserted directly into Eq. (12.20) to yield the same result as was obtained in Ex. 12.2.

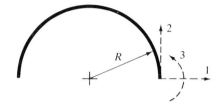

(a) Structure and nodes
for part (a)

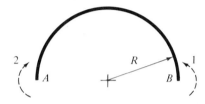

(b) Nodes for part (b)

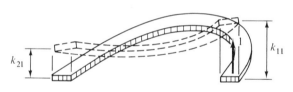

(c) Displaced shape
of analogous
foundation

Fig. E12.5

b. Following the interpretation of the stiffness matrix, Eq. (12.20), as relations
between applied unit loads and resulting deformations of an analogous
foundation on a subsoil of unit modulus, we apply a unit load along node 1
at point *B*, and seek to determine the resulting displacement along node 2
at point *A*; the displaced analogous foundation is shown in part (c), and
the desired displacement is computed as

$$\Delta_A = k_{21} = \frac{p_A}{E} = \frac{1}{E}\left[\frac{1}{A} + \frac{1 \cdot \bar{v}}{I_u} v_A - \frac{1 \cdot \bar{u}}{I_v} u_A\right]$$

$$= \frac{EI}{R}\left[\frac{1}{3.14} + \frac{.637R}{.297R^2}.637R - \frac{R}{1.57R^2}R\right] = \underline{1.05\frac{EI}{R}}$$

as already computed by a lengthier method in Ex. 11.4.

Example 12.6: Find the (4×4) stiffness matrix $[k]$ for the tapering beam
shown in part (a) of the figure, with linearly varying moment of inertia.
Consider only flexural distortions.

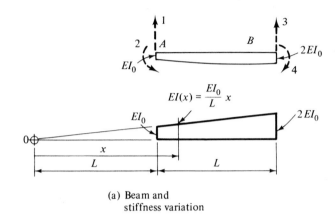

(a) Beam and
    stiffness variation

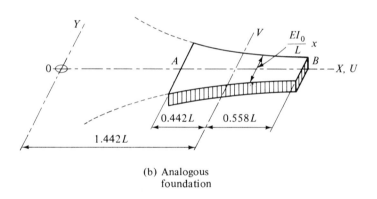

(b) Analogous
    foundation

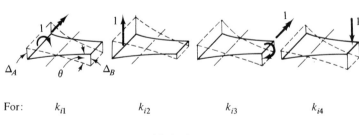

For: k_{i1} k_{i2} k_{i3} k_{i4}

(c) Analogous
    loads and
    displacements

Fig. E12.6

Solution: The computations are carried out most conveniently by placing the origin of the variable x so that the denominator $EI(x)$ can be expressed in one term, as shown in part (a) of the figure:

$$EI(x) = EI_0 \frac{x}{L}$$

The analogous foundation, of width $1/[EI(x)] = L/EI_0 x$, is shown in part (b); we determine its properties, starting with the centroidal distance

$$\bar{x} = \frac{\int_A x\, dA}{\int_A dA}$$

where

$$A = \int_A dA = \int_{x=L}^{2L} \frac{L\, dx}{EI_0 x} = \frac{L}{EI_0} \ln \frac{2L}{L} = .693 \frac{L}{EI_0}$$

and

$$\int_A x\, dA = \int_{x=L}^{2L} \frac{Lx\, dx}{EI_0 x} = \frac{L^2}{EI_0}$$

so that

$$\bar{x} = \frac{L}{.693} = 1.442L \qquad \text{(measured from point } O\text{)}$$

The moment of inertia about the centroidal axis V is

$$I_v = \int_A x^2\, dA - A \cdot \bar{x}^2$$

$$= \frac{L}{EI_0} \left[\int_{x=L}^{2L} \frac{x^2\, dx}{x} - .693(1.442L)^2 \right]$$

$$= [1.500 - 1.442] \frac{L^3}{EI_0} = .058 \frac{L^3}{EI_0}$$

Note that precision is in order here because of the small difference of two large numbers.

We next represent the unit displacements along nodes 1 to 4 of the real structure by the corresponding unit loads along these nodes of the analogous structure, as shown, one at a time, in part (c). The analogous structure rests on a subsoil of modulus E of value unity. The desired stiffnesses are the deformations also shown in part (c), so we have to find these by elementary beam methods: For k_{i1}, apply moment $M_A = 1$. Then

$$\theta_A = \theta_B = k_{11} = -k_{31} = \frac{M}{EI_v} = \frac{1}{.058(L^3/EI_0)} = 17.24 \frac{EI_0}{L^3} = -k_{13}$$

$$\Delta_A = k_{21} = \theta \cdot u_A = 17.24 \frac{EI_0}{L^3} .442L = 7.63 \frac{EI_0}{L^2} = k_{12}$$

$$\Delta_B = k_{41} = \theta \cdot u_B = 17.24 \frac{EI_0}{L^3}(-.558L) = -9.62 \frac{EI_0}{L^2} = k_{14}$$

For k_{i2}, apply force $P_A = 1$. Then

$$\Delta_A = k_{22} = \frac{P}{AE} + \frac{M \cdot u_A}{EI_v} = \frac{1}{.693(L/EI_0)} + \frac{(1 \cdot 442L).442L}{.058(L^3/EI_0)} = 4.82\frac{EI_0}{L}$$

$$\theta_B = k_{32} = \frac{M}{EI_v} = \frac{(1 \cdot 442L)}{.058(L^3/EI_0)} = 7.63\frac{EI_0}{L^2} = k_{23}$$

$$\Delta_B = k_{42} = \frac{P}{AE} - \frac{Mu_B}{EI_v} = \frac{1}{.693(L/EI_0)} - \frac{(1 \cdot 442L).558L}{.058(L^3/EI_0)}$$

$$= -2.81\frac{EI_0}{L} = k_{24}$$

For k_{i3}, the applied moment $M_B = 1$ is the negative of that applied for determination of k_{i1}; therefore, all displacements $k_{i3} = -k_{i1}$.
For k_{i4}, apply force $P_B = 1$. Then

$$\Delta_B = k_{44} = \frac{P}{AE} + \frac{Mu_B}{EI_v} = \frac{1}{.693(L/EI_0)} + \frac{(1 \cdot .558L).558L}{.058(L^3/EI_0)}$$

$$= 6.80\frac{EI_0}{L}$$

We have now all elements of the stiffness matrix, and insert them in the proper array:

$$[k] = \frac{EI_0}{L^3}\begin{bmatrix} 17.24 & & \text{sym} & \\ 7.63L & 4.82L^2 & & \\ -17.24 & -7.63L & 17.24 & \\ -9.62L & -2.81L^2 & 9.62L & 6.80L^2 \end{bmatrix}$$

This is the desired stiffness matrix.

12.4. Influence Lines by the Force Method

It will be recalled from Chapter 6 that influence lines plot the variation of a force as a function of the location of the unit load which causes this force. Thus, the problem of determining influence lines for indeterminate structures consists of a statically indeterminate analysis for multiple loading conditions, a task that is suitable for computer analysis. In general, this involves a large number of forces (for instance, the moments at all $\frac{1}{10}$ points of each span of a continuous beam due to a unit load at each of these points). The size of the matrices therefore may become quite large, and some care is required in organizing the work.

Here, we shall confine the matrix analysis to the determination of redundants, and compute the remaining required internal forces by statics. Rather than talk in generalities, we shall illustrate the procedure by the following example problem.

Example 12.7: For the two-span two-times indeterminate beam of part (a) of the figure determine and plot the influence lines for the support moments M_1 and M_2 and for the span moments M_4 and M_7.

Solution: This beam was already analyzed for other loading conditions in Exs. 10.2 and 10.3, where we find the needed flexibilities $[F]$. We use the continuity conditions in the form of Eq. (10.4) to solve for the two unknown redundant moments M_1 and M_2 at points B and C from the equations

$$\frac{L}{6EI}\begin{bmatrix} 4 & 1 \\ 1 & 2 \end{bmatrix}\begin{bmatrix} M_{11} & M_{12} & \cdots \\ M_{21} & M_{22} & \cdots \end{bmatrix} + \begin{bmatrix} \Delta^\circ_{11} & \Delta^\circ_{12} & \cdots \\ \Delta^\circ_{21} & \Delta^\circ_{22} & \cdots \end{bmatrix} = \begin{bmatrix} 0 & 0 & \cdots \\ 0 & 0 & \cdots \end{bmatrix}$$

(12.21)

The $[M]$, $[\Delta^\circ]$, and $[0]$ matrices contain as many columns as points at which the unit loads are considered, that is, one for each loading condition. In this case, we shall determine influence ordinates at the $\frac{1}{5}$ points of each span, numbered as in part (a), for a total of eight loading conditions.

The $[\Delta^\circ]$ matrix contains the angle changes at the cuts of the primary structure shown in part (b) due to unit loads applied at these points, one at a time. These angle changes, equal to the end rotations of a simply supported beam subject to a concentrated unit load acting at a point x_1 from the left, x_2 from the right end, as shown in part (c), can be calculated by any method (for instance, the curvature-area method) as

$$\theta_1 = \frac{L^2}{6EI}\left[\left(\frac{x_1}{L}\right) - \left(\frac{x_1}{L}\right)^3\right]$$

$$\theta_2 = \frac{L^2}{6EI}\left[\left(\frac{x_2}{L}\right) - \left(\frac{x_2}{L}\right)^3\right]$$

(12.22)

For the unit load between points A and B of the beam, that is, for the first four loading conditions, $\Delta^\circ_1 = \theta_2$ and $\Delta^\circ_2 = 0$. For the load between points B and C, that is, for the next four loading conditions, $\Delta^\circ_1 = \theta_1$ and $\Delta^\circ_2 = \theta_2$. Substituting the appropriate values of x/L into Eqs. (12.22), and solving Eqs. (12.21) for the redundant moments by inverting the $[F]$ matrix, we obtain

$$\begin{bmatrix} M_{13} & M_{14} & \cdots \\ M_{23} & M_{24} & \cdots \end{bmatrix}$$

$$= -\frac{6EI}{7L}\begin{bmatrix} 2 & -1 \\ -1 & 4 \end{bmatrix}\begin{bmatrix} .192 & .336 & .384 & .288 & .288 & .384 & .336 & 192 \\ 0 & 0 & 0 & 0 & .192 & .336 & .384 & .288 \end{bmatrix}$$

$$= -\begin{bmatrix} .0549 & .0960 & .1098 & .0823 & .0549 & .0618 & .0411 & .0138 \\ -.0274 & -.0480 & -.0549 & -.0411 & .0687 & .1371 & .1716 & .1372 \end{bmatrix}L$$

Each row contains the influence ordinates for the moment designated by its first subscript due to a unit load at the point of the beam indicated by the second subscript. For longhand calculation, a tabular form is indicated for such computations.

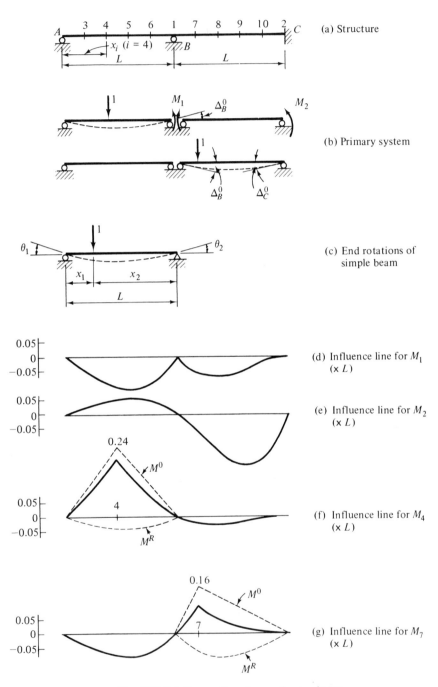

(a) Structure

(b) Primary system

(c) End rotations of simple beam

(d) Influence line for M_1 ($\times L$)

(e) Influence line for M_2 ($\times L$)

(f) Influence line for M_4 ($\times L$)

(g) Influence line for M_7 ($\times L$)

Fig. E12.7 Influence lines for continuous beam.

Parts (d) and (e) plot the influence lines for the redundant moments M_1 and M_2. In contrast with the statically determinate influence lines obtained in Chapter 6, these statically indeterminate influence lines consist always of curved line segments. Furthermore, the shape of these influence lines should remind us of the validity of the theorem of Mueller–Breslau discussed in Sec. 6.1, according to which all influence lines can be interpreted as the deflection curve generated by the introduction of a unit deformation at the point and in the sense of the force to be determined; in our case, a unit kink at point B of the beam would cause the deflections shown in part (d), and a unit kink at point C would cause the deflections shown in part (e).

The influence lines for any other force in the structure can now be found by statics from the influence lines for the redundants, using analytical, numerical, or graphical means. Here, we shall use superposition of the moment due to the redundants, M^R, and the statically determinate simple beam moment, M^0, at any section i for which the influence line is desired.

In the left span AB of the continuous beam, for instance, the moments due to the redundant M_1 vary linearly from zero at the left end A $(x_i = 0)$ to M_1 at the right end B $(x_i = L)$; so the moment M_i^R at any section i, located at x_i from the left end, is

$$M_i^R = \frac{x_i}{L} M_1$$

The statically determinate portion, M_i^0 is obtained from the simple-span influence line for M_i, shown dashed in part (f), so that the total influence ordinate is

$$M_i = M_i^R + M_i^0 = \frac{x_i}{L} M_1 + M_i^0$$

For section 4, for instance, $x_i/L = .4$. The statically determinate and the statically indeterminate parts of the influence line are shown dashed in part (f), and added graphically to obtain the influence line for the moment at secton 4 of the beam.

For any point in the right-hand span BC of the beam, the indeterminate portion of the moment is due to the two redundant end moments M_1 and M_2. Between these member ends the moment variation is linear, so that the redundant moment at a section j, at a distance x_j from point B, is

$$M_j^R = \left(\frac{L - x_j}{L} M_1 + \frac{x_j}{L} M_2 \right)$$

To this must be added the statically determinate influence line for the moment at this section. In part (g), the statically determinate and indeterminate portions of the influence line for section 7, $(x_j/L) = .2$, are shown and superposed to arrive at the influence line for the moment M_7.

Influence lines for other forces, such as shears or reactions, could be determined in an analogous manner.

PROBLEMS

As in all analysis problems, first attempt to draw the expected deformed shape of the structure and review it in the light of the results of the analysis.

1. a. The group redundants $\bar{X}_1$ and $\bar{X}_2$ each consist of the two reactions applied simultaneously, as shown in part (b) of the figure. Determine the values of the coefficients k_1 and k_2 such that $\bar{F}_{ij}$ $(i \neq j)$ vanish; then, find the values of $\bar{F}_{ij}$ $(i = j)$. Write the diagonalized (2×2) flexibility matrix $[\bar{F}]$.
   b. Using the matrix $[\bar{F}]$ determined in Part a, analyze the continuous beam under uniform load shown in part (a) of the figure.

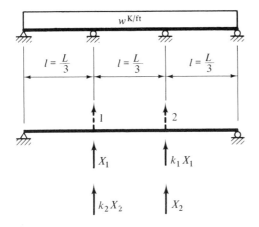

(a) Continuous beam
   EI constant

(b) Primary system with group redundants:

   $= \bar{X}_1$

   $= \bar{X}_2$

Prob. 1

2. a. Locate the elastic center and conjugate axes for the rigid frame.
   b. Determine the flexibilities $\bar{F}_{ij}$ for redundant nodes at the elastic center; verify that the F_{ij} $(i \neq j)$ are zero.
   c. Select a suitable primary system, write the statically determinate moment expression M_0, and compute the displacements at the elastic center due to the applied load.

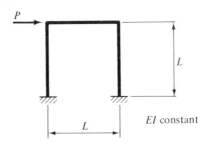

Prob. 2

    d. Compute and plot the moments in the structure; from the moments, draw the shear diagram and compute the reactions by statics.

3. a. Locate the elastic center, conjugate axes, and determine the diagonalized flexibilities $\bar{F}_{ij}$ for the closed ring.

    b. Select a suitable primary system (*hint:* preserve symmetry), compute the displacements of the elastic center due to the loads, and determine and plot the moment variation due to the diametrically opposed point loads.

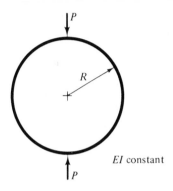

Prob. 3

4. a. Locate the elastic center and conjugate axes, and determine the diagonalized flexibilities $\bar{F}_{ij}$ for the square box culvert shown.

    b. Select a suitable primary system (*hint:* preserve symmetry), compute the displacements of the elastic center due to the applied loads, and determine and plot moments and shears in the structure.

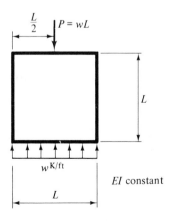

Prob. 4

5. Rework Prob. 2 by the method of column analogy; try to avoid formal integrations, but compare computations to those of Prob. 2.

6. Rework Prob. 3 by the method of column analogy; use any useful information from Prob. 3.

7. Rework Prob. 4 by the method of column analogy; compute properties of, and loads on, the analogous foundation in the simplest, most straightforward way. Compare computations to those of Prob. 4.

Work the following problems by any diagonalizing procedure that avoids the solution of simultaneous equations.

8. a. Compute the fixed-end moments at points A and B due to the concentrated load P. Compare results to those of Ex. 11.1.

b. Draw the moment diagram for the beam.

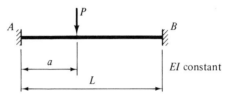

Prob. 8

9. Repeat Prob. 8 for a uniform load of intensity w kips/ft.

10. Compute the fixed-end moments and plot the moment diagram for the variable stiffness beam under uniform load w.

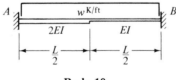

Prob. 10

11. Compute the fixed-end moments at points A and B of the linearly tapering two-flange beam shown due to a uniform load w. [*Hint:* Computations can be

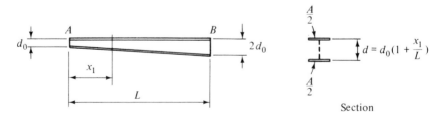

Section

Prob. 11

simplified by assuming the origin so that $d = d_0(x/L)$. Compare calculations with those of Prob. 11.5.]

12. Compute the fixed-end moments of points A and B of the haunched beam due to the concentrated load P. (*Hint:* Information from Probs. 8 and 10 can be used here.)

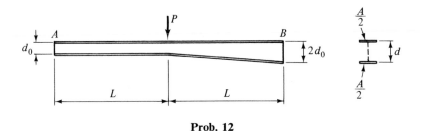

Prob. 12

13. Analyze the rigid frame under uniform load, and draw moment and shear diagrams.

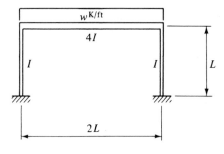

Prob. 13

14. Analyze the closed box-type structure with walls of unequal stiffness. Draw

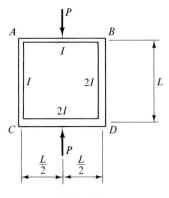

Prob. 14

moment diagram. (*Note:* The horizontal and vertical axes are *not* principal axes here. Find these by a transformation procedure such as is discussed in books on mechanics of rigid bodies, for instance. Verify your result by realizing that an axis through *A* and *D* is an axis of symmetry, and therefore the inclination of the principal axes must be at $\pm 45°$ with the horizontal.

15. Compute the stiffness matrix, of order (4 × 4), for the prismatic beam shown. Use information from Prob. 8 if available.

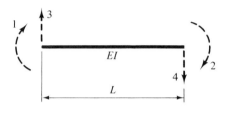

Prob. 15

16. Compute the stiffness matrix, of order (4 × 4), for the numbered nodes of the nonprismatic beam shown; information from Prob. 10 is useful.

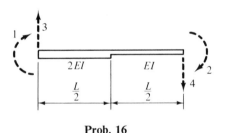

Prob. 16

17. Compute the (4 × 4) stiffness matrix for the numbered nodes of the wedge beam. (*Hint:* Use information from Prob. 11, and compare effort and results to those of Prob. 11.)

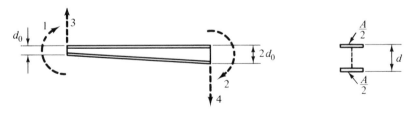

Prob. 17

18. Compute the (4 × 4) stiffness matrix for the numbered nodes of the rigid frame shown. Information from Prob. 13 is helpful here.

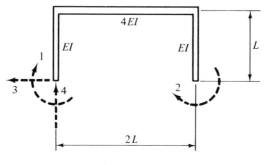

Prob. 18

19. Develop an analogous foundation for the *pin-ended* frame shown, and use it to analyze the structure due to a transverse uniform load acting on member *BC*. (*Hint:* Represent the infinite flexibility of the pins at *A* and *C* by a section of foundation of infinite width $1/EI = 1/0 = \infty$; the neutral, or conjugate, axis of the foundation will then have to pass through these points.

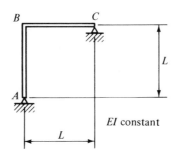

Prob. 19

20. You may recall that, contrary to reality, we are assuming framing members

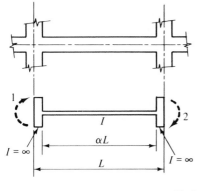

(a) Portion of building frame

(b) Beam with infinitely stiff haunches

Prob. 20

to end at the intersection of member centerlines, as shown in part (a) of the figure, whereas actually they should be calculated with ends of increased stiffness to represent the joint areas, as shown in part (b). This effect is to be investigated. Assuming a joint area of infinite stiffness, of length $(L/2)(1 - \alpha)$, at each end, compute the (2×2) stiffness matrix for the numbered nodes as a function of the ratio α. Plot the value of k_{11} for values of α ranging from .80 to 1.00, and draw engineering conclusions.

21. Compute and draw influence lines for
   a. M_B.
   b. R_B.
   c. V_{B^-}.
   d. M_D.
   Calculate influence ordinates at $\frac{1}{5}$ points of span, and use symmetry.

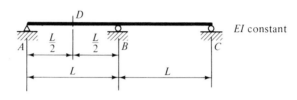

Prob. 21

22. Compute and draw the influence line for
   a. Horizontal reaction at E.
   b. Moment at D.
   c. Shear at C.
   Calculate influence ordinates for $\frac{1}{5}$ points of span BD, and use symmetry.

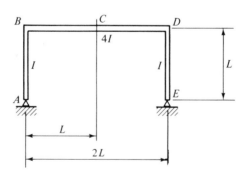

Prob. 22

V

The Displacement Method

In Part IV the analysis of statically indeterminate structures was carried out by considering the redundant forces as unknowns. In this part an analogous method will be discussed, which considers the displacements as unknowns. Whether one or the other of these methods is preferable will depend on the nature of the structure and its support conditions, as well as the availability of automatic computing equipment and programs.

The ideas and formulation of the displacement method are presented in Chapter 13; the emphasis here is on a generally valid matrix scheme suitable for computer operations. Chapter 14 deals with two older longhand versions of the displacement method.

In Chapter 15 we intend to reinforce some of the basic thinking by discussing and correlating different energy formulations of structural theory. Finally, Chapter 16 generalizes the displacement method by presenting the finite-element method, which unifies and extends the thought process to the analysis of any type of structure, including those previously accessible only by means of higher theory.

CHAPTER 13

The Displacement Method

The displacement method is entirely analogous to the force method; whereas in the latter the forces were unknowns, here the displacements are unknowns.

Section 13.1 presents basic ideas and conditions under which the displacement method may be preferred. The concept of stiffness matrices, which forms an important building block of the method, has already been introduced in Chapter 11 and is further covered in Secs. 13.2 and 13.3 and a general matrix formulation is contained in Sec. 13.4. Additional tools for the application of the displacement method are explained in Secs. 13.5 and 13.6; and Sec. 13.7, finally, implements the method by means of a computer program suitable for analyzing three-dimensional framed structures.

13.1. Basic Ideas of the Displacement Method

In the force method discussed in Part IV, the redundant forces are unknowns. A set of equations of geometry, equal to the number of unknown forces (the degree of static indeterminacy r), has to be solved to determine these forces.

In the displacement method, the nodal displacements are unknowns. A set of equations of equilibrium equal to the number of unknown displacements (the degree of *kinematic* indeterminacy k) has to be solved to determine these displacements.

A decision as to the relative advantage of one method or the other for any particular problem might be based on the number of unknowns contained in either formulation. If the number of unknown displacements (also called the *degrees of freedom*), k, is less than the number of redundant forces, r, the displacement method is advantageous. If, on the other hand, $r < k$, the force method might be preferred.

Figure 13.1 illustrates this concept. Figure 13.1(a) shows a three-member truss of one degree of static indeterminacy; that is, $r = 1$. Its common joint

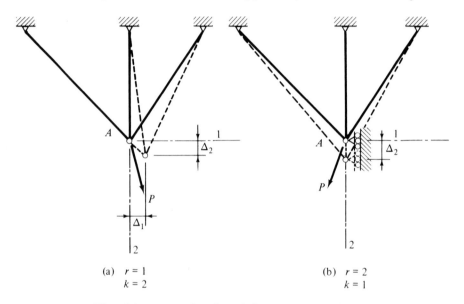

$$(a) \quad r = 1$$
$$k = 2$$

$$(b) \quad r = 2$$
$$k = 1$$

Fig. 13.1 Degrees of static and kinematic indeterminacy.

A has two unknown displacement components, Δ_1 and Δ_2, so that $k = 2$. Thus, the force method requires one equation of geometry, but the displacement method requires two equations of statics.

We now modify the three-member truss by adding a horizontal support at joint A, as shown in Fig. 13.1(b). Now the degree of static indeterminacy $r = 2$, but only one unknown displacement, Δ_2, remains, so that $k = 1$. For this case, the displacement method may be preferred. We see that from this viewpoint the choice of method depends on the nature of the structure and its support conditions. However, we shall soon recognize that the displacement method in its matrix form lends itself to more effective automatization than the force method and may thus be preferred even for cases in which $k > r$. Most computer routines for structural analysis are in fact based on the displacement method.

The basic building blocks of structural mechanics are of course used in this as in any other rational method; only the sequence of application is somewhat different. In the force method we start with a primary system that satisfies statics at all stages; geometry is then satisfied by later correction. In the displacement method we start with a primary system that satisfies geometry at all stages; statics is then satisfied by later correction.

The following examples are intended to convey the ideas underlying the displacement method in a simple fashion. The sequence of operations is here carried out in the following order:

1. Computation of the force–displacement relations, that is, the stiffnesses.

2. Establishment of the geometrical relations.

3. Establishment of the equilibrium equations to determine the unknown displacements.

4. Calculation of forces by substituting the displacements found in step 3 into the force–displacement relations of step 1.

We shall follow these steps in several examples.

Example 13.1: Calculate the moments in the structure shown in part (a) of the figure due to the cranked-in moment M.

Solution: This structure is statically indeterminate to the fourth degree, but its kinematic degree of indeterminacy can be considered as one, since all forces can be determined in terms of the single unknown rotation θ of joint A. We proceed along the outlined steps:

1. The rotational stiffnesses (that is, the end moments associated with a unit rotation at end A) are found for all members as outlined, and as shown in part (b).

MEMBER	REMARKS	STIFFNESS
AB	End B is pinned, member is statically determinate	$k^1 = \dfrac{M_{AB}}{\theta_{AB}} = \dfrac{3EI}{2L}$
AC	Calculate spring force due to unit rotation of the rigid arm; compute moment M_{AC} by statics	$k^2 = \dfrac{M_{AC}}{\theta_{AC}} = \dfrac{EI}{L}$
AD	Same as member AB	$k^3 = \dfrac{M_{AD}}{\theta_{AD}} = 3\dfrac{EI}{L}$
AE	Far end E fixed; use stiffness k_{11} from Eq. (11.4)	$k^4 = \dfrac{M_{AE}}{\theta_{AE}} = 4\dfrac{EI}{L}$

2. Since joint A is rigid, all member ends turn through the same angle θ:

$$\theta_{AB} = \theta_{AC} = \theta_{AD} = \theta_{AE} = \theta$$

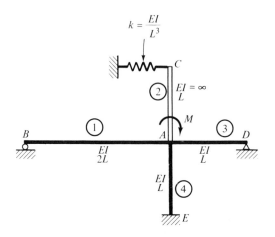

(a) Structure

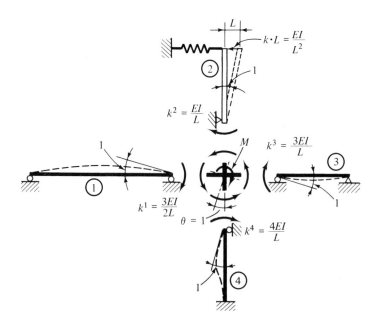

(b) Stiffness

Fig. E13.1

3. The single equilibrium equation needed sums the moments on joint A:

$$M_A = 0: \qquad M_{AB} + M_{AC} + M_{AD} + M_{AE} = M$$

Using the stiffnesses from step 1 and the geometrical relation of step 2, we express the moments in terms of the angle θ to obtain the equilibrium equation in terms of displacements:

$$\frac{EI}{L}\left(\frac{3}{2} + 1 + 3 + 4\right)\theta = M$$

or

$$\sum k^i \cdot \theta = M$$

from which we can solve for the unknown displacement:

$$\theta = \frac{M}{\sum k^i} = \frac{2}{19}\frac{L}{EI}M$$

Note that the total stiffness K against rotation of the structure at joint A, defined by the moment M required to achieve a unit rotation,

$$K = \frac{M}{\theta} = \sum k^i$$

is obtained by adding the member stiffnesses at joint A.

4. To obtain the member end moments at A, we substitute the value of the unknown displacement θ found in step 3 into the stiffness relations of step 1:

$$M_{AB} = \tfrac{3}{19}M; \qquad M_{AC} = \tfrac{2}{19}M; \qquad M_{AD} = \tfrac{6}{19}M; \qquad M_{AE} = \tfrac{8}{19}M$$

Note that the applied moment M is distributed to the adjacent members in direct proportion to their stiffness. The fraction of the total moment resisted by member i is called the *distribution factor* (DF) and is equal to the ratio of the rotational stiffness of member i to the sum K of the stiffnesses of all members entering the joint:

$$DF_i = \frac{k^i}{\sum\limits_m k^i} = \frac{k^i}{K}$$

Example 13.2: Analyze the three-member truss shown in part (a) of the figure by the displacement method. All members have identical stiffness AE.

Solution: We identify the structure as possessing two degrees of translational freedom, the horizontal and vertical displacements at joint A, and accordingly label them as nodes 1 and 2. We now follow the basic steps outlined earlier.

The stiffnesses k_{ij}^m for a typical member m will now be computed for the general case of a two-force member inclined at an angle α with the horizontal,

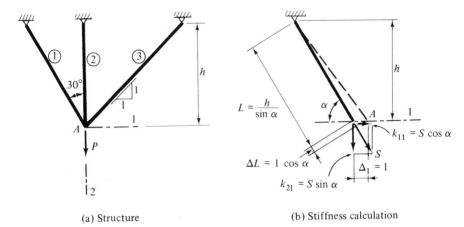

(a) Structure   (b) Stiffness calculation

Fig. E13.2

shown in part (b). Proceeding in the most straightforward manner, we first visualize end A displaced an amount $\Delta_1 = 1$ along node 1; the corresponding member elongation $\Delta L = 1 \cdot \cos \alpha$, and the resulting axial force

$$S = \frac{\Delta L \cdot AE}{L} = \frac{AE \cdot \cos \alpha}{h/\sin \alpha}$$

The horizontal and vertical components of this force are the desired stiffnesses:

$$k_{11}^m = \frac{AE}{h} \sin \alpha \cos^2 \alpha; \qquad k_{21}^m = \frac{AE}{h} \sin^2 \alpha \cos \alpha$$

In a similar manner, we displace the member end A through $\Delta_2 = 1$, and compute:

$$k_{12}^m = \frac{AE}{h} \sin^2 \alpha \cos \alpha; \qquad k_{22}^m = \frac{AE}{h} \sin^3 \alpha$$

Substituting appropriate values for α for each member, we obtain the following stiffnesses [all values $\cdot (AE/h)$]:

MEMBER	α	$k_{11} = \dfrac{S_1}{\Delta_1}$	$k_{12} = \dfrac{S_1}{\Delta_2} = k_{21} = \dfrac{S_2}{\Delta_1}$	$k_{22} = \dfrac{S_2}{\Delta_2}$
1	60°	.216	.375	.650
2	90°	0	0	1.000
3	135°	.354	−.354	.354

The geometrical conditions simply state the interconnectivity of all member ends at joint A:

$$\Delta_1^1 = \Delta_1^2 = \Delta_1^3 = \Delta_1$$

$$\Delta_2^1 = \Delta_2^2 = \Delta_2^3 = \Delta_2$$

and the equilibrium equations state the balance of forces acting on joint A along nodes 1 and 2:

$$\sum F_1 = 0: \qquad S_1^1 + S_1^2 + S_1^3 = 0$$
$$\sum F_2 = 0: \qquad S_2^1 + S_2^2 + S_2^3 = P$$

or, in terms of the displacements Δ_1 and Δ_2, by substituting the stiffnesses,

$$(.216 + 0 + .354)\frac{AE}{h}\Delta_1 + (.375 + 0 - .354)\frac{AE}{h}\Delta_2 = 0$$

$$(.375 + 0 - .354)\frac{AE}{h}\Delta_1 + (.650 + 1.000 + .354)\frac{AE}{h}\Delta_2 = P$$

from which the unknown displacements are

$$\Delta_1 = -.0184\frac{Ph}{AE}; \qquad \Delta_2 = .500\frac{Ph}{AE}$$

The components along nodes 1 and 2 of the member forces are now obtained by resubstituting these values into the force–displacement relations, which for member m are

$$S_i^m = k_{i1}^m\Delta_1 + k_{i2}^m\Delta_2 = \sum_{j=1}^{2} k_{ij}^m\Delta_j \qquad (13.3)$$

The S_1 and S_2 components for all members, as well as their resultant forces S, are given in the table:

MEMBER	S_1	S_2	S
1	.183P	.317P	.365P
2	0	.500P	.500P
3	−.183P	.183P	.259P

This concludes the example.

We note that this type of computation may be suitable for matrix formulation. Such an approach will be stressed in Sec. 13.4; prior to doing this, however, we shall in the following sections take another look at the stiffness matrices.

13.2. Element Stiffness Matrix

As already discussed in Sec. 11.2, the *element stiffness matrix* $[k]$ contains the stiffnesses, k_{ij}, defined as the forces that must be applied to the structural element at node i to ensure a unit displacement at node j, all others zero.

The calculation of the stiffnesses for beam elements by the force method was outlined in Sec. 11.2, and Eq. (11.4) gives the flexural stiffness matrix for

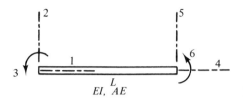

Fig. 13.2 Plane beam element.

a planar, prismatic, straight, elastic beam element. If axial displacements are to be considered as well, it is necessary to add nodes at both member ends in the axial direction, as shown in Fig. 13.2; the corresponding stiffness matrix is then given by

$$[k] = \begin{array}{c} \\ 1 \\ 2 \\ 3 \\ 4 \\ 5 \\ 6 \end{array}
\begin{array}{cccccc}
\ \ 1 & \ \ 2 & \ \ 3 & \ \ 4 & \ \ 5 & \ \ 6 \\
\left[\begin{array}{cccccc}
\dfrac{AE}{L} & & & & & \\[2mm]
0 & \dfrac{12EI}{L^3} & & & \text{sym} & \\[2mm]
0 & \dfrac{6EI}{L^2} & \dfrac{4EI}{L} & & & \\[2mm]
-\dfrac{AE}{L} & 0 & 0 & \dfrac{AE}{L} & & \\[2mm]
0 & -\dfrac{12EI}{L^3} & -\dfrac{6EI}{L^2} & 0 & \dfrac{12EI}{L^3} & \\[2mm]
0 & \dfrac{6EI}{L^2} & \dfrac{2EI}{L} & 0 & -\dfrac{6EI}{L^2} & \dfrac{4EI}{L}
\end{array}\right]
\end{array}$$

(13.4)

For the more general case of a three-dimensional prismatic beam element, we must consider the twelve nodes shown in Fig. 13.3, that is, axial and transverse translatory nodes along three axes at each member end and rotatory nodes about each of these axes. With these nodes, axial forces and dis-

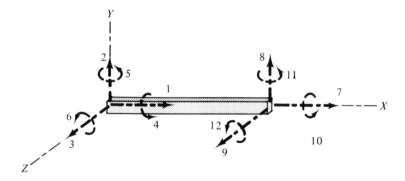

Fig. 13.3 Three-dimensional beam element.

$$[k] = \begin{bmatrix}
\dfrac{EA}{L} & & & & & & & & & & & \\[6pt]
& \dfrac{12EI_z}{L^3} & & & & & & & & & & \\[6pt]
& & \dfrac{12EI_y}{L^3} & & & & & & & & & \\[6pt]
& & & \dfrac{GC}{L} & & & & & & & & \\[6pt]
& & -\dfrac{6EI_y}{L^2} & & \dfrac{4EI_y}{L} & & & & & & & \\[6pt]
& \dfrac{6EI_z}{L^2} & & & & \dfrac{4EI_z}{L} & & & & & & \\[6pt]
-\dfrac{EA}{L} & & & & & & \dfrac{EA}{L} & & & & & \\[6pt]
& -\dfrac{12EI_z}{L^3} & & & & -\dfrac{6EI_z}{L^2} & & \dfrac{12EI_z}{L^3} & & & & \\[6pt]
& & -\dfrac{12EI_y}{L^3} & & \dfrac{6EI_y}{L^2} & & & & \dfrac{12EI_y}{L^3} & & & \\[6pt]
& & & -\dfrac{GC}{L} & & & & & & \dfrac{GC}{L} & & \\[6pt]
& & -\dfrac{6EI_y}{L^2} & & \dfrac{2EI_y}{L} & & & & \dfrac{6EI_y}{L^2} & & \dfrac{4EI_y}{L} & \\[6pt]
& \dfrac{6EI_z}{L^2} & & & & \dfrac{2EI_z}{L} & & -\dfrac{6EI_z}{L^2} & & & & \dfrac{4EI_z}{L}
\end{bmatrix}$$

sym

(13.5)

327

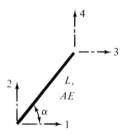

Fig. 13.4 Truss element.

placements, transverse shears and corresponding translations, twisting, and bending about two axes can be accounted for. With the numbering of Fig. 13.3, elementary beam and torsion theory leads to the stiffness matrix of Eq. (13.5) on page 327.

Another standard element stiffness matrix is the one for the inclined planar truss member with the nodal numbering of Fig. 13.4. Calculations of the type already performed in Ex. 13.2 lead to the stiffness matrix

$$
[k] = \begin{matrix} & & 1 & 2 & 3 & 4 \\ & 1 \\ & 2 \\ & 3 \\ & 4 \end{matrix} \frac{AE}{L} \begin{bmatrix} \cos^2\alpha & & & \text{sym} \\ \sin\alpha\cos\alpha & \sin^2\alpha & & \\ -\cos^2\alpha & -\sin\alpha\cos\alpha & \cos^2\alpha & \\ -\sin\alpha\cos\alpha & -\sin^2\alpha & \sin\alpha\cos\alpha & \sin^2\alpha \end{bmatrix} \quad (13.6)
$$

A catalogue of stiffness matrices of the type shown in Eqs. (13.4) to (13.6) is necessary to perform efficient computer analysis by the displacement method. Although computer programs available for production runs have matrices such as Eq. (13.5) built in, more specialized work may require stiffnesses of other members, such as tapering or curved members, or more general structural elements such as in the finite-element method. Otherwise, it will be necessary to model these elements as an assembly of straight prismatic members to which Eq. (13.5) can be applied.

13.3. Structure Stiffness Matrix

The *structure stiffness matrix* $[K]$ relates the forces and displacements of a structure composed of elements; the force X_i at node i is linearly related to the displacements Δ_j:

$$
X_i = K_{i1}\Delta_1 + K_{i2}\Delta_2 + \cdots + K_{ij}\Delta_j + \cdots
$$
$$
\{X\} = [K]\{\Delta\} \quad (13.7)
$$

Each element K_{ij} of this stiffness matrix is defined as the force that must be applied to the complete structure at node i to ensure a unit displacement at node j, all others zero.

We recall from Exs. 13.1 and 13.2 that the total, or structure, stiffness at a node is obtained by adding the stiffnesses of all members at this node. This summing of element stiffnesses results from expressing the equilibrium equations in terms of the compatible displacements, as will be shown for a general plane case with reference to Fig. 13.5(a); we consider one single joint connect-

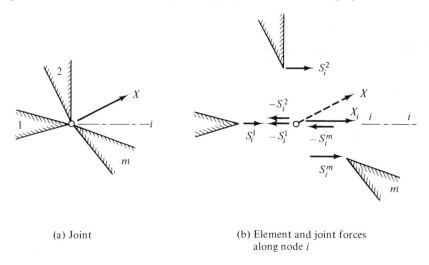

(a) Joint

(b) Element and joint forces along node i

Fig. 13.5

ing several elements $1, 2, \ldots, m, \ldots$, and with an external force X, of components X_1, X_2 acting on it.

Figure 13.5(b) shows the internal element forces S_i^m of element m along node i, as well as their equal and opposite reactions on the joint. Each of these element forces is related to the element displacements at node j by the element stiffness k_{ij}^m; for instance, the ith force on the mth element is

$$S_i^m = k_{i1}^m \Delta_1 + k_{i2}^m \Delta_2 + \cdots + k_{ij}^m \Delta_j + \cdots \qquad (13.8)$$

where the j extends over all the nodes attached to the element. The equilibrium equation for the forces along node i acting on the joint is

$$\sum F_i = 0: \qquad -S_i^1 - S_i^2 - \cdots - S_i^m - \cdots + X_i = 0$$

from which, substituting Eq. (13.8), we get

$$-(k_{i1}^1 \Delta_1 + k_{i2}^1 \Delta_2 + \cdots) - (k_{i1}^2 \Delta_1 + k_{i2}^2 \Delta_2 + \cdots) - \cdots$$
$$- (k_{i1}^m \Delta_1 + k_{i2}^m \Delta_2 + \cdots) - \cdots + X_i = 0$$

Combining coefficients of identical Δ's, we obtain

$$(k_{i1}^1 + k_{i1}^2 + \cdots + k_{i1}^m + \cdots)\Delta_1 + (k_{i2}^1 + k_{i2}^2 + \cdots + k_{i2}^m + \cdots)\Delta_2$$
$$+ \cdots = X_i \qquad (13.9)$$

By comparing Eqs. (13.7 and 13.9), we can now equate in each case

$$K_{ij} = k^1_{ij} = k^2_{ij} + \cdots + k^m_{ij} + \cdots = \sum_m k^m_{ij} \qquad (13.10)$$

Thus, the structure stiffness K_{ij} is obtained by superposition of the element stiffnesses k^m_{ij}. It is essential that each node of the structure be carefully labeled, and that the nodal numbering of each element correspond to that of the structure. The element stiffness matrices for all members are then written and superposed, or *assembled*, to form the structure stiffness matrix as shown in the following example. We should always recall that the summing of the element stiffnesses to obtain the structure stiffness amounts to the satisfaction of equilibrium equations expressed in terms of displacements.

> **Example 13.3:** Assemble the stiffness matrix for the structure and nodes shown in part (a) of the figure.

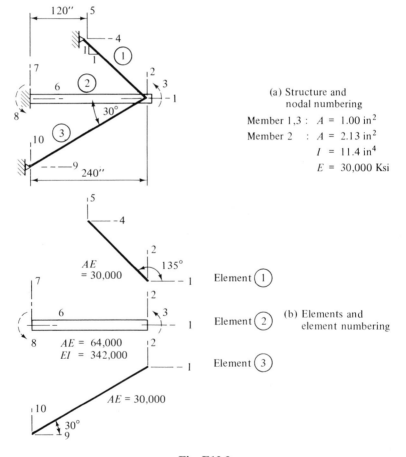

(a) Structure and
    nodal numbering

Member 1,3 : $A = 1.00$ in^2
Member 2   : $A = 2.13$ in^2
              $I = 11.4$ in^4
              $E = 30,000$ Ksi

Element ①
Element ②   (b) Elements and
              element numbering
Element ③

Fig. E13.3

Solution: We first write out the element stiffness matrices, member by member, taking care to adhere to the established numbering shown in part (a). These matrices follow Eq. (13.6) for the two-force members 1 and 3, and Eq. (13.4) for the planar beam element 2; accordingly, using the structure properties given in part (a), we obtain

$$
[k^1] = \begin{array}{c c} & \begin{array}{cccc} 1 & 2 & 4 & 5 \end{array} \\ \begin{array}{c} 1 \\ 2 \\ 4 \\ 5 \end{array} & \left[\begin{array}{cccc} 89 & & & \text{sym} \\ -89 & 89 & & \\ -89 & 89 & 89 & \\ 89 & -89 & -89 & 89 \end{array}\right] \end{array}
$$

$$
[k^2] = \begin{array}{c c} & \begin{array}{cccccc} 6 & 7 & 8 & 1 & 2 & 3 \end{array} \\ \begin{array}{c} 6 \\ 7 \\ 8 \\ 1 \\ 2 \\ 3 \end{array} & \left[\begin{array}{cccccc} 267 & & & & & \text{sym} \\ 0 & .298 & & & & \\ 0 & 35.8 & 5{,}700 & & & \\ -267 & 0 & 0 & 267 & & \\ 0 & -.298 & -35.8 & 0 & .298 & \\ 0 & 35.8 & 2{,}850 & 0 & -35.8 & 5{,}700 \end{array}\right] \end{array}
$$

$$
[k^3] = \begin{array}{c c} & \begin{array}{cccc} 1 & 2 & 9 & 10 \end{array} \\ \begin{array}{c} 1 \\ 2 \\ 9 \\ 10 \end{array} & \left[\begin{array}{cccc} 81 & & & \text{sym} \\ 47 & 27 & & \\ -81 & -47 & 81 & \\ -47 & -27 & 47 & 27 \end{array}\right] \end{array}
$$

Assembling these element stiffnesses into the locations of the (10 × 10) structure stiffness matrix as indicated by the numbering of the rows and columns, we obtain

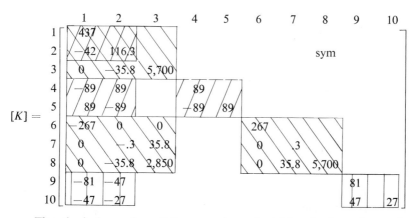

The physical meaning of these numbers should be clearly understood; element K_{11}, for instance, is about four times the value of K_{22}, indicating that the horizontal force required to stretch the structure a specified amount

horizontally is four times as much as a vertical force at the same point causing the same amount of vertical displacement. The contribution of the beam member, k_{22}^2, is only a small fraction of the total value of the stiffness K_{22}, indicating that this stiffness is mainly due to the inclined truss members. Considerations of this type can be of great value to the designer in assessing the effectiveness of a structural configuration.

13.4. Matrix Formulation of the Displacement Method

The matrix displacement method, if used in conjunction with appropriate computing equipment, is the most powerful of the various methods of analysis. In its generalization as the *finite-element method* it is capable of analyzing any solid body; in this section, however, its application will be restricted to the analysis of framed structures.

The *displacement method* relates the forces to the displacements by means of the structure stiffness matrix $[K]$:

$$[X] = [K][\Delta] \tag{13.11}$$

Each node will have a force and a corresponding displacement, one of which is known, the other unknown. For a structure with k degrees of kinematic indeterminacy (or degrees of freedom), we collect the k nodes with unknown displacements and known forces (loads) at the top, the remaining f nodes corresponding to the known support displacements and unknown forces (reactions) at the bottom, identify the former nodal values by the subscript α, the latter by β, and partition the matrix equation:

$$
\begin{array}{c}
\text{Known forces} \\
\text{(loads)} \\
\text{Unknown forces} \\
\text{(reactions)}
\end{array}
\,
\begin{array}{c}
k \\
f
\end{array}
\left\{
\begin{bmatrix} X_\alpha \\ \hline X_\beta \end{bmatrix}
\right.
=
\begin{bmatrix} K_{\alpha\alpha} & K_{\alpha\beta} \\ \hline K_{\beta\alpha} & K_{\beta\beta} \end{bmatrix}
\underbrace{\begin{bmatrix} \Delta_\alpha \\ \hline \Delta_\beta \end{bmatrix}}_{k \quad f}
\left.
\begin{array}{c}
\text{Unknown displacements} \\
\text{at load points} \\
\text{Known (specified)} \\
\text{displacements}
\end{array}
\right.
\tag{13.12}
$$

The solution is obtained in two steps: first, the first k equations are solved for the unknown displacements:

$$[X_\alpha] = [K_{\alpha\alpha}][\Delta_\alpha] + [K_{\alpha\beta}][\Delta_\beta]$$

from which

$$[\Delta_\alpha] = [K_{\alpha\alpha}]^{-1}\big[[X_\alpha] - [K_{\alpha\beta}][\Delta_\beta]\big] \tag{13.13}$$

Note that a $(k \times k)$ square matrix has to be inverted, corresponding to a solution of k simultaneous equations.

The displacements $[\Delta_\alpha]$ thus found are then substituted in the f remaining equations to solve for the unknown forces:

$$[X_\beta] = [K_{\beta\alpha}][\Delta_\alpha] + [K_{\beta\beta}][\Delta_\beta] \tag{13.14}$$

With these basic concepts in mind, we outline the steps necessary for the formulation of the matrix displacement method:

1. Identify the separate elements of the structure. The interconnections between these elements are called *joints*.

2. At each joint, identify and number the nodes for which forces and corresponding displacements exist. In the most general three-dimensional case, there may be six nodes (three forces and corresponding translations, and three moments and corresponding rotations) at each joint. In other cases, there may be less. Number the nodes with unknown displacements first.

3. Calculate and write the stiffness matrix for each element, adhering to the numbering established in step 2.

4. Assemble the structure stiffness matrix by superposition of the element stiffness matrices.

5. Write the matrix equation $[X] = [K][\Delta]$, substitute known values of forces and displacements, partition into Eqs. (13.13) and (13.14), and solve for the unknown displacements $[\Delta_\alpha]$ and the unknown reactions $[X_\beta]$.

6. To find the element (internal) forces, $[S]$, use the force–displacement relations for each element m:

$$[S^m] = [k^m][\Delta] \tag{13.15}$$

This concludes the analysis by the displacement method.

Example 13.4: Analyze the structure already considered in Ex. 13.3, and shown again in the figure, by the matrix displacement method due to the applied load of 10 kips. Members 1 and 3 are two force; member 2 is subject to flexural and axial deformations.

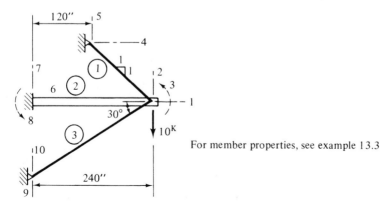

For member properties, see example 13.3

Fig. E13.4 For member properties, see Fig. E13.3

Solution: The nodes are identified and numbered as shown, remembering to number first the three nodes for which the displacements are unknown. The element stiffnesses were already written and assembled into the appropriate structure stiffness matrix in Ex. 13.3. The given values of the applied loads, X_1 to X_3, and the specified support displacements, Δ_4 to Δ_{10} (all zero), are now inserted into the force–displacement Eqs. (13.12) to yield

$$
\begin{pmatrix} 0 \\ -10 \\ 0 \\ X_4 \\ X_5 \\ X_6 \\ X_7 \\ X_8 \\ X_9 \\ X_{10} \end{pmatrix}
=
\begin{bmatrix}
437 & & K_{\alpha\alpha} & & & K_{\alpha\beta} & & & \\
-42 & 116.3 & & & & & & & \\
0 & -35.8 & 5{,}700 & & & & & & \\
\hline
-89 & 89 & & 89 & & & & & \\
89 & -89 & & -89 & 89 & & & & \\
-267 & 0 & 0 & & & 267 & & & \\
0 & -.3 & 35.8 & & & 0 & .3 & & \\
0 & -35.8 & 2{,}850 & & & 0 & 35.8 & 5{,}700 & \\
-81 & -47 & K_{\beta\alpha} & & & K_{\beta\beta} & & & 81 \\
-47 & -27 & & & & & & & 47 & 27
\end{bmatrix}
\begin{pmatrix} \Delta_1 \\ \Delta_2 \\ \Delta_3 \\ 0 \\ 0 \\ 0 \\ 0 \\ 0 \\ 0 \\ 0 \end{pmatrix}
$$

Equations (13.13) and (13.14) can now be solved to yield the matrix of unknown displacements and the matrix of unknown reactions:

$$
\begin{Bmatrix} \Delta_1 \\ \Delta_2 \\ \Delta_3 \end{Bmatrix} = \begin{Bmatrix} -.0086 \\ -.0893 \\ -.0006 \end{Bmatrix} \text{ in.}
$$

$$
\begin{Bmatrix} X_4 \\ X_5 \\ X_6 \\ X_7 \\ X_8 \\ X_9 \\ X_{10} \end{Bmatrix} = \begin{Bmatrix} -7.1802 \\ 7.1802 \\ 2.2904 \\ .0067 \\ 1.5977 \\ 4.8898 \\ 2.8131 \end{Bmatrix} \text{ kips, except } X_8 \text{ in kip-inches}
$$

Note that these reactions satisfy the equilibrium conditions. The interior element forces, that is, the end forces and moments of members 1, 2, and 3, can now be found by applying Eq. (13.15) to each of the members; for example, member 1: $\{S^1\} = [k^1]\{\Delta\}$

$$
\begin{Bmatrix} S_1^1 \\ S_2^1 \\ S_3^1 \\ S_4^1 \end{Bmatrix}
=
\begin{bmatrix}
89 & & & \text{sym} \\
-89 & 89 & & \\
-89 & 89 & 89 & \\
89 & -89 & -89 & 89
\end{bmatrix}
\begin{Bmatrix} -.0086 \\ -.0893 \\ 0 \\ 0 \end{Bmatrix}
=
\begin{Bmatrix} 7.18 \\ -7.18 \\ -7.18 \\ 7.18 \end{Bmatrix} \text{ kips}
$$

Note that only the horizontal and vertical components of the two-force members, rather than the axial force, are obtained. This latter force, which is needed for design purposes, can be determined by simple statics as

$$N^1 = \frac{1}{\sqrt{2}}7.18 + \frac{1}{\sqrt{2}}7.18 = 10.02 \text{ kips}$$

Section 13.5 will show how such design forces can be determined from automatic matrix calculations.

13.5. Transformations

For practical automatic analysis of general structures, the matrix displacement method outlined in Sec. 13.4 is still incomplete; the procedure shows the following important deficiencies:

1. The two truss bars of Ex. 13.4 are similar, and should therefore have similar stiffness properties. But the element stiffness matrix, Eq. (13.6), furnishes different values because of its dependence on the member inclination. In this section we will redefine the element stiffness matrix so as to indicate the intrinsic properties of the element, independent of its orientation.

2. When two beam elements are inclined to each other, as shown in Fig. 13.6, the element stiffness matrices, Eqs. (13.4) and (13.5), are written with respect to different member axes and cannot be added without first being referred to a common system of axes.

3. The internal forces $\{S\}$ in Ex. 13.4 were obtained in an inconvenient

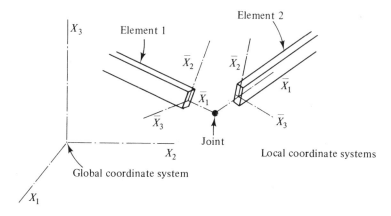

Fig. 13.6 Global and local coordinate systems.

coordinate system; an automatic routine is necessary for the determination of member design forces.

Because of these considerations we still need the following:

1. A local, or element, coordinate system with respect to which the element stiffness will be calculated.

2. A common global, or structure, coordinate system with respect to which all element stiffnesses must be written before assembly. Such a system is, for instance, indicated in Fig. 13.6.

3. A simple system of transformation between the two axis systems. Such a transformation can be devised by use of vectors, as will now be discussed.

We consider a vector V, as in Fig. 13.7, represented by its components X_1, X_2, and X_3 in the set of orthogonal axes X_1, X_2, and X_3. To represent the same vector by its coordinates $\bar{X}_1$, $\bar{X}_2$, and $\bar{X}_3$ along the new set of axes $\bar{X}_1$,

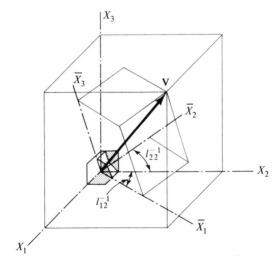

Fig. 13.7 Vector in two coordinate systems.

$\bar{X}_2$, and $\bar{X}_3$, we write the vector transformation

$$\begin{Bmatrix} \bar{X}_1 \\ \bar{X}_2 \\ \bar{X}_3 \end{Bmatrix} = \begin{bmatrix} l_{11} & l_{12} & l_{13} \\ l_{21} & l_{22} & l_{23} \\ l_{31} & l_{32} & l_{33} \end{bmatrix} \begin{Bmatrix} X_1 \\ X_2 \\ X_3 \end{Bmatrix}$$

or

$$\{\bar{\mathbf{X}}\} = [\lambda]\{\mathbf{X}\} \qquad (13.16)$$

wherein l_{ij} represents the direction cosine between the new (barred) $\bar{X}_i$ axis and the old (unbarred) X_j axis.

We can also transform inversely from the barred to the unbarred system; from Eq. (13.16)

$$\{\mathbf{X}\} = [\lambda]^{-1}\{\bar{\mathbf{X}}\} \qquad (13.17)$$

If both sets of reference axes are orthogonal (that is, all axes in one set are normal to each other), then

$$[\lambda]^{-1} = [\lambda]^T \qquad (13.18)$$

which saves calculation, because the transpose of a matrix is easier to obtain than its inverse.

In plane cases, the third row and third column of Eq. (13.16) can be deleted, so that in this case, shown in Fig. 13.8,

$$\begin{Bmatrix} \bar{X}_1 \\ \bar{X}_2 \end{Bmatrix} = \begin{bmatrix} \cos \alpha & \sin \alpha \\ -\sin \alpha & \cos \alpha \end{bmatrix} \begin{Bmatrix} X_1 \\ X_2 \end{Bmatrix} \qquad (13.19)$$

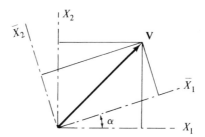

Fig. 13.8 Plane-vector transformation.

Example 13.5 The beam is oriented as shown in the figure. An analysis has resulted in the element forces of 5 kips and 10 kips, respectively, along axes X_1 and X_2. Calculate the axial force $\bar{S}_1$ and the shear force $\bar{S}_2$ in the beam.

Solution: We recognize this as a transformation of a force vector from the unbarred X_1, X_2 to the barred $\bar{X}_1, \bar{X}_2$ axial and transverse axis system. Accordingly, Eq. (13.19) becomes

$$\begin{Bmatrix} \bar{S}_1 \\ \bar{S}_2 \end{Bmatrix} = \begin{bmatrix} \cos 60° & \sin 60° \\ -\sin 60° & \cos 60° \end{bmatrix} \begin{Bmatrix} 5 \text{ kips} \\ 10 \text{ kips} \end{Bmatrix} = \begin{Bmatrix} 11.16 \text{ kips} \\ .67 \text{ kips} \end{Bmatrix}$$

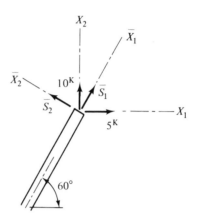

Fig. E13.5

Since forces, moments, displacements, and rotations are all vector quantities, the transformation, Eq. (13.16), is applicable to all these quantities. To transform the forces **S** and moments **M** at both ends of the inclined member shown in Fig. 13.9 from the global (unbarred) axis system to the principal member axes (the barred system), we can transform the four vectors (one force and one moment vector at each of two member ends) all at the same time

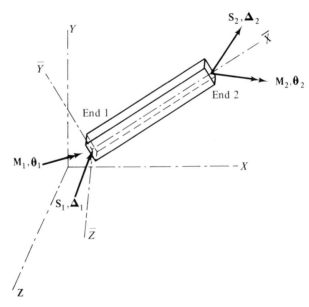

Fig. 13.9 Element-force and displacement vectors.

by the following transformation:

$$
\begin{Bmatrix} \bar{X}_1 \\ \bar{Y}_1 \\ \bar{Z}_1 \\ \bar{M}_{X1} \\ \bar{M}_{Y1} \\ \bar{M}_{Z1} \\ \bar{X}_2 \\ \bar{Y}_2 \\ \bar{Z}_2 \\ \bar{M}_{X2} \\ \bar{M}_{Y2} \\ \bar{M}_{Z2} \end{Bmatrix}
=
\begin{bmatrix} \lambda & 0 & 0 & 0 \\ 0 & \lambda & 0 & 0 \\ 0 & 0 & \lambda & 0 \\ 0 & 0 & 0 & \lambda \end{bmatrix}
\begin{Bmatrix} X_1 \\ Y_1 \\ Z_1 \\ M_{X1} \\ M_{Y1} \\ M_{Z1} \\ X_2 \\ Y_2 \\ Z_2 \\ M_{X2} \\ M_{Y2} \\ M_{Z2} \end{Bmatrix}
\tag{13.20}
$$

or

$$\{\bar{S}\} = [T]\{S\}$$

Similarly, we can transform the four corresponding displacement vectors Δ, consisting of one translation vector Δ and one rotation vector θ at each of two member ends, from the global to the local axes by the transformation

$$\{\bar{\Delta}\} = [T]\{\Delta\} \tag{13.21}$$

To transform the entire stiffness matrix, we define the *element stiffness matrix in local coordinates* $[\bar{k}]$ by the force–displacement relation in the barred system:

$$\{\bar{S}\} = [\bar{k}]\{\bar{\Delta}\} \tag{13.22}$$

To express this relation in global coordinates, we substitute Eqs. (13.20) and (13.21) into Eq. (13.22) and obtain

$$[T]\{S\} = [\bar{k}][T]\{\Delta\}$$

from which we solve for the forces in global coordinates; making use of Eq. (13.18),

$$\{S\} = [T]^T[\bar{k}][T]\{\Delta\}$$

If we now define the *element stiffness matrix in global coordinates* $[k]$ as

$$[k] = [T]^T[\bar{k}][T] \tag{13.23}$$

we can write the force–displacement relation in the global (unbarred) system as

$$\{S\} = [k]\{\Delta\} \tag{13.24}$$

The element stiffness matrices in local coordinates must be transformed into the common global coordinate system by Eq. (13.23) before they can be assembled into the structure stiffness matrix according to Eq. (13.10). (Note that in Ex. 13.4 no such transformation was necessary because all element stiffness matrices were already written in a common axis system.)

Example 13.6 For the plane elastic two-force member shown, calculate
a. The element stiffness matrix $[\bar{k}]$ in the local $\bar{X}_1$, $\bar{X}_2$ coordinate system.
b. The matrix $[T]$ for transformation of the local vectors into a global X_1, X_2 coordinate system as shown.
c. The element stiffness matrix $[k]$ in global coordinates.

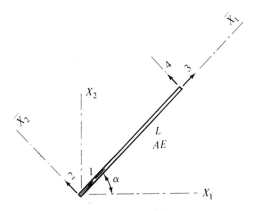

Fig. E13.6

Solution:
a. The nodal numbering of end forces and displacements in local coordinates is shown in the figure. By elementary considerations, the stiffness matrix in terms of the longitudinal and transverse member axes is

$$[\bar{k}] = \begin{bmatrix} \dfrac{AE}{L} & 0 & -\dfrac{AE}{L} & 0 \\ 0 & 0 & 0 & 0 \\ -\dfrac{AE}{L} & 0 & \dfrac{AE}{L} & 0 \\ 0 & 0 & 0 & 0 \end{bmatrix} = \dfrac{AE}{L} \begin{bmatrix} 1 & 0 & -1 & 0 \\ 0 & 0 & 0 & 0 \\ -1 & 0 & 1 & 0 \\ 0 & 0 & 0 & 0 \end{bmatrix}$$

b. The transformation matrix $[T]$ must transform two vectors, one at each end, simultaneously; accordingly, for this planar structure we obtain $[\lambda]$ by Eq. (13.19), and write it twice along the diagonal of the (4×4) $[T]$ matrix:

$$[T] = \begin{bmatrix} \lambda & 0 \\ 0 & \lambda \end{bmatrix} = \begin{bmatrix} \cos\alpha & \sin\alpha & & \\ -\sin\alpha & \cos\alpha & & 0 \\ & & \cos\alpha & \sin\alpha \\ 0 & & -\sin\alpha & \cos\alpha \end{bmatrix}$$

c. To transform the $[\bar{k}]$ matrix to global coordinates, we use Eq. (13.23); performing the two indicated matrix multiplications, one after the other, we obtain

$$[k] = [T]^T[\bar{k}][T] = \frac{AE}{L} \begin{bmatrix} \cos^2\alpha & & & \text{sym} \\ \sin\alpha\cos\alpha & \sin^2\alpha & & \\ -\cos^2\alpha & -\sin\alpha\cos\alpha & \cos^2\alpha & \\ -\sin\alpha\cos\alpha & -\sin^2\alpha & \sin\alpha\cos\alpha & \sin^2\alpha \end{bmatrix}$$

This element stiffness matrix in global coordinates was already presented in Eq. (13.6).

It remains to determine the direction cosines l_{ij} in the transformation matrix $[\lambda]$ for a general linear member. This is a relatively simple matter for the case of members of a planar structure; considering the forces (and corresponding displacements, which are analogous) acting on one end of the beam lying in plane 1–2 of Fig. 13.10, we number the axial and shear forces S_1 and

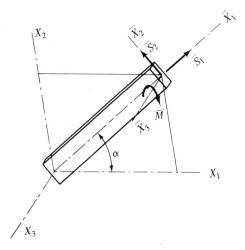

Fig. 13.10 Transformation, plane case.

S_2, and add the end moment M as force S_3. Since the moment does not change under rotation of the beam in its plane, it follows that the transformation of the end forces from the global X to the local $\bar{X}$ axes results by simple extension of Eq. (13.19):

$$\left\{\begin{array}{c} \bar{X} \\ \bar{Y} \\ \bar{M} \end{array}\right\} = \left\{\begin{array}{c} \bar{S}_1 \\ \bar{S}_2 \\ \bar{S}_3 \end{array}\right\} = \left[\begin{array}{ccc} \cos\alpha & \sin\alpha & 0 \\ -\sin\alpha & \cos\alpha & 0 \\ 0 & 0 & 1 \end{array}\right] \left\{\begin{array}{c} S_1 \\ S_2 \\ S_3 \end{array}\right\} \qquad (13.25)$$

or

$$\{\bar{S}\} = [\lambda]\{S\}$$

and analogously,

$$\{\bar{\Delta}\} = [\lambda]\{\Delta\}$$

For a three-dimensional case the situation is somewhat more involved. We consider the general member of Fig. 13.11 of length L, with the principal member axes shown as $\bar{\bar{X}}_1$, $\bar{\bar{X}}_2$, and $\bar{\bar{X}}_3$ and the global axes as X_1, X_2, and X_3 with unit vectors $\mathbf{i}$, $\mathbf{j}$, and $\mathbf{k}$. The member end coordinates are x_1^1, x_2^1, and x_3^1 at end 1, and x_1^2, x_2^2, and x_3^2 at end 2.

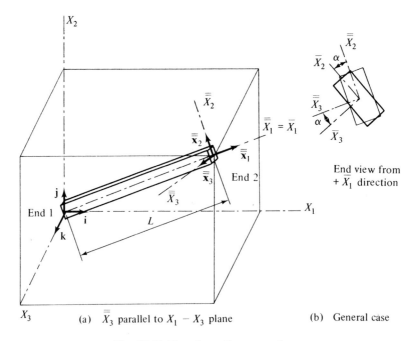

(a) $\bar{\bar{X}}_3$ parallel to $X_1 - X_3$ plane   (b) General case

Fig. 13.11 Transformation, general case.

The direction cosines of the longitudinal axis $\bar{X}_1$ of the member are determined by the geometry of Fig. 13.11 as

$$l_{11} = \frac{1}{L}(x_1^2 - x_1^1); \qquad l_{12} = \frac{1}{L}(x_2^2 - x_2^1); \qquad l_{13} = \frac{1}{L}(x_3^2 - x_3^1)$$

The major principal cross-sectional (or strong-bending) axis is in many cases horizontal, that is, normal to the vertical global axis X_2; in this case, a unit vector $\bar{\mathbf{x}}_3$ along this cross-sectional axis, denoted $\bar{X}_3$ in Fig. 13.11(a), is determined according to the rules of vector analysis already used in Sec. 3.4, by Eq. (3.5); since it is normal to both vectors $\mathbf{j}$ and $\bar{\mathbf{x}}_1$,

$$\bar{\mathbf{x}}_3 = \frac{\bar{\mathbf{x}}_1 \times \mathbf{j}}{|\bar{\mathbf{x}}_1 \times \mathbf{j}|} = \frac{1}{(l_{11}^2 + l_{13}^2)^{1/2}} \begin{vmatrix} \mathbf{i} & \mathbf{j} & \mathbf{k} \\ l_{11} & l_{12} & l_{13} \\ 0 & 1 & 0 \end{vmatrix} = \frac{-\mathbf{i} \cdot l_{13} + \mathbf{k} \cdot l_{11}}{(l_{11}^2 + l_{13}^2)^{1/2}}$$

The direction cosines l_{3j} are observed to be the unit vector components along the global axes; calling the denominator quantity $(l_{11}^2 + l_{13}^2)^{1/2} \equiv L_1$,

$$l_{31} = -\frac{l_{13}}{L_1}; \qquad l_{32} = 0; \qquad l_{33} = \frac{l_{11}}{L_1}$$

The weak-bending principal axis $\bar{X}_2$ is normal to the unit vectors along $\bar{X}_1$ and $\bar{X}_3$; by Eq. (3.6) its unit vector $\bar{\mathbf{x}}_2$ is therefore obtained by

$$\bar{\mathbf{x}}_2 = \bar{\mathbf{x}}_3 \times \bar{\mathbf{x}}_1 = \frac{1}{L_1} \begin{vmatrix} \mathbf{i} & \mathbf{j} & \mathbf{k} \\ -l_{13} & 0 & l_{11} \\ l_{11} & l_{12} & l_{13} \end{vmatrix}$$

$$= \frac{1}{L_1}[\mathbf{i}(-l_{11}l_{12}) - \mathbf{j}(-l_{13}^2 - l_{11}^2) + \mathbf{k}(-l_{12}l_{13})]$$

The direction cosines are then equal to the components of the unit vector:

$$l_{21} = -\frac{1}{L_1}(l_{11}l_{12}); \qquad l_{22} = \frac{1}{L_1}(l_{13}^2 + l_{11}^2); \qquad l_{23} = -\frac{1}{L_1}(l_{12}l_{13})$$

These direction cosines are now assembled into the $[\lambda]$ matrix for the case in which the $\bar{X}_3$ principal member axis is normal to the X_2 global axis:

$$\begin{Bmatrix} \bar{X}_1 \\ \bar{X}_2 \\ \bar{X}_3 \end{Bmatrix} = \begin{bmatrix} l_{11} & l_{12} & l_{13} \\ -\dfrac{1}{L_1}l_{11}l_{12} & \dfrac{1}{L_1}(l_{11}^2 + l_{13}^2) & -\dfrac{1}{L_1}l_{12}l_{13} \\ -\dfrac{1}{L_1}l_{13} & 0 & \dfrac{1}{L_1}l_{11} \end{bmatrix} \begin{Bmatrix} X_1 \\ X_2 \\ X_3 \end{Bmatrix} \qquad (13.26)$$

or

$$\{\bar{X}\} = [\lambda_1]\{X\}$$

where

$$l_{1i} = \frac{x_i^2 - x_i^1}{L}; \qquad L_1 = (l_{11}^2 + l_{13}^2)^{1/2}$$

In the most general case shown in Fig. 13.11(b), the principal cross-sectional axes $\bar{X}_2$, $\bar{X}_3$ are entirely arbitrary, making an angle α with the previously defined $\bar{\bar{X}}_2$, $\bar{\bar{X}}_3$ axes. The longitudinal member axis $\bar{X}_1$ is identical with the $\bar{\bar{X}}_1$ axis.

We now transform additionally from the $\bar{\bar{X}}$ to the $\bar{X}$ axes:

$$\begin{Bmatrix} \bar{X}_1 \\ \bar{X}_2 \\ \bar{X}_3 \end{Bmatrix} = \begin{bmatrix} 1 & 0 & 0 \\ 0 & \cos\alpha & \sin\alpha \\ 0 & -\sin\alpha & \cos\alpha \end{bmatrix} \begin{Bmatrix} \bar{\bar{X}}_1 \\ \bar{\bar{X}}_2 \\ \bar{\bar{X}}_3 \end{Bmatrix}$$

or

$$\{\bar{X}\} = [\lambda_2]\{\bar{\bar{X}}\} \tag{13.27}$$

The resultant transformation now consists of two successive transformations, first from the X to the $\bar{\bar{X}}$ axes by Eq. (13.26), then from the $\bar{\bar{X}}$ to the $\bar{X}$ axes by Eq. (13.27):

$$\{\bar{X}\} = [\lambda_2][\lambda_1]\{X\} \equiv [\lambda]\{X\} \tag{13.28}$$

and, analogously,

$$\{\bar{\Delta}\} = [\lambda]\{\Delta\}$$

where

$$[\lambda] = [\lambda_2][\lambda_1]$$

Note that when the angle α is zero, the $[\lambda_2]$ matrix degenerates into a unit matrix and can be dropped, and for the case when the member lies in the X_1X_2 plane, that is, the planar case shown in Fig. 13.10, $l_{13} = l_{23} = l_{31} = l_{32} = 0$, $l_{33} = 1$, and the general transformation, Eq. (13.16), degenerates into Eq. (13.25). The general transformation matrix, Eq. (13.28), can thus be used for all cases, except the following:

In the special case when the member axis is vertical, this transformation breaks down; since the direction cosines l_{11} and l_{13} are both zero in this case, the denominator quantity L_1 in Eq. (13.26) becomes zero, and some of the terms in the transformation matrix $[\lambda]$ become indeterminate.

We can, however, redefine the angle α in this case as the angle in the horizontal X_1X_3 plane between the global X_3 and the local $\bar{X}_3$ axes, positive when turning from the global X_3 to the global X_1 axis, as shown in Fig. 13.12.

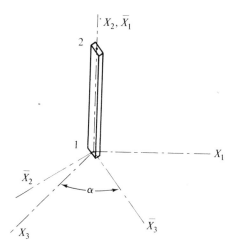

Fig. 13.12 Transformation for vertical member.

Going then through the same procedure as before, we can determine the appropriate transformation matrix as

$$[\lambda] = \begin{bmatrix} 0 & l_{12} & 0 \\ -l_{12}\cos\alpha & 0 & l_{12}\sin\alpha \\ \sin\alpha & 0 & \cos\alpha \end{bmatrix} \tag{13.29}$$

in which $l_{12} = 1$ for this case.

    We can now also discuss the calculation of the internal element forces $\{\bar{S}\}$. The procedure of Sec. 13.4, Eq. (13.15), provided for the computation of element forces in global coordinates as linear functions of the nodal displacements in global coordinates:

$$[S] = [k][\Delta] \tag{13.30}$$

To obtain the desired design forces, in terms of the member axes, we apply the transformation, Eq. (13.20), to the element forces in global coordinates for each element:

$$\{\bar{S}\} = [T]\{S\} = [T][k][\Delta] \tag{13.31}$$

This enables us to determine the member forces, element by element, in a form convenient for plotting shear, moment, and axial force diagrams, necessary for design of the members.

    With the basic steps of Sec. 3.4 and the transformation procedure of this section, we can now provide a general framework of operations for the displacement method:

    1. Identify elements and joints, and define a set of global coordinate axes; number all nodes (free nodes first).

2. For each element, define an appropriate set of local coordinates and calculate the element stiffness matrix in local coordinates.

3. For each element, calculate the transformation matrix from global to local coordinates.

4. By use of the transformation matrix obtained in step 3, transform the element stiffness matrix in local coordinates for each element computed in step 2 into an element stiffness matrix in global coordinates.

5. Superimpose the element stiffness matrices in global coordinates obtained in step 4 to assemble the structure stiffness matrix, making sure to adhere to consistent nodal numbering.

6. Following the operations of Sec. 13.4, insert the known boundary displacements and forces into Eq. (13.12), and solve for the unknown displacements and forces by Eqs. (13.13) and (13.14). These quantities will be in global coordinates.

7. For each element, compute the element design forces $\{\bar{S}\}$ in local coordinates by Eq. (13.31). This usually concludes the analysis process.

This outline will have made it fairly obvious that the entire procedure is intended for automatic computer routines. However, to gain the insight necessary for efficient programming of the operations, it is useful to attempt at least a simple problem in longhand. The following example will therefore trace through the operations in terms of a planar two-member structure.

Example 13.7. Analyze the planar structure shown by the matrix displacement method. Joint B is a pin.

Solution: We define a global coordinate system X_1, X_2 and number the nodes (free nodes first), as shown in part (a) of the figure. Next, we identify the elements, attach suitable local coordinates, and label the nodes. For element 1, the sequence of nodal numbering follows Fig. 13.2, and Eq. (13.4) gives the stiffness matrix for this prismatic plane beam element:

$$
[\bar{k}^1] = \begin{array}{c} \\ 6 \\ 7 \\ 8 \\ 1 \\ 2 \\ 3 \end{array}
\begin{array}{c}
\begin{array}{cccccc} 6 & 7 & 8 & 1 & 2 & 3 \end{array} \\
\left[\begin{array}{cccccc}
29{,}462 & & & & & \\
0 & 246 & & & \text{sym} & \\
0 & 20{,}833 & 2{,}357{,}020 & & & \\
-29{,}462 & 0 & 0 & 29{,}462 & & \\
0 & -246 & -20{,}833 & 0 & 246 & \\
0 & 20{,}833 & 1{,}178{,}510 & 0 & -20{,}833 & 2{,}358{,}020
\end{array} \right]
\end{array} \times 10^{-6} E
$$

The transformation matrix from local to global coordinates is obtained from

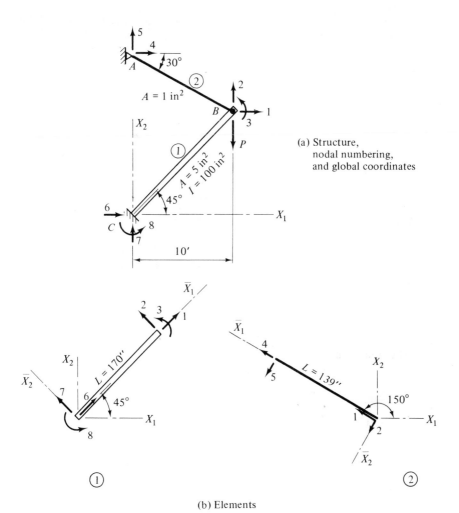

(a) Structure,
nodal numbering,
and global coordinates

(b) Elements

Fig. E13.7

Eq. (13.25) for a planar beam, with $\alpha = 45°$:

$$[T^1] = \begin{array}{c} \\ 6 \\ 7 \\ 8 \\ 1 \\ 2 \\ 3 \end{array} \begin{array}{cccccc} 6 & 7 & 8 & 1 & 2 & 3 \\ \left[\begin{array}{ccc|ccc} .707 & .707 & 0 & & & \\ -.707 & .707 & 0 & & 0 & \\ 0 & 0 & 1 & & & \\ \hline & & & .707 & .707 & 0 \\ & 0 & & -.707 & .707 & 0 \\ & & & 0 & 0 & 1 \end{array}\right] \end{array}$$

and transforming into the global X_1, X_2 coordinate system by use of Eq. (13.23), we obtain the stiffness matrix for element 1 in global coordinates:

$$[k^1] = \begin{array}{c} \\ 6 \\ 7 \\ 8 \\ 1 \\ 2 \\ 3 \end{array} \begin{array}{cccccc} 6 & 7 & 8 & 1 & 2 & 3 \\ \left[\begin{array}{cccccc} 14{,}854 & & & & & \\ 14{,}609 & 14{,}854 & & & \text{sym} & \\ -14{,}731 & 14{,}731 & 2{,}357{,}000 & & & \\ -14{,}854 & -14{,}609 & 14{,}731 & 14{,}854 & & \\ -14{,}609 & -14{,}854 & -14{,}731 & 14{,}609 & 14{,}854 & \\ -14{,}731 & 14{,}731 & 1{,}178{,}510 & 14{,}731 & -14{,}731 & 2{,}357{,}000 \end{array}\right] \end{array} \times 10^{-6}E$$

Proceeding similarly with element 2, we write the stiffness matrix in the local coordinate system of part (b) and, following Ex. 13.6, write

$$[\bar{k}^2] = \begin{array}{c} \\ 1 \\ 2 \\ 4 \\ 5 \end{array} \begin{array}{cccc} 1 & 2 & 4 & 5 \\ \left[\begin{array}{cccc} 7{,}216 & 0 & -7{,}216 & 0 \\ 0 & 0 & 0 & 0 \\ -7{,}216 & 0 & 7{,}216 & 0 \\ 0 & 0 & 0 & 0 \end{array}\right] \end{array} \times 10^{-6}E$$

The transformation matrix is the same used in Ex. 13.6, with $\alpha = 150°$:

$$[T^2] = \begin{array}{c} \\ 1 \\ 2 \\ 4 \\ 5 \end{array} \begin{array}{cccc} 1 & 2 & 4 & 5 \\ \left[\begin{array}{cc|cc} -.866 & .500 & & \\ -.600 & -.866 & & 0 \\ \hline & & -.866 & .500 \\ & 0 & -.500 & -.866 \end{array}\right] \end{array}$$

Using Eq. (13.23), we obtain the stiffness matrix for element 2 in global coordinates:

$$[k^2] = \begin{array}{c} \\ 1 \\ 2 \\ 4 \\ 5 \end{array} \begin{array}{cccc} 1 & 2 & 4 & 5 \\ \left[\begin{array}{cccc} 5{,}412 & & & \text{sym} \\ -3{,}125 & 1{,}805 & & \\ -5{,}412 & 3{,}125 & 5{,}412 & \\ 3{,}125 & -1{,}805 & -3{,}125 & 1{,}805 \end{array}\right] \end{array} \times 10^{-6}E$$

Now that we have all element stiffnesses referred to a common axis system, we can add them to obtain the structure stiffness matrix on page 349.
We now proceed with the displacement analysis by relating forces and displacements according to Eq. (13.12), in which the specified loads and specified

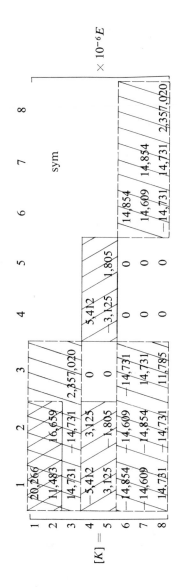

$$[K] = \begin{bmatrix}
20{,}266 & 11{,}483 & -14{,}731 & -5{,}412 & 3{,}125 & -14{,}854 & -14{,}609 & 14{,}731 \\
 & 16{,}659 & -14{,}731 & 3{,}125 & -1{,}805 & -14{,}609 & -14{,}854 & -14{,}731 \\
 & & 2{,}357{,}020 & 0 & 0 & -14{,}731 & 14{,}731 & 11{,}785 \\
 & & & 5{,}412 & -3{,}125 & 0 & 0 & 0 \\
 & & & & 1{,}805 & 0 & 0 & 0 \\
 & & & & & 14{,}854 & 14{,}609 & -14{,}731 \\
 & \text{sym} & & & & & 14{,}854 & 14{,}731 \\
 & & & & & & & 2{,}357{,}020
\end{bmatrix} \times 10^{-6} E$$

support displacements are

$$\{X_\alpha\} = \begin{Bmatrix} X_1 \\ X_2 \\ X_3 \end{Bmatrix} = \begin{Bmatrix} 0 \\ -P \\ 0 \end{Bmatrix}; \qquad \{\Delta_\beta\} = \begin{Bmatrix} \Delta_4 \\ \Delta_5 \\ \Delta_6 \\ \Delta_7 \\ \Delta_8 \end{Bmatrix} = \begin{Bmatrix} 0 \\ 0 \\ 0 \\ 0 \\ 0 \end{Bmatrix}$$

Solving according to Eqs. (13.13) and (13.14), we obtain the nodal displacements and support reactions

$$\{\Delta_\alpha\} = \begin{Bmatrix} \Delta_1 \\ \Delta_2 \\ \Delta_3 \end{Bmatrix} = \begin{Bmatrix} 57.813 \\ -100.756 \\ -.991 \end{Bmatrix}\frac{P}{E}; \qquad \{X_\beta\} = \begin{Bmatrix} X_4 \\ X_5 \\ X_6 \\ X_7 \\ X_8 \end{Bmatrix} = \begin{Bmatrix} -.6277 \\ .3625 \\ .6277 \\ .6375 \\ 1.1680 \end{Bmatrix}P$$

The member end forces can now be obtained by Eq. (13.31) from the displacements $\{\Delta\}$ to conclude the problem.

13.6. Matrix Displacement Method for Distributed Loads

The formulation of the matrix displacement method so far presented has been restricted to the case of specified loads on, and specified deformations of, the joints. It is now desirable to extend the method to allow analysis due to loads applied between joints.

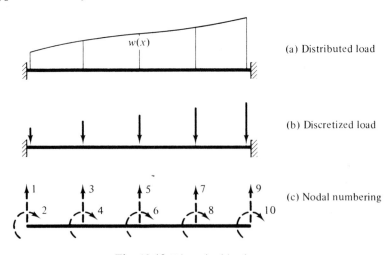

(a) Distributed load

(b) Discretized load

(c) Nodal numbering

Fig. 13.13 Discretized load system.

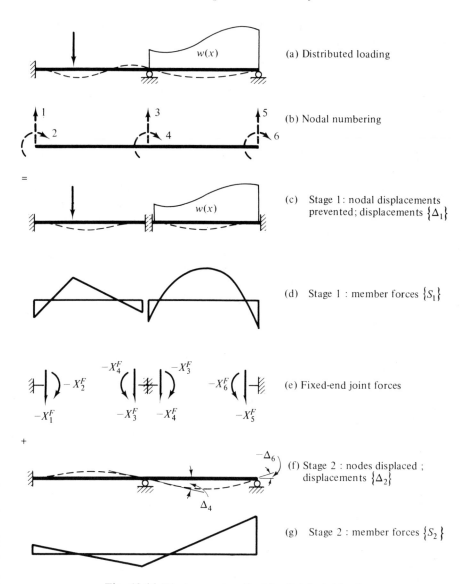

(a) Distributed loading

(b) Nodal numbering

(c) Stage 1: nodal displacements prevented; displacements $\{\Delta_1\}$

(d) Stage 1 : member forces $\{S_1\}$

(e) Fixed-end joint forces

(f) Stage 2 : nodes displaced ; displacements $\{\Delta_2\}$

(g) Stage 2 : member forces $\{S_2\}$

Fig. 13.14 Displacement method for distributed loads.

We understand that any load system could be approximated by an appropriate set of nodal concentrated loads, as discussed in Secs. 3.6 and 7.5. But this process requires the introduction of additional nodes for each of these discretized loads, thus increasing the degree of kinematic indeterminacy considerable. If, for instance, we were to simulate the distributed load on the plane beam of Fig. 13.13(a) by the concentrated loads of Fig. 13.13(b), we would have to provide the nodes shown in Fig. 13.13(c); it is apparent that in addition to the error involved in the discretization, computations of unduly large proportions would result.

We shall now discuss an alternative procedure in terms of the beam of Fig. 13.14(a) under arbitrary loading. We shall introduce nodes only at the member ends, as shown in Fig. 13.14(b), and consider the response in two parts:

1. The beam under the effect of the applied loads when all nodal displacements are prevented, as shown in Fig. 13.14(c), where all joints are locked against rotation and translation. The resulting forces acting on the member ends in this condition are identified as the fixed-end nodal forces $\{S^F\}$, whose calculation was already discussed in Sec. 11.1. The equal and opposite forces *on the joints* will be $\{X^F\} = -\{S^F\}$. The internal member forces in the loaded members, with joints constrained against displacements [shown by the fixed-end moment diagram of Fig. 13.14(d)] will be designated by the symbol $\{S_1\}$; the corresponding member deformations are $\{\Delta_1\}$.

2. We now allow the joints to displace into their equilibrium positions, as indicated by the rotations $\{\Delta\}$ of the joints of Fig. 13.14(f). The nodal member forces due to all displacements $\{\Delta\}$ are, by the definition of the structure stiffness matrix $[K]$, given by $\{S_2\} = [K]\{\Delta\}$; the equal and opposite joint forces are then $\{X_2\} = -[K]\{\Delta\}$. The internal member forces due to these joint displacements will be called $\{S_2\}$, and are symbolized by the moment diagram, Fig. 13.14(g). The displacements at intermediate points of the members due to the joint displacements, shown in Fig. 13.14(f), will be called $\{\Delta_2\}$.

The sum of all joint forces due to stages 1 and 2 and the actual external joint forces $\{X\}$ applied to the joints, which may be specified loads or reactions, must be in equilibrium:

$$\{X^F\} - [K]\{\Delta\} + \{X\} = 0$$

or

$$\{X\} + \{X^F\} = [K]\{\Delta\} \tag{13.32}$$

Note that this equation is similar to Eq. (13.11), with the addition of the fixed-end *joint* forces $\{X^F\}$. The signs of the forces follow the sense of the joint nodes.

The further solution is similar to the procedure of Sec. 13.4; we list the α nodes corresponding to the unknown joint displacements first, followed by

β nodes corresponding to the unknown forces, and write

$$\begin{bmatrix} X_\alpha \\ \hline X_\beta \end{bmatrix} + \begin{bmatrix} X_\alpha^F \\ \hline X_\beta^F \end{bmatrix} = \begin{bmatrix} K_{\alpha\alpha} & \vdots & K_{\alpha\beta} \\ \hline K_{\beta\alpha} & \vdots & K_{\beta\beta} \end{bmatrix} \begin{bmatrix} \Delta_\alpha \\ \hline \Delta_\beta \end{bmatrix} \tag{13.33}$$

from which we solve first for the unknown displacements:

$$\{\Delta_\alpha\} = [K_{\alpha\alpha}]^{-1}[[X_\alpha] + [X_\alpha^F] - [K_{\alpha\beta}][\Delta_\beta]] \tag{13.34}$$

followed by the solution for the unknown reactions:

$$\{X_\beta\} = [K_{\beta\alpha}][\Delta_\alpha] + [K_{\beta\beta}][\Delta_\beta] - [X_\beta^F] \tag{13.35}$$

The member forces are determined by superposition of the fixed-end member forces and those due to the joint displacements:

$$[S] = [S_1] + [S_2] \tag{13.36}$$

Since the nodal displacements of stage 1 are zero, the nodal displacements from Eq. (13.34) furnish the correct values; but deformations at intermediate points of the members must again be calculated by superposition of parts 1 and 2:

$$[\Delta] = [\Delta_1] + [\Delta_2] \tag{13.37}$$

The effects of self-strains due to temperature changes, prestressing, or similar causes can be handled in a similar manner. It is only necessary to compute the fixed-end forces on the joints, $[X^F]$, and to proceed in the outlined manner.

Example 13.8 The structure of part (a) of the figure (which was already considered in Ex. 13.4) is subjected to two different loading conditions:
a. A uniform load of $w = 1$ kip/ft on the horizontal beam.
b. An axial shortening of 0.10 in. of the inclined bracing member 3.
For each case, compute nodal forces and deformations and draw the moment diagram for the beam.

Solution: We compute the fixed-end forces for both members due to both loading conditions by the methods presented in Chapter 11. In loading condition 1, only member 2 is subjected to load; for this beam, the procedure of Sec. 11.1 leads to the fixed-end moments and shears shown in part (b). In loading condition 2, the shortening of member 3 leads to axial fixed-end forces, which are proportional to the member stiffnesses: $S^F = \Delta L \cdot (AE/L)$ $= 10.8$ kips tension; this fixed-end force has to be expressed in terms of its components parallel to the nodal axes, as shown in part (c).
The fixed-end forces acting on the joints for both loading conditions are

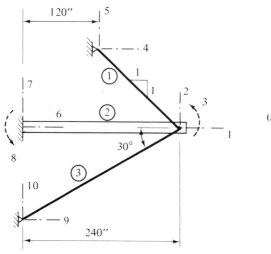

(a) Structure

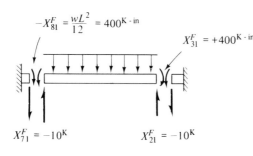

$$-X_{81}^F = \frac{wL^2}{12} = 400^{\text{K} \cdot \text{in}}$$

$$X_{31}^F = +400^{\text{K} \cdot \text{in}}$$

(b) Loading condition 1

Fixed-end forces

$$X_{71}^F = -10^K \qquad X_{21}^F = -10^K$$

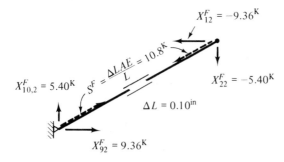

$$X_{12}^F = -9.36^K$$

$$X_{22}^F = -5.40^K$$

$$X_{10,2}^F = 5.40^K$$

$$S^F = \frac{\Delta L A E}{L} = 10.8^K$$

$$\Delta L = 0.10^{\text{in}}$$

(c) Loading condition 2

$$X_{92}^F = 9.36^K$$

Fig. E13.8

now assembled in the fixed-end force matrix $[X^F]$, adhering to the nodal numbering system of part (a):

$$
[X^F] =
\begin{array}{c}
 \\
1 \\
2 \\
3 \\
4 \\
5 \\
6 \\
7 \\
8 \\
9 \\
10
\end{array}
\begin{array}{c}
1 \qquad\qquad 2 \\
\begin{bmatrix}
0 & -9.36 \\
-10 & -5.40 \\
400 & 0 \\
0 & 0 \\
0 & 0 \\
0 & 0 \\
-10 & 0 \\
-400 & 0 \\
0 & 9.36 \\
0 & 5.40
\end{bmatrix}
\end{array}
$$

The other matrices in Eq. (13.33) for this problem were already calculated and presented in Ex. 13.4. Inserting all matrices into Eq. (13.33), and proceeding through the solution steps (13.34) and (13.35), we obtain the desired nodal displacements and reactive nodal forces:

$$
\begin{bmatrix}
\Delta_{11} & \Delta_{12} \\
\Delta_{21} & \Delta_{22} \\
\Delta_{31} & \Delta_{23}
\end{bmatrix}
=
\begin{bmatrix}
-.0064 & -.0268 \\
-.0672 & -.0564 \\
.0698 & -.0004
\end{bmatrix}
\begin{array}{l}
\text{in.} \\
\text{in.} \\
\text{Rad.}
\end{array}
$$

$$
\begin{bmatrix}
X_{41} & X_{42} \\
X_{51} & X_{52} \\
X_{61} & X_{62} \\
X_{71} & X_{72} \\
X_{81} & X_{82} \\
X_{91} & X_{92} \\
X_{10,1} & X_{10,2}
\end{bmatrix}
=
\begin{bmatrix}
-5.376 & -2.612 \\
5.376 & 2.612 \\
1.705 & 7.144 \\
12.505 & .004 \\
601.20 & 1.004 \\
3.671 & -4.533 \\
2.119 & -2.616
\end{bmatrix}
\text{All kips except } X_8 \text{ in kip-in.}
$$

Member forces can now be calculated by the use of Eq. (13.36) to conclude the problem.

13.7. Computer Program for the Analysis of Framed Structures by the Displacement Method

The procedures of this chapter could hardly be applied in a meaningful way without an appropriate computer program. For this reason, program FRAME for the analysis of arbitrary two- or three-dimensional framed structures, which is listed in Appendix A, is discussed and demonstrated in this section.

The computations performed by this program closely follow the steps outlined in Secs. 13.4 to 13.6. The program can accommodate the effects of joint loads, joint displacements, and fixed-end forces due to member loads. Two-force members with pin connections can also be handled by appropriate manipulation. In contrast to the rules governing nodal numbering, which have been described in Sec. 13.4, only numbering of the joints is required by the program, along with an indication for each joint node as to whether it is free or constrained; the program incorporates a routine that automatically orders the nodes in the appropriate sequence in accordance with this "tagging."

Figure 13.15 represents a simple flow chart that lists the sequence of operations contained in program FRAME, and keys it to the presentation of the theory of Secs. 13.4 to 13.6. The program listing in Appendix A contains further explanatory statements, as well as information for the required data input. This latter requires the following steps:

1. Preparation of structure: Introduce a global set of axes X_1, X_2, and X_3 of which the X_2 axis should be vertical. Number each joint and each member.

2. General structure information: One card, indicating number of elements, joints, and loading conditions.

3. Joint information: For each joint, one card, specifying number of joint, three joint coordinates, six restraint numbers indicating whether the joint nodes are free or constrained [the sequence of nodes here follows that of the global coordinate system, with the three translational degrees of freedom first, followed by the three rotational degrees of freedom, as shown in Fig. 13.16(a)], and two numbers indicating whether or not the joint is subject to applied loads, or specified nonzero displacements. If either applied joint load or displacement is indicated, one additional card is required for each loading condition, specifying the values of the load or displacement components in the same nodal order outlined above.

4. Member information: For each member, one card specifying member number, member incidences, prismatic member properties, and one number indicating whether or not the member is loaded. If an applied member load is indicated, two cards are required for each loading condition, specifying the fixed-end member forces in the sequence of member end node numbering shown in Fig. 13.16(b).

This concludes the data input. The output consists of all joint displacements and joint forces in the global coordinate system, as well as the member end forces in the local coordinate system shown in Fig. 13.16; thus, these forces may be identified as axial and shear forces, and torsional and bending moments, respectively, and used for design purposes.

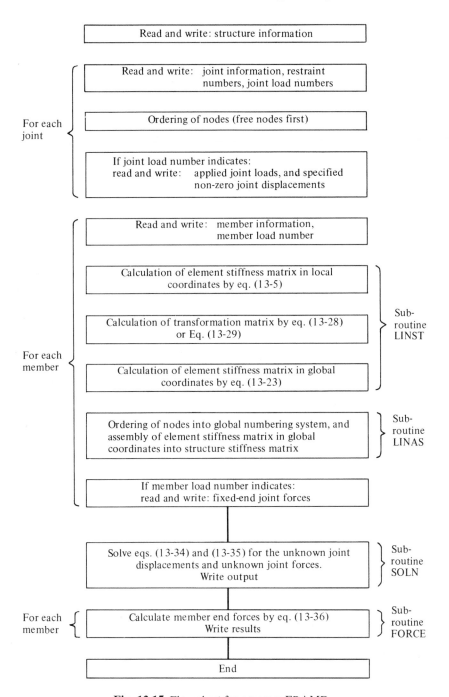

Fig. 13.15 Flow chart for program FRAME.

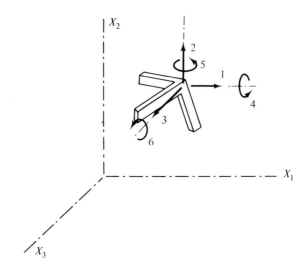

(a) Numbering of joint nodes

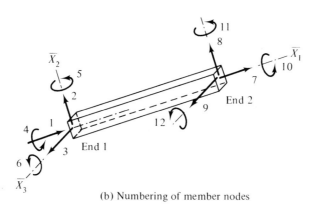

(b) Numbering of member nodes

Fig. 13.16 Nodal numbering for program FRAME.

To determine the design forces along the individual members, the statically determinate member forces must be superimposed upon those due to the member end forces to obtain the final design forces in the member. This last operation must be done longhand by the analyst.

Example 13.9 The fixed-ended three-dimensional rigid frame shown in part (a) of the figure is to be analyzed by use of program FRAME for the following three loading conditions:

a. A uniform vertical load, of intensity *w*, on member 1.

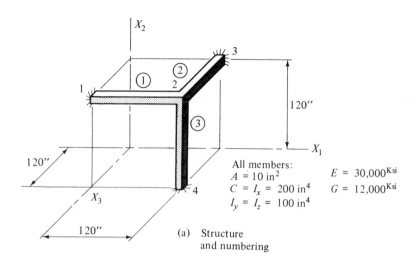

All members:
$A = 10$ in^2 $E = 30,000^{Ksi}$
$C = I_x = 200$ in^4 $G = 12,000^{Ksi}$
$I_y = I_z = 100$ in^4

(a) Structure and numbering

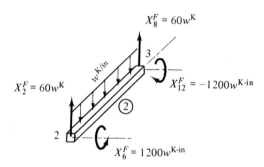

$X_8^F = 60w^K$

$X_2^F = 60w^K$ $w^{K/in}$ $X_{12}^F = -1200w^{K \cdot in}$

$X_6^F = 1200w^{K \cdot in}$

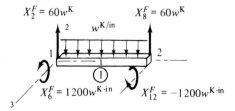

$X_2^F = 60w^K$ $X_8^F = 60w^K$

$w^{K/in}$

$X_6^F = 1200w^{K \cdot in}$ $X_{12}^F = -1200w^{K \cdot in}$

(b) Fixed-End Forces on Loaded Members

Fig. E13.9(a) and (b)

359

NUMBER OF ELEMENTS, JOINTS, LOADING CONDITIONS

3   4   3

X,Y,AND Z COORDINATES AND RESTRAINT NUMBERS FOR JOINT   1

0.0000   120.0000   120.0000   1   1   1   1

X,Y,AND Z COORDINATES AND RESTRAINT NUMBERS FOR JOINT   2

120.0000   120.0000   120.0000   0   0   0   0

X,Y,AND Z COORDINATES AND RESTRAINT NUMBERS FOR JOINT   3

120.0000   120.0000   0.0000   1   1   1   1

X,Y,AND Z COORDINATES AND RESTRAINT NUMBERS FOR JOINT   4

120.0000   0.0000   120.0000   1   1   1   1

SPECIFIED DISPLACEMENTS IN STRUCTURE COORDINATES OF JOINT   4

ROW/COLUMN	1	2	3
1	-0.	-0.	-0.
2	-0.	-0.	-1.00000E+00
3	-0.	-0.	-0.
4	-0.	-0.	-0.
5	-0.	-0.	-0.
6	-0.	-0.	-0.

$\times \Delta$, in.

360

MEMBER ENDS,E,G,AREA,IX,IY,IZ,ALPHA FOR ELEMENT 2

2	3	30000	12000	10.00	200.00	100.00	100.00	0.00000

FIXED SUPPORT FORCES DUE TO LOADS ON MEMBER 2(IN MEMBER COORDINATES)

ROW/COLUMN	1	2	3
1	-0.	-0.	-0.
2	-0.	6.00000E+01	-0.
3	-0.	-0.	-0.
4	-0.	-0.	-0.
5	-0.	-0.	-0.
6	-0.	1.20000E+03	-0.
7	-0.	-0.	-0.
8	-0.	6.00000E+01	-0.
9	-0.	-0.	-0.
10	-0.	-0.	-0.
11	-0.	-0.	-0.
12	-0.	-1.20000E+03	-0.

x W, Kips or Kip-in.

MEMBER ENDS,E,G,AREA,IX,IY,IZ,ALPHA FOR ELEMENT 3

4	2	30000	12000	10.00	200.00	100.00	100.00	0.00000

MEMBER ENDS,E,G,AREA,IX,IY,IZ,ALPHA FOR ELEMENT 1

1	2	30000	12000	10.00	200.00	100.00	100.00	0.00000

FIXED SUPPORT FORCES DUE TO LOADS ON MEMBER 1(IN MEMBER COORDINATES)

ROW/COLUMN	1	2	3
1	-0.	-0.	-0.
2	6.00000E+01	6.00000E+01	-0.
3	-0.	-0.	-0.
4	-0.	-0.	-0.
5	-0.	-0.	-0.
6	1.20000E+03	1.20000E+03	-0.
7	-0.	-0.	-0.
8	6.00000E+01	6.00000E+01	-0.
9	-0.	-0.	-0.
10	-0.	-0.	-0.
11	-0.	-0.	-0.
12	-1.20000E+03	-1.20000E+03	-0.

x W, Kips or Kip-in.

Fig. E13.9(c) Computer input.

b. A uniform vertical load, of intensity w, on both members 1 and 2.

c. A vertical support settlement of amount Δ at point 4.

Solution: A global set of axes is introduced as in part (a), with axis X_2 vertical.

FINAL JOINT DISPLACEMENTS(IN STRUCTURE COORDINATES)

0.0000		0.0000		0.0000	
0.0000		0.0000		0.0000	
0.0000		0.0000		0.0000	
0.0000		0.0000		0.0000	
0.0000		0.0000		0.0000	
0.0000		0.0000		0.0000	
-.0026		-.0026		.0028	
-.0210		-.0421		-.9892	
.0001		-.0026		.0028	
.0001		-.0052		.0056	
-.0000		0.0000		.0000	
.0054	× W	.0052	× W	-.0056	× Δ, in. or rad.
0.0000		0.0000		0.0000	
0.0000		0.0000		0.0000	
0.0000		0.0000		0.0000	
0.0000		0.0000		0.0000	
0.0000		0.0000		0.0000	
0.0000		0.0000		0.0000	
-0.0000		-0.0000		-0.0000	
-0.0000		-0.0000		-1.0000	
-0.0000		-0.0000		-0.0000	
-0.0000		-0.0000		-0.0000	
-0.0000		-0.0000		-0.0000	
-0.0000		-0.0000		-0.0000	

FINAL JOINT FORCES(IN STRUCTURE COORDINATES)

6.5968		6.4305		-6.9294	
67.1257		67.4141		13.5624	
.0178		.0536		-.0577	
-2.3978		104.6027		-112.7184	
-.6854		-3.2152		3.4647	
1493.7942		1514.0926		954.6416	
0.0000		0.0000		0.0000	
0.0000		0.0000		0.0000	
0.0000		0.0000		0.0000	
0.0000	× W	0.0000	× W	0.0000	× Δ, Kips or
0.0000		0.0000		0.0000	Kip-in.
0.0000		0.0000		0.0000	
.0358		.0536		-.0577	
.2884		67.4141		13.5624	
-.1663		6.4305		-6.9294	
-20.2984		-1514.0926		-954.6416	
2.5299		3.2152		-3.4647	
-107.0005		-104.6027		112.7184	
-6.6326		-6.4841		6.9872	
52.5859		105.1718		-27.1248	
.1485		-6.4841		6.9872	
5.9114		-258.2914		278.3313	
.3074		0.0000		-.0000	
264.2028		258.2914		-278.3313	

Fig. E13.9(d) Computer output.

FINAL MEMBER END FORCES FOR ELEMENT   1 (MEMBER COORDINATES)

ROW/COLUMN	1	2	3
1	6.59680E+00	6.43049E+00	-6.92941E+00
2	6.71257E+01	6.74141E+01	1.35624E+01
3	1.78275E-02	5.35874E-02	-5.77451E-02
4	-2.39783E+00	1.04603E+02	-1.12718E+02
5	-6.85382E-01	-3.21525E+00	3.46470E+00
6	1.49379E+03	1.51409E+03	9.54642E+02
7	-6.59680E+00	-6.43049E+00	6.92941E+00
8	5.28743E+01	5.25859E+01	-1.35624E+01
9	-1.78275E-02	-5.35874E-02	5.77451E-02
10	2.39783E+00	-1.04603E+02	1.12718E+02
11	-1.45392E+00	-3.21525E+00	3.46470E+00
12	-6.38705E+02	-6.24401E+02	6.72846E+02

(columns: 1 × W, 2 × W, 3 × Δ)

FINAL MEMBER END FORCES FOR ELEMENT   2 (MEMBER COORDINATES)

ROW/COLUMN	1	2	3
1	-1.66306E-01	6.43049E+00	-6.92941E+00
2	-2.88352E-01	5.25859E+01	-1.35624E+01
3	-3.57600E-02	-5.35874E-02	5.77451E-02
4	-1.07000E+02	-1.04603E+02	1.12718E+02
5	1.76133E+00	3.21525E+00	-3.46470E+00
6	-1.43038E+01	6.24401E+02	-6.72846E+02
7	1.66306E-01	-6.43049E+00	6.92941E+00
8	2.88352E-01	6.74141E+01	1.35624E+01
9	3.57600E-02	5.35874E-02	-5.77451E-02
10	1.07000E+02	1.04603E+02	-1.12718E+02
11	2.52986E+00	3.21525E+00	-3.46470E+00
12	-2.02984E+01	-1.51409E+03	-9.54642E+02

(columns: 1 × W, 2 × W, 3 × Δ)

FINAL MEMBER END FORCES FOR ELEMENT   3 (MEMBER COORDINATES)

ROW/COLUMN	1	2	3
1	5.25859E+01	1.05172E+02	-2.71248E+01
2	6.63256E+00	6.48408E+00	-6.98715E+00
3	1.48478E-01	-6.48408E+00	6.98715E+00
4	3.07414E-01	0.	-1.32349E-15
5	-5.91142E+00	2.58291E+02	-2.78331E+02
6	2.64203E+02	2.58291E+02	-2.78331E+02
7	-5.25859E+01	-1.05172E+02	2.71248E+01
8	-6.63256E+00	-6.48408E+00	6.98715E+00
9	-1.48478E-01	6.48408E+00	-6.98715E+00
10	-3.07414E-01	0.	1.32349E-15
11	-1.19060E+01	5.19798E+02	-5.60127E+02
12	5.31704E+02	5.19798E+02	-5.60127E+02

(columns: 1 × W, 2 × W, 3 × Δ)

Fig. E13.9(d) Computer output. (*cont.*)

The joints and members of the frame are numbered as shown, and the joint coordinates are computed.

The structure information is input, followed by the joint information; note that all nodes at joints 1, 3, and 4 are constrained, while all nodes at joint 2 are free to displace. This information is listed in the input data of part (c).

All joint loads are zero, but loading condition 3 involves a specified nonzero displacement of joint 4; therefore, the joint 4 card is followed by three cards, one for each loading condition, of which the first two are blank and the third lists the appropriate specified nodal displacement in accordance with the nodal sequence of Fig. 13.16(a).

The member information is furnished next; the horizontal members 1 and 2 are subject to an applied load in conditions 1 and 2, so each member card is followed by six load cards listing the fixed-end forces for each of the three loading conditions. These fixed-end forces are shown in part (b) and are entered on the load cards in accordance with the nodal sequence of Fig. 13.16(b). Since loading condition 3 does not involve any member forces, the last two cards are blank in each case.

Member 3 is vertical; thus the angle is defined with respect to the X_3, rather than the vertical, axis, as per Eq. (13.29). No load cards are required for this unloaded member. This concludes the data input.

Part (d) shows the listing of the computer output, consisting of joint forces and displacements in the global system and member end forces in the

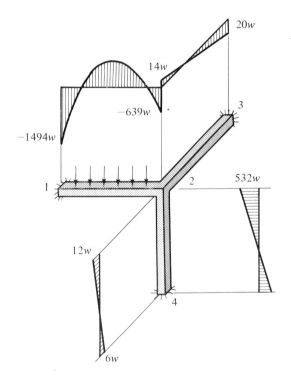

(e) Partial Moment Diagram, Loading Condition 1

Fig. E13.9(e)

local system. From the latter, the total member design forces are obtained by superposition of the statically determinate forces, as shown by the partial moment diagrams of part (e).

This concludes the example.

PROBLEMS

Note: Problems 1 to 18 are suitable for hand computation. Problems 21 to 33 are intended for computer analysis.

1. The systems shown consist of linear elastic springs and perfectly rigid bars. For each one, specify the number of degrees of static indeterminancy, r, and the degree of kinematic indeterminacy, k. Take advantage of symmetry where appropriate.

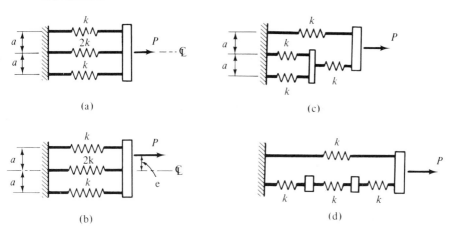

Prob. 1

2. To each of the beams shown, attach a nodal numbering system capable of representing joint actions and displacements. Then specify for each the degrees of static and of kinematic indeterminacy.

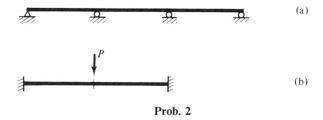

Prob. 2

3. The system shown consists of elastic springs of specified moduli and a perfectly rigid bar AB, hinged at point A. Analyze the system by the displacement

method. Find all spring forces. (*Hint:* All deformations can be expressed in terms of one displacement, for instance, the rotation of the bar *AB*.)

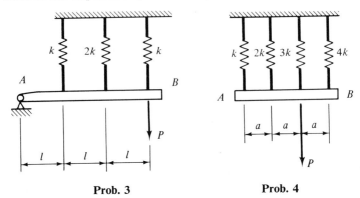

Prob. 3   **Prob. 4**

4. Analyze the system shown, consisting of elastic springs of given stiffnesses and a perfectly rigid bar *AB* by the displacement method. Find forces in all springs. (*Hint:* All displacements can be expressed in terms of two degrees of freedom.)

5. Analyze the two-span prismatic beam under applied moment by the displacement method. Draw shear and moment diagrams.

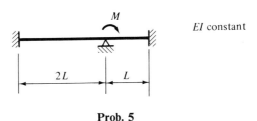

Prob. 5

6. The two beams are each fixed at one end and connected by a common pin joint at point *B*; this joint rests on an elastic spring of given modulus. Analyze the structure by the displacement method due to an applied load *P*. Draw shear and moment diagrams.

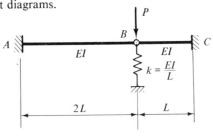

Prob. 6

7. Analyze the rigid frame shown by the displacement method for the effects of the load *P* applied to the overhang. Neglect axial distortions. Draw the deflected

shape, and shear and moment diagrams. (*Hint:* The load can be represented as a cranked-in moment on joint *B*.)

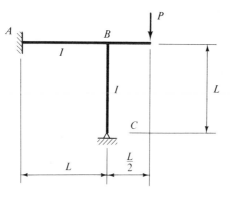

Prob. 7

8. Analyze the plane three-member truss shown due to the applied load by use of the displacement method. All members are of equal length and cross section.

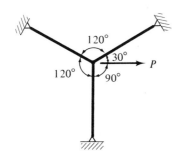

Prob. 8

9. Derive the element stiffness matrix, Eq. (13.6), by
   a. Geometric methods.
   b. By the method of virtual displacements.
   (*Hint:* Consider a real force system resulting from a real unit displacement along node *j*; then consider a virtual displacement system consisting of a unit displacement along node *i* and the resulting internal strains, and satisfy the equation of virtual work.)

10. Derive the elements k_{i2} of the second column of the stiffness matrix for a planar beam, Eq. (13.4), by solution of the differential equation for an unloaded prismatic beam segment:

$$\frac{d^4v}{dx^4} = 0$$

The solution of this ordinary, homogeneous differential equation is

$$v = a + bx + cx^2 + dx^3$$

where the constants can be determined by appropriate boundary conditions, and the end moments, k_{32} and k_{62}, and the end shears, k_{22} and k_{52}, can be calculated by appropriate differentiation of the displacements:

$$M = EI\frac{d^2v}{dx^2}; \qquad V = EI\frac{d^3v}{dx^3}$$

11. Repeat the analysis of Ex. 13.2, using a formal matrix approach as in Sec. 13.4; attach suitable nodes to all truss joints (numbering the free nodes first), and use Eqs. (13.13) and (13.14) to determine displacements and reactive components.

12. The fixed-ended, nonuniform beam is to be analyzed by the displacement method, following the steps outlined in Sec. 13.4. Remember to number the two unknown nodes, the translation and rotation at point B, first. Neglect axial deformations, so that no displacements parallel to the X axis need be considered. Draw the deformed shape of the beam (indicating nodal displacements), as well as shear and moment diagrams.

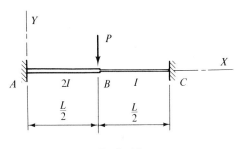

Prob. 12

13. The propped cantilever beam with flexible connection is to be analyzed by the displacement method. Identify the two elastic elements, attach suitable nodes, and go through a formal matrix solution for unknown displacements and actions. [*Hint:* A total of either three or five nodes could be considered here, depending on whether the v components of displacement (which are zero) are to be deleted or included; in the former case, appropriate rows and columns should be deleted from the beam stiffness matrix, Eq. (13.4). In either case, a rotational node should be placed on either side of the coil spring, of which the right is one of the two unknown nodes, the left is zero. An appropriate (2×2) stiffness matrix must be written for the coil spring.]

Compare the method and results with those of Prob. 3 of Chapter 10, which was done by the force method.

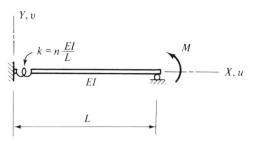

Prob. 13

14. Analyze the structure shown by the matrix displacement method, following the outline of Sec. 13.4. The support at point C allows translation u parallel to the X axis, but no rotation. Neglect axial deformations in the members, so that the v displacement of joint B can be neglected, and the u displacements of points B and C are equal. There are thus two degrees of freedom, θ_B and u_B, which should be numbered first, followed by the remaining four nodes. Draw the deformed shape, indicating the nodal displacements, as well as the shear and moment diagrams.

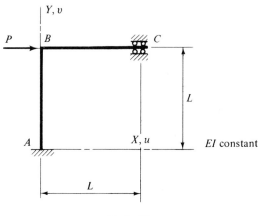

Prob. 14

15. Analyze the structure shown by the displacement method. Identify the two free nodes and number them first, followed by five constrained nodes.

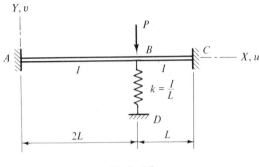

Prob. 15

16. Analyze the rigid frame by the displacement method. Identify and number the three free nodes first (axial elongations can be neglected, so that the u displacements at joints B and C are identical, and their v displacements are zero). Draw the deflected shape, and shear and moment diagrams.

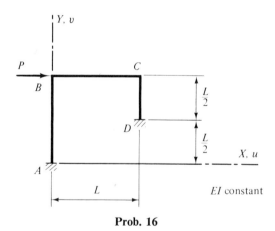

Prob. 16

17. Analyze the two-element fixed arch by the matrix displacement method,
a. Neglecting axial distortions.
b. Including axial distortions; axial stiffness $AE = n(EI/L^2)$.
Provide a solution for the moment at point B as a function of the stiffness ratio n, and discuss the effect of axial shortening (or "rib shortening") on stresses in arches.

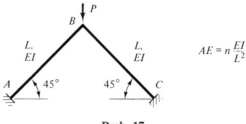

$$AE = n\frac{EI}{L^2}$$

Prob. 17

18. Analyze the bent pipe structure by the displacement method. Axial deformations may be neglected; this reduces the required nodes to a total of nine, of which three are unknown. Write the stiffness matrices in accordance with the adopted nodal numbering, and solve the problem by the method of Sec. 13.4. Draw plots of bending and torsional moments.

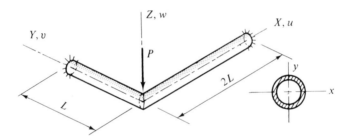

Typical section
$$EI_x = EI_y = GC$$

Prob. 18

19. Write a complete set of instructions for program FRAME in Appendix A. Describe, in concise step-by-step fashion, the indication of axes and numbering of the structure to be analyzed, and the required sequence, information, and format of the data cards. Also describe the expected output information from the program.

20. Discuss the use of program FRAME for the following special cases:
   a. Planar structures.
   b. Trusses and articulated structures.
   c. Neglect of axial distortions.

21. Repeat Prob. 16 by use of program FRAME for the following cases:
    a. Neglecting axial distortions.
    b. $AE = 1,000(EI/L^2)$.

22. Repeat Prob. 17 by use of program FRAME, for the case
    a. $n = 400$.
    b. $n = 4,000$.
    c. Comment on the effect of axial distortions on the internal forces in the structure.

23. Repeat Prob. 18 by use of program FRAME.

24. The (6×6) stiffness matrix $[k]$ of the tapering plane beam of linearly varying moment of inertia is to be determined approximately by considering it as an assembly of four prismatic subelements, each of constant I equal to that of the actual beam at the segment midpoint. Adopt a suitable nodal system, write and assemble the element stiffness matrices, and set up and solve the force–deformation equations (13.12); note that you will have to consider six loading cases, each consisting of a specified unit end displacement, with all others equal to zero. The corresponding end forces are the desired stiffnesses. Compare the results with the exact ones of Ex. 12.6.

25. Repeat Prob. 24, this time taking six subelements; compare the results of Probs. 24 and 25 and draw conclusions regarding the convergence of the results to the exact solution with an increasing number of elements.

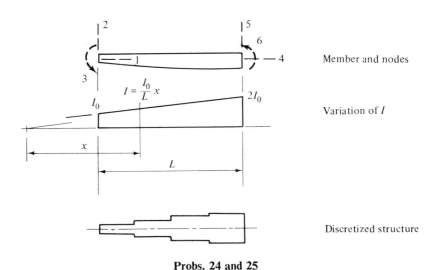

Probs. 24 and 25

26. The rigid-jointed space frame is to be analyzed due to
    a. A load P parallel to the X axis, applied to joint E.

b. A support settlement $w = w_0$ of joint A.

Axial deformations may be neglected. Compare forces and deformations in the different portions of the structure, and discuss the feasibility of analyzing the structure as a series of individual plane frames.

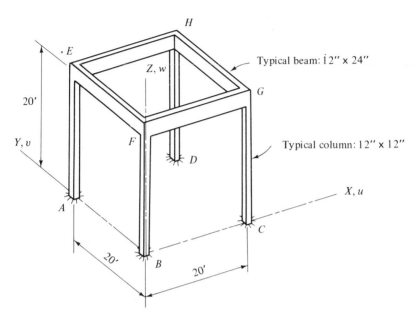

Typical beam: 12" x 24"

Typical column: 12" x 12"

Prob. 26

27. Now assume that there is a floor slab at the upper level *EFGH* of the frame of Prob. 26 which is so rigid in its plane that the horizontal displacements *u* and *v* of joints *E*, *F*, *G*, and *H* are constrained to deform together. Consider the effects of these constraints, introduce an appropriate nodal system, and analyze by the matrix displacement method for the same loads in Prob. 21. Compare results with those of Prob. 21, and discuss the effect of rigid floor diaphragms on the response of building frames.

28. Analyze the equal-legged tripod structure shown. All legs are of the circular thin-walled cross section shown, of material with $G = \frac{1}{2}E$. Joint A is rigidly welded, and joints B, C, and D are fixed supports. Loads are
   a. A load P, parallel to the X axis, applied to joint A.
   b. A settlement w_0 of support D, parallel to the Z axis.
   c. Now, assume all joints are pin connected, analyze the structure as an ideal truss, and compare results. Discuss appropriate simplifications for such structures, and discuss the effect of structural parameters on the results.

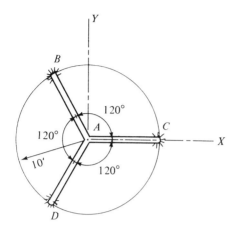

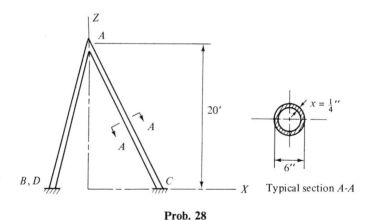

Typical section *A-A*

Prob. 28

29. Analyze the structure of Prob. 15 due to a uniform load of intensity *w* on beam *ABC* by means of the matrix displacement method.

30. Analyze the plane rigid-jointed frame shown due to
    a. A vertical, uniform load *w* kips/ft on member *AB*, as shown.
    b. A prestressing force of uniform magnitude *P* kips, with parabolically varying eccentricity, as shown.
    c. Calculate an optimum magnitude of the prestressing force *P* of of part b of the problem to counteract the moments due to the applied load *w* of part a.

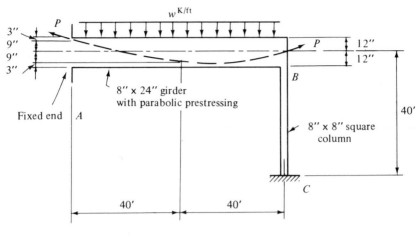

Prob. 30

31. Analyze the structure of Prob. 16 due to
    a. A uniform vertical load of intensity w_1 on member *BC*.
    b. A uniform horizontal load of intensity w_2 on member *AB*.
    c. A transverse temperature gradient ΔT in member *BC*. The coefficient of thermal expansion is α and the member depth is *d*.

32. Analyze the structure of Prob. 18 due to
    a. Its deadweight, of intensity *w* kips/ft.
    b. A uniform shrinkage of amount ϵ in./in.

33. Analyze the space frame of Prob. 26 due to:
    a. Its deadweight; members are of concrete weighing 150 lb/ft³.
    b. A prestressing force *P* applied to each of the four horizontal members, as shown.

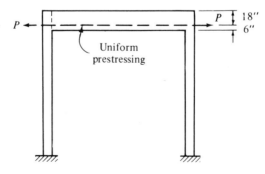

Prob. 33b

CHAPTER 14

Two Longhand Displacement Methods

Because the structural engineer should be capable of analyzing structures even without the help of a computer, we present in this chapter two displacement methods suitable for longhand use.

The *slope-deflection method*, covered in Sec. 14.1, was the first systematic displacement method in use; however, its vaster implications were not realized until much later. With the advent of matrix–computer stiffness methods, this technique is becoming obsolete, but its basic steps should still be known to the well-educated engineer.

The *moment-distribution method*, covered in Secs. 14.2. and 14.3, is an iterative technique, which can be carried to any desired degree of accuracy without the need to solve simultaneous equations, and for this reason will continue to be used for preliminary and exact longhand analysis. Because of the lack of abstraction underlying its thought process, it conveys a good physical feel for structural behavior and can therefore be of great help in improving the judgment of the designer.

14.1. Slope-Deflection Method

The general matrix displacement method outlined in Chapter 13 is suitable only for use with appropriate computing facilities. A simpler version of the displacement method, which is particularly appropriate for longhand calculation, will be outlined in this section. This method is called the slope-deflection method.

The slope-deflection method was the first application of the displacement method. In its classic formulation, as presented here, it is applicable to flexural planar structures with rigid joints in which axial and shear distortions are neglected. However, if the basic ideas of the displacement method are well understood, it could well be extended to include more complicated cases.

The slope-deflection method follows the typical steps of the displacement method outlined in Sec. 13.1:

1. Write the stiffness relations for each member; these relations are expressed in the slope-deflection equation, which relates member end moments to the applied loads and to the member end rotations and translations.

2. The geometrical relations are satisfied by writing the displacements common to several members in terms of an appropriate notation, and by inserting the given support conditions.

3. Equilibrium is satisfied by writing one equilibrium equation corresponding to each unknown displacement. By substitution of the slope-deflection equations, these equilibrium equations are expressed in terms of the unknown rotations and translations, and solved for these displacements.

4. The displacements found in step 3 are resubstituted into the slope-deflection equations to find the member end moments. All other desired forces may then be determined by statics.

The method of nodal numbering used in the matrix displacement method could be used here; however, we shall follow classical notation. The joints of the structure are designated by letters, and each member is designated by the double-letter subscript corresponding to the member ends. The member end moments, displacements, and rotations are also identified by the letters representing the member ends; thus, referring to the beam *AB* of Fig. 14.1,

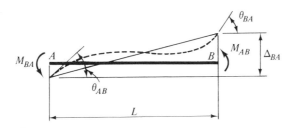

Fig. 14.1 Slope-deflection convention.

the member end moments at ends A and B, respectively, are designated M_{AB} and M_{BA}, the end rotations as θ_{AB} and θ_{BA}, and the relative member end translation as Δ_{BA}. Counterclockwise member end moments and end rotations, and relative member end translations associated with counterclockwise chord rotations, are considered positive.

The slope-deflection equations are force–displacement relations and can be written in terms of the stiffnesses. The relations between member end moments and end rotations, for instance, are given in matrix form as

$$\begin{Bmatrix} M_{AB} \\ M_{BA} \end{Bmatrix} = \begin{bmatrix} k_{AA} & k_{AB} \\ k_{BA} & k_{BB} \end{bmatrix} \begin{Bmatrix} \theta_{AB} \\ \theta_{BA} \end{Bmatrix} \tag{14.1}$$

where k_{AB}, for instance, denotes the moment at end A due to a unit rotation at end B, all other member end displacements zero.

The member end moments due to relative joint translation can be obtained by the superposition procedure of Fig. 14.2. Figure 14.2(a) shows beam AB subjected to a relative joint translation Δ_{BA}; this deformed shape can be constructed by superposition of the three shapes of Figs. 14.2(b) to (d), of which Fig. 14.2(b) shows a rigid-body rotation, which does not cause any moments, and Figs. 14.2(c) and (d) show a rotation of magnitude Δ_{BA}/L applied, respectively, at ends A and B, with the other end fixed. Thus, the end

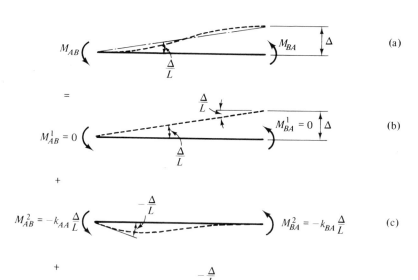

Fig. 14.2 Effect of joint translation.

moments can be written in terms of the rotational stiffnesses as shown. Adding the moments at each end due to the conditions of Figs. 14.2(b) to (d), we find the end moments due to the relative end translation Δ_{BA}:

$$M_{AB} = -(k_{AA} + k_{AB})\frac{\Delta_{BA}}{L}$$

$$M_{BA} = -(k_{BA} + k_{BB})\frac{\Delta_{BA}}{L}$$

(14.2)

Adding these moments to those of Eqs. (14.1) due to the end rotations, and to the fixed-end moments M_{AB}^F and M_{BA}^F due to loads on the beam, we obtain the total member end moments in terms of the end displacements as

$$\begin{Bmatrix} M_{AB} \\ M_{BA} \end{Bmatrix} = \begin{Bmatrix} M_{AB}^F \\ M_{BA}^F \end{Bmatrix} + \begin{bmatrix} k_{AA} & k_{AB} & -\frac{1}{L}(k_{AA} + k_{AB}) \\ k_{BA} & k_{BB} & -\frac{1}{L}(k_{AB} + k_{BB}) \end{bmatrix} \begin{Bmatrix} \theta_{AB} \\ \theta_{BA} \\ \Delta_{BA} \end{Bmatrix}$$

or, in non-matrix formulation, the slope-deflection equations are

$$M_{AB} = M_{AB}^F + k_{AA}\theta_{AB} + k_{AB}\theta_{BA} - \frac{1}{L}(k_{AA} + k_{AB})\Delta_{BA}$$

$$M_{BA} = M_{BA}^F + k_{BA}\theta_{AB} + k_{BB}\theta_{BA} - \frac{1}{L}(k_{BA} + k_{BB})\Delta_{BA}$$

(14.3)

For the specific case of a straight prismatic member of length L and flexural stiffness EI, the rotational end stiffnesses are

$$k_{AA} = k_{BB} = \frac{4EI}{L}; \qquad k_{AB} = k_{BA} = \frac{2EI}{L}$$

so that the slope-deflection equations for this case become

$$M_{AB} = M_{AB}^F + \frac{2EI}{L}\left(2\theta_{AB} + \theta_{BA} - 3\frac{\Delta_{BA}}{L}\right)$$

$$M_{BA} = M_{BA}^F + \frac{2EI}{L}\left(\theta_{AB} + 2\theta_{BA} - 3\frac{\Delta_{BA}}{L}\right)$$

(14.4)

The signs of the fixed-end moments must conform to the earlier stated convention; that is, counterclockwise moments on the member ends are positive.

The use of Eq. (14.3) or (14.4) in analysis is demonstrated in the following problems, in which the typical steps of the displacement method are followed.

Example 14.1: Analyze the two-span prismatic beam of part (a) of the figure due to the concentrated load *P*.

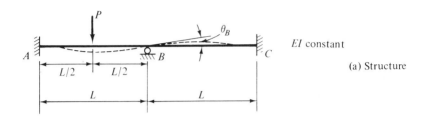

EI constant

(a) Structure

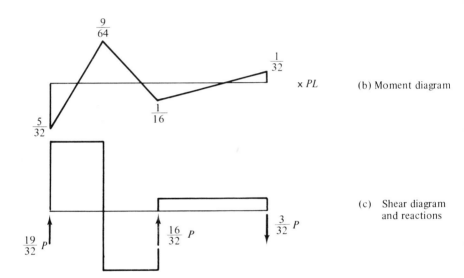

× *PL*   (b) Moment diagram

(c)   Shear diagram and reactions

Fig. E14.1

Solution: Note that this structure is three times statically indeterminate, so that three simultaneous equations would be required by the force method. By the displacement method, only one unknown rotation, θ_B, has to be determined.

We follow the steps of the displacement method:

1. *Force–deformation relations*

     Member *AB*:

$$\text{fixed end moments:} \qquad M^F_{AB} = +\frac{PL}{8}; \qquad M^F_{AB} = -\frac{PL}{8}$$

     (force-method analysis of fixed beam)

Slope-deflection equations

Member AB:

$$M_{AB} = \frac{PL}{8} + \frac{EI}{L}\left[4\theta_{AB} + 2\theta_{BA} - 6\frac{\Delta_{BA}}{L}\right]$$

$$M_{BA} = -\frac{PL}{8} + \frac{EI}{L}\left[2\theta_{AB} + 4\theta_{BA} - 6\frac{\Delta_{BA}}{L}\right]$$

Member BC:

$$M_{BC} = 0 + \frac{EI}{L}\left[4\theta_{BC} + 2\theta_{CB} - 6\frac{\Delta_{CB}}{L}\right]$$

$$M_{CB} = 0 + \frac{EI}{L}\left[2\theta_{BC} + 4\theta_{CB} - 6\frac{\Delta_{CB}}{L}\right]$$

2. *Geometry.* We substitute values of specified support displacements, and indicate the equality of member end rotations at joint B.

$$\theta_{AB} = \theta_{CB} = \Delta_{BA} = \Delta_{CB} = 0$$
$$\theta_{BA} = \theta_{BC} = \theta_{B}$$

Substituting these values, the slope-deflection equations become then

$$M_{AB} = \frac{PL}{8} + \frac{EI}{L}\cdot 2\theta_{B}$$

$$M_{BA} = -\frac{PL}{8} + \frac{EI}{L}\cdot 4\theta_{B}$$

$$M_{BC} = \frac{EI}{L}\cdot 4\theta_{B}$$

$$M_{CB} = \frac{EI}{L}\cdot 2\theta_{B}$$

3. *Statics.* We write a moment equilibrium equation of joint B, corresponding to the unknown joint rotation θ_{B}:

$$\sum \overset{+}{\overgroup{M_{B}}} = 0: \qquad M_{BA} + M_{BC} = 0$$

Substituting the slope-deflection equations, we obtain the equilibrium equation in terms of θ_{B}:

$$\left(-\frac{PL}{8} + \frac{EI}{L}\cdot 4\theta_{B}\right) + \left(\frac{EI}{L}\cdot 4\theta_{B}\right) = 0$$

Solving for θ_{B}:

$$\theta_{B} = \frac{PL^{2}}{64EI}$$

4. Resubstituting θ_B into the slope-deflection equations, we solve for the end moments of the members:

$$M_{AB} = \frac{PL}{8} + \frac{1}{32}PL = \frac{5}{32}PL$$

$$M_{BA} = \frac{PL}{8} + \frac{1}{16}PL = -\frac{1}{16}PL$$

$$M_{BC} = \frac{1}{16}PL = \frac{1}{16}PL$$

$$M_{CB} = \frac{1}{32}PL = \frac{1}{32}PL$$

Note that a positive end moment is counterclockwise on the member; the reactions, shear, and moment diagrams can now be determined by statics, as shown in parts (b) and (c). The moment diagram is, as usual, drawn on the compression side of the beam; its signs are not necessarily identical with those of the slope-deflection solution.

The slope-deflection method can also be used to analyze the effects of specified joint displacements; in this case, the applied member loads, and therefore the fixed-end moments, will be zero. The appropriate known joint rotations or displacements must be specified, as in the following example; special care with proper signs must be observed here.

Example 14.2: Analyze the two-span prismatic beam of part (a) of the figure due to a support settlement of point C of amount $\boldsymbol{\Delta}$.

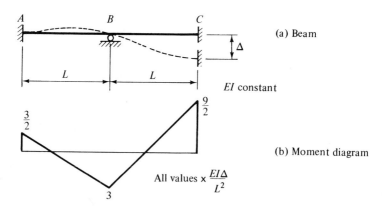

Fig. E14.2

Solution: This beam is identical to that analyzed in Ex. 14.1; we therefore carry over as much information as possible. In particular, we use the slope-deflection equations of that example, but with all fixed-end moments equal to zero. The conditions of geometry are

$$\theta_{AB} = \theta_{CB} = 0; \quad \theta_{BA} = \theta_{BC} = \theta_B; \quad \Delta_{BA} = 0; \quad \Delta_{CB} = -\Delta$$

Substituting these values into the slope-deflection equations and inserting these into the moment equilibrium equation of joint B, as in the previous example, we get

$$\sum \overset{+}{M_B} = 0: \quad M_{BA} + M_{BC} = 0: \quad \frac{EI}{L}[4\theta_B] + \frac{EI}{L}\left[4\theta_B + 6\frac{\Delta}{L}\right] = 0$$

from which

$$\theta_B = -\frac{3}{4}\frac{\Delta}{L}$$

Inserting the values of joint displacements into the slope-deflection equations, we obtain the moment values:

$$M_{AB} = \frac{EI}{L}\left[\; 0 \quad -\frac{3}{2}\frac{\Delta}{L} - \; 0 \; \right] = -\frac{3}{2}\frac{EI\Delta}{L^2}$$

$$M_{BA} = \frac{EI}{L}\left[\; 0 \quad -3\frac{\Delta}{L} - \; 0 \; \right] = -3\frac{EI\Delta}{L^2}$$

$$M_{BC} = \frac{EI}{L}\left[-3\frac{\Delta}{L} + \; 0 \quad +6\frac{\Delta}{L} \right] = 3\frac{EI\Delta}{L^2}$$

$$M_{CB} = \frac{EI}{L}\left[-\frac{3}{2}\frac{\Delta}{L} + \; 0 \quad +6\frac{\Delta}{L} \right] = \frac{9}{2}\frac{EI\Delta}{L^2}$$

The moment diagram is plotted in part (b).

In general, the number of equilibrium equations must conform to the number of unknown degrees of freedom. For each unknown joint rotation, a corresponding moment equilibrium equation must be set up. Similarly, the equilibrium equation corresponding to an unknown joint translation can be written for forces parallel to this translation.

A case involving joint translation is analyzed in the following example.

Example 14.3: Analyze the single-bay single-story rigid frame under lateral load shown in part (a) of the figure.

Solution

1. *Slope-deflection equations and geometrical relations.* Fixed-end moments (counterclockwise moments on member ends are positive, as shown in Fig. 14.1):

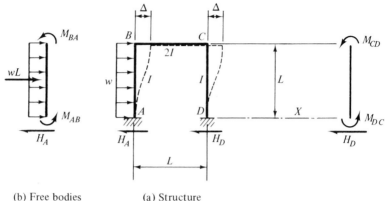

(b) Free bodies   (a) Structure

Fig. E14.3

$$M_{AB}^F = \frac{wL^2}{12} = -M_{BA}^F, \qquad \text{all others zero}$$

$$M_{AB} = \frac{wL^2}{12} + \frac{EI}{L}\left[2\theta_B \qquad -\frac{6}{L}\Delta\right]$$

$$M_{BA} = -\frac{wL^2}{12} + \frac{EI}{L}\left[4\theta_B \qquad -\frac{6}{L}\Delta\right]$$

$$M_{BC} = \frac{2EI}{L}[4\theta_B + 2\theta_C]$$

$$M_{CB} = \frac{2EI}{L}[2\theta_B + 4\theta_C]$$

$$M_{CD} = \frac{EI}{L}\left[4\theta_C - \frac{6}{L}\Delta\right]$$

$$M_{DC} = \frac{EI}{L}\left[2\theta_C - \frac{6}{L}\Delta\right]$$

2. *Equilibrium equations*

$$\sum M_B = 0: \quad M_{BA} + M_{BC} = 0: \quad 6\theta_B + 2\theta_C - \frac{3}{L}\Delta = \frac{wL^3}{24EI} \qquad \text{(a)}$$

$$\sum M_C = 0: \quad M_{CB} + M_{CD} = 0: \quad 2\theta_B + 6\theta_C - \frac{3}{L}\Delta = 0 \qquad \text{(b)}$$

$$\sum F_X = 0: \quad -H_A - H_D + wL = 0 \qquad \text{(c)}$$

Referring to free bodies, part (b), and expressing shears in terms of member end moments,

$$H_A = \frac{wL}{2} + \frac{M_{AB} + M_{BA}}{L}$$

$$H_D = \frac{M_{CD} + M_{DC}}{L}$$

Expressing the moments in terms of displacements, and substituting into Eq. (c),

$$\overset{+}{\sum F_X} = 0: \quad -\frac{wL}{2} - \frac{1}{L}\left\{\left[\frac{wL^2}{12} - \frac{EI}{L}\left(2\theta_B - \frac{6}{L}\Delta\right)\right]\right.$$
$$+ \left[-\frac{wL^2}{12} + \frac{EI}{L}\left(4\theta_B - \frac{6}{L}\Delta\right)\right]$$
$$\left. + \left[\frac{EI}{L}\left(4\theta_C - \frac{6}{L}\Delta\right) + \frac{EI}{L}\left(2\theta_C - \frac{6}{L}\Delta\right)\right]\right\} + wL$$
$$= 0$$

or, simplifying,

$$\theta_B + \theta_C - \frac{4}{L}\Delta = \frac{wL^3}{12EI} \qquad \qquad \text{(d)}$$

Equations (a), (b), and (d) are solved for the unknown displacements

$$\theta_B = -.0012\frac{wL^3}{EI}; \quad \theta_C = -.0115\frac{wL^3}{EI}; \quad \Delta = -.0241\frac{wL^4}{EI}$$

and these values are substituted into the slope-deflection equations to yield

$$M_{AB} = .225wL^2; \quad M_{BA} = -M_{BC} = .056wL^2$$
$$M_{CB} = -M_{CD} = -.097wL^2; \quad M_{DC} = .121wL^2$$
$$H_A = .782wL; \quad H_D = .218wL$$

From these values, shear and moment diagrams can be drawn by statics. Note that in Eqs. (a), (b), and (d) of this example the fixed-end moments (which depend only on the loading condition) are in each case transposed to the right-hand side of the equation, and the stiffness terms (which are properties of the structure) remain on the left-hand side. Thus, different loading conditions can be accommodated by changing only the load terms on the right-hand side of the equations.

We note that the establishment of the force-equilibrium equation in the preceding example was somewhat complicated because the forces had to be expressed in terms of the end moments before they could be written in terms of the displacements. A more convenient and more systematic method for writing equilibrium conditions directly in terms of the end moments relies on the theorem of virtual displacements.

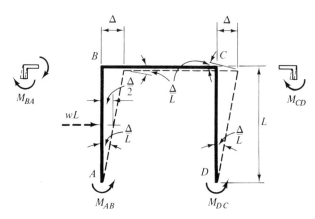

Fig. 14.3 Equilibrium by theorem of virtual displacement.

The virtual displacement will consist of a displacement corresponding to the unknown degree of freedom of the structure so articulated that the real member end moments can do internal work. For the frame of Ex. 14.3, for instance, we would visualize the virtually displaced shape due to a sway Δ, shown in Fig. 14.3. With the moment convention as before, the virtual work equation is then

$$EVW = IVW$$

$$wL \cdot \frac{\Delta}{2} - (M_{AB} + M_{DC})\frac{\Delta}{L} = (M_{BA} + M_{CD})\frac{\Delta}{L}$$

or

$$M_{AB} + M_{BA} + M_{CD} + M_{DC} = \frac{wL^2}{2}$$

This equilibrium equation is identical to that leading to Eq. (d) of Ex. 14.3.

For problems such as the gabled frame of Fig. 14.4(a), which involves two degrees of translational freedom, the method of virtual displacements is particularly useful for writing the corresponding equilibrium equations. The two virtual displacements shown in Fig. 14.4(b) and (c) can be interpreted as rigid-body rotations of the appropriately articulated structure about the instantaneous centers shown and the virtual rotations at the joints, and the displacements of the external real loads can be computed by methods found in books on rigid-body kinematics.

Similarly, in multistory framed structures each story will have one degree of translational freedom, or sway, and the corresponding equilibrium equa-

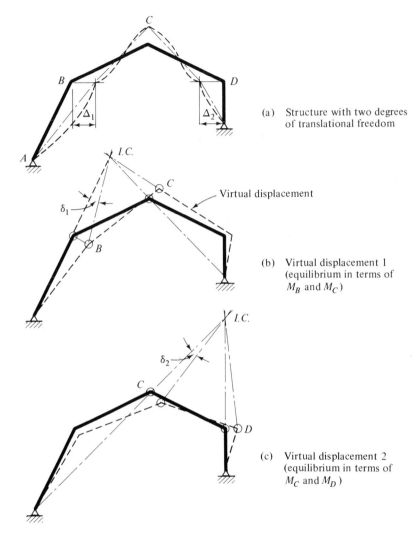

(a) Structure with two degrees
     of translational freedom

(b) Virtual displacement 1
     (equilibrium in terms of
     M_B and M_C)

(c) Virtual displacement 2
     (equilibrium in terms of
     M_C and M_D)

Fig. 14.4 Sway conditions.

tion can be written by equating the external and internal virtual work due to virtual displacements as shown in Figs. 14.5(a) and (b).

The application of a longhand method such as presented here is always restricted by the size of the set of equations to be solved. In most cases it is probable that preference will be given to the matrix displacement method when appropriate computing facilities are available.

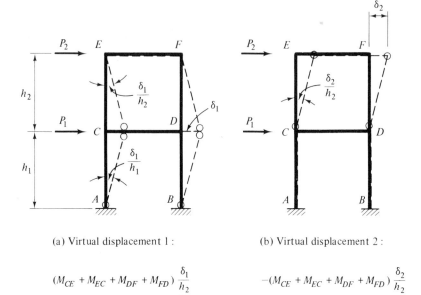

(a) Virtual displacement 1 :   (b) Virtual displacement 2 :

$$(M_{CE} + M_{EC} + M_{DF} + M_{FD}) \frac{\delta_1}{h_2}$$

$$-(M_{AC} + M_{CA} + M_{BD} + M_{DB}) \frac{\delta_1}{h_1}$$

$$= P_1 \cdot \delta_1$$

$$-(M_{CE} + M_{EC} + M_{DF} + M_{FD}) \frac{\delta_2}{h_2}$$

$$= P_2 \cdot \delta_2$$

Fig. 14.5 Sway equilibrium equations.

14.2. Moment-Distribution Method

The method of moment distribution is a numerical application of the displacement method in which the desired quantities are determined by a method of successive approximation that is suitable for longhand calculations. Because the procedure lends itself to simple physical interpretation, it can be used for quick approximate as well as exact solutions, and the thought process involved can guide the designer's judgment, because it can give a definite feel for the structural action.

Before outlining the procedure, we review some preliminary concepts:

1. Fixed-end moment: The member end moments due to loads when the member ends are prevented from rotating, shown in Fig. 14.6(a). The calculation of fixed-end moments was discussed in Sec. 11.1.

2. Member end stiffness [Fig. 14.6(b)]: The member end moment necessary to cause unit rotation of the end when all other supports are prevented from movement. Their calculation was discussed in Sec. 11.2, and it was shown there that, in particular, the rotational stiffness

(a) Fixed-end moments

(b) Rotational stiffness
and carry-over factor

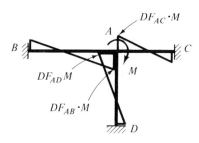

(c) Distribution and carry-
over factors

Fig. 14.6 Basic concepts for moment distribution.

    of a prismatic member is proportional to the member characteristic EI/L.

3. Distribution factor [Fig. 14.6(c)]: Defines the member end moment at a joint as a fraction of the total unbalanced moment applied to the joint. In Sec. 13.1 this factor was shown to be equal to the ratio of the member end stiffness to the sum of stiffnesses of all members at the joint; that is, for member i,

$$\text{DF}_i = \frac{k_i}{\sum\limits_{i} k}$$

    The sum of the distribution factors for any joint equals 1.

4. Carry-over factor C_{AB} [Fig. 14.6(b)]: The ratio of the moment at the far, fixed end B of a member to the cranked-in moment at the near end A of the member, that is, the ratio of the rotational end stiffnesses: $C_{AB} = k_{BA}/k_{AA}$; for a prismatic beam, this value is $\frac{1}{2}$; for haunched sections, the stiffnesses can be calculated by the force method.

5. Sign convention: It is convenient to use the same convention for moments as in the slope-deflection method shown in Fig. 14.1; that

is, moments are positive when acting counterclockwise on the member or clockwise on the joint.

The moment-distribution procedure begins with the moments due to loads on the geometrically determinate structure, that is, all joints prevented from movement. These are the fixed-end moments. The structure next is gradually eased into its final deformed shape by allowing one joint at a time to rotate. Each time a joint is released in this fashion, the unbalanced moment on the joint is distributed to the adjacent members (whose far ends are fixed at this stage) in accordance with their distribution factor; a fraction of the moment thus distributed to the near member end, equal to the carry-over factor, is carried over to the far, fixed member end. This carry-over moment will in turn become an unbalanced moment to be distributed to the adjacent members upon release of that joint. As the joints are successively released, the residual unbalanced moments become smaller and smaller, allowing the total moments, obtained by addition of all incremental moments, to converge to the correct solution. In general, the procedure converges rapidly; furthermore, premature termination of the scheme will result in approximate results that may be useful for various design purposes.

The successful application of the procedure depends on an efficient tabular scheme, as shown in the following example; in this scheme, joints are released in the balancing cycle designated "Bal" in column 2, following which the residual moments are carried over to the far member ends (which are considered fixed at this stage) in the carry-over cycle designated "CO" This process is repeated until the residual moments have become sufficiently small, at which time all moments are added to obtain the final moments.

We shall consider first a structure whose joints are fixed against translations.

Example 14.4: Analyze the rigid-jointed frame of part (a) of the figure by moment distribution.

Solution: We assume all joints locked against rotation, and calculate the fixed-end moments (FEM):

$$M_{AB}^F = -M_{BA}^F = \frac{wL^2}{12} = 50 \text{ kip-ft}$$

We next set up the table shown, grouping together all member ends with a common joint. The rotational stiffnesses are entered (note that relative values are sufficient), and the distribution factors are determined. The distribution factors of value 0 at the fixed supports A and D may be justified by considering the rigid support as a member of $k = \infty$, so that, for support A, $k_{AB}/(k_{AB} + \infty) = 0$, and any unbalanced moment is resisted entirely by the rigid support. In the alternative case of a pinned end, a support member of rota-

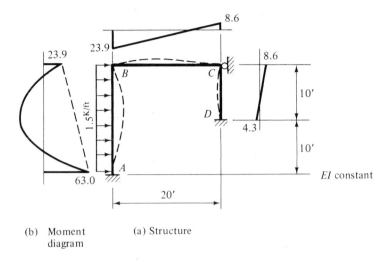

(b) Moment   (a) Structure
     diagram

Fig. E14.4

tional stiffness 0 could be visualized, leading to a distribution factor of value 1.

The fixed-end moments are written, and the distribution process started. The analyst should be able to visualize the current constraints on the structure at all stages. In the tabular solution, the stage numbered 1 in the first column considers the moments due to the applied loads with all joints locked against rotation. At stage 2, joint *B* is released; all other joints remain locked. At stage 3, joint *B* has been locked again, and joint *C* is released, and so on.

	JOINT	A	B		C		D		
	MEMBER	*AB*	*BA*	*BC*	*CB*	*CD*	*DC*		COMMENTS
	k	1	1	1	1	2	2	$\times \dfrac{EI}{5}$	
	DF	0	.5	.5	.33	.67	0		
1	FEM	+50.0	−50.0						kip-ft All joints locked
2	{Bal	0	+25.0	+25.0					}Joint *B* released
	{CO	+12.5	0	0	+12.5				
3	{Bal	0	0	0	−4.2	−8.3			}*B* locked,
	{CO			−2.1	0	0	−4.1		}*C* released
	.								
	.								
	.								
	Final moment	+63.0	−23.9	+23.9	+8.6	−8.6	−4.3		kip-ft Sum of all moments

All reactions and internal forces can be obtained by statics from these end moments, and the moment diagram drawn as shown in part (b).

In more general loading cases, such as that of the following example, the unlocking of one joint at a time would lead to unduly lengthy tabulations; in such cases, all joints can be released simultaneously in the balancing cycle, following which all distributed moments are carried over to the far member ends simultaneously in the carry-over cycle.

When a structure is to be analyzed by moment distribution for the effects of a given support displacement, or support rotation, the fixed-end moments are determined as those of the structure displaced accordingly, with all other joint displacements and rotations prevented. For instance, the continuous

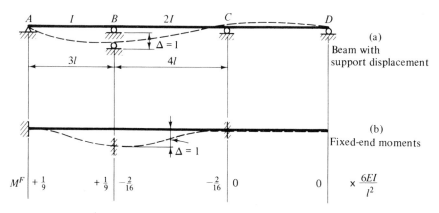

Fig. 14.7 Fixed-end moments due to support settlement.

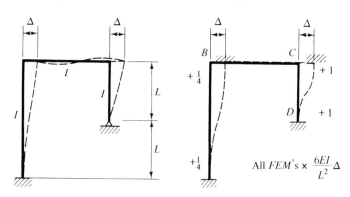

(a) Frame and load   (b) Fixed-end moments

Fig. 14.8 Fixed-end moments due to sway.

beam of Fig. 14.7, whose support B is displaced an amount $\Delta = 1$, suffers the fixed-end moments associated with the displaced shape of Fig. 14.7(b), which can be identified as the translatory stiffnesses k_{13} and k_{23} of Eq. (11.4) ($= 6EI/L^2$). Their values are indicated at appropriate places below the structure. It is particularly important to observe signs of these moments.

When a rigid frame is subjected to a specified sway, as shown in Fig. 14.8(a), the fixed-end moments are those caused by the translation Δ, which causes the fixed-end moments shown in Fig. 14.8(b). Subsequent joint rotations are allowed while additional sway is prevented, thereby easing the joints into their final slopes. The following example demonstrates this analysis for the frame of the previous example.

Example 14.5: Analyze the rigid frame shown for a horizontal displacement Δ of support C to the right.

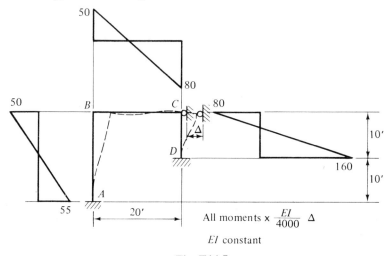

Fig. E14.5

Solution: The fixed-end moments due to the specified sway occur in the legs of the frame:

$$M_{AB}^F = M_{BA}^F = \frac{6EI\Delta}{20^2} = \frac{60}{4,000}EI\Delta$$

$$M_{CD}^F = M_{DC}^F = \frac{6EI}{10^2}\Delta = \frac{240}{4,000}EI\Delta$$

We introduce these values into the table already used in Ex. 14.4; to obtain moment values of convenient magnitude for the operations, we factor out a constant of appropriate magnitude and relax the joints by distributing the moments:

JOINT	A	B		C		D	
MEMBER	AB	BA	BC	CB	CD	DC	
DF	0	.5	.5	.33	.67	0	
FEM	+60	+60			+240	+240	$\times \dfrac{EI}{4,000}\Delta$
Bal	0	−30	−30	−80	−160	0	
CO	−15	0	−40	−15	0	−80	
Bal	0	+20	+20	+5	+10	0	
.							
.							
.							
M_{sway}	+55	+50	−50	−80	+80	+160	$\times \dfrac{EI}{4,000}\Delta$

From these member end moments, the moment diagram can be drawn as shown and all other forces determined.

In frames with joints free to sway, the moment distribution can be carried out in the following steps:

1. Restrain all joints against translation, and carry out the moment distribution as before; this leads to a set of moments M_L.

2. Corresponding to each translatory degree of freedom of the structure, introduce a unit displacement, calculate the resulting fixed-end moments, and distribute them in accordance with the method outlined earlier. This involves a number of separate moment distributions equal to the number of translatory degrees of freedom, each one resulting in moments M_i due to a unit displacement. The moments due to an actual (but as yet unknown) displacement Δ_i are $\Delta_i M_i$.

3. Corresponding to each translatory degree of freedom, introduce a force equilibrium equation (the shear equation of the slope-deflection method, Sec. 14.1). These equations are written in terms of the member end moments and will therefore contain the unknown Δ_i's. The equations are solved for the Δ_i's.

4. The moments M_i of step 2 are multiplied by the appropriate translations Δ_i, and added to the moments M_L of step 1 to obtain the final moments. The remaining forces can then be found by statics.

The method is used to solve the following example.

Example 14.6: Analyze the single-story single-bay rigid frame of part (a) of the figure.

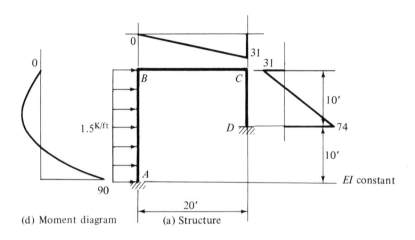

(d) Moment diagram   (a) Structure

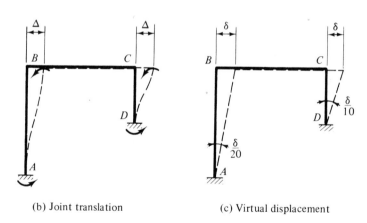

(b) Joint translation   (c) Virtual displacement

Fig. E14.6

Solution: This structure has one degree of translatory freedom (neglecting axial deformations), the sway of the top joints. We first restrain the top against this sway by introducing an imaginary restraint, and perform a moment distribution. This is identical to Ex. 14.4, and we shall therefore only list here the resulting moments M_L:

JOINT	A	B		C		D
MEMBER	AB	BA	BC	CB	CD	DC
DF	0	.50	.50	.33	.67	0
M_L	+63.0	−23.9	+23.9	+8.6	−8.6	−4.3	kip-ft

Next, we displace the top story of the frame an assumed amount Δ, as shown in part (b), and analyze the moments in the frame due to this sway. This calculation was carried out in Ex. 14.5, and we only tabulate the results of that analysis here:

MEMBER	AB	BA	BC	CB	CD	DC
M_{sway}	+55	+50	−50	−80	+80	+160	$\times \dfrac{EI}{4,000}\Delta \equiv \Delta'$

The actual value of the sway is determined so as to satisfy the equilibrium of horizontal forces; we use the theorem of virtual displacements, as discussed in Sec. 14.1 and shown in part (c). Equating internal and external virtual work, we obtain

$$(M_{AB} + M_{BA})\frac{\delta}{20} + (M_{CD} + M_{DC})\frac{\delta}{10} = (1.5 \times 20)\left(\frac{\delta}{2}\right) \tag{a}$$

where, from the two moment distributions,

$$M_{AB} = (63.0 + 55\Delta'); \qquad M_{BA} = (-23.9 + 50\Delta')$$
$$M_{CD} = (-8.6 + 80\Delta'); \qquad M_{DC} = (-4.3 + 160\Delta')$$

We substitute these values into the equilibrium equation (a), cancel the virtual displacement value δ, and solve for the sway:

$$\Delta' = .489$$

We continue the tabular calculations by prorating the sway moments, and add them to the moments due to the loads to determine the final moments:

JOINT	A	B		C		D
MEMBER	AB	BA	BC	CB	CD	DC
M_L	+63	−24	+24	+8	−8	−4
M_{sway}	+27	+24	−24	−39	+39	+78
M_{total}	+90	0	0	−31	+31	+74	kip-ft

All remaining forces can be found by statics. The final moment diagram is
shown in part (d).

PROBLEMS

1. a. Analyze the two-span continuous beam for the concentrated load *P*.
   b. Analyze the beam for a support settlement of amount Δ_B at point *B*. [*Hint:*
      Here, the last term in the slope-deflection equations, Eq. (14.4), for each of
      members *AB* and *BC* is known; watch signs.]

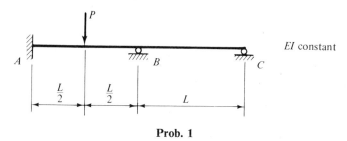

Prob. 1

2. A tank wall with dimensions and supports as shown is subjected to water
   pressure of the full tank. Analyze a 1-ft-wide vertical strip, and draw shear
   and moment diagrams as well as its deformed shape.

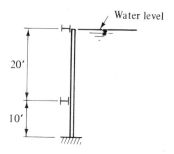

Prob. 2

3. Analyze the rigid frame shown due to
   a. The uniform vertical load *w* on *BC*.
   b. A vertical support settlement Δ_D of column foot *D*. [*Hint:* This will cause
      the last term in the slope-deflection equations, Eq. (14.4), for members *AB*
      and *BC* to be known. Watch your signs.]
   c. A shortening of member *BC* of amount ΔL. (*Hint:* This will induce a known
      translation in member *CE*.)
   For each case, draw moment diagrams.

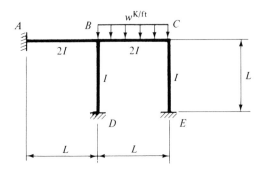

Prob. 3

4. Analyze the structure shown by the slope-deflection method for
   a. A uniform load *w* on beam *ABC*.
   b. A specified rotation of joint *D* of amount θ_D. (*Hint:* Substitute this known rotation in the appropriate term of the slope-deflection equation.)

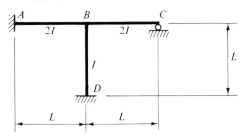

Prob. 4

5. Analyze the rigid frame, using the slope-deflection method, for
   a. An applied vertical load *w* on member *BC*.
   b. A horizontal load *P* acting at joint *B* to the right.
   c. A support rotation θ_A of joint *A*.
   Draw the moment diagram in each case. Also, sketch the deformed shape, making use of the calculated joint displacements.

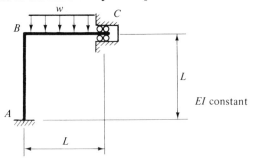

Prob. 5

6. Analyze the rigid frame shown due to
   a. A vertical uniform load *w* on member *BC*.
   b. A horizontal load *P* applied at joint *B* to the right.
   c. A support settlement Δ_A at point *A*.
   Draw deformed shape and moment diagrams for each case.

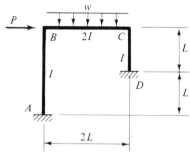

Prob. 6

7. Analyze the rigid frame with sloping legs by the slope-deflection method.

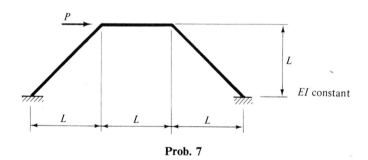

Prob. 7

8. Analyze the rigid frame shown for
   a. A uniform vertical load *w* acting on member *BC*.
   b. A concentrated horizontal load *P* to the right acting on joint *B*.
      (*Hint:* This is not a symmetrical problem; watch your signs.)

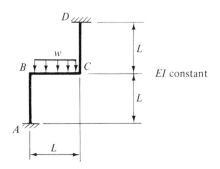

Prob. 8

9. Analyze the two-story frame shown due to
   a. A vertical load w_1 acting on member CD.
   b. Horizontal loads to the right of magnitude P on joint A and $2P$ on joint C.
   c. A uniform lateral load of intensity w_2 on members ACE.

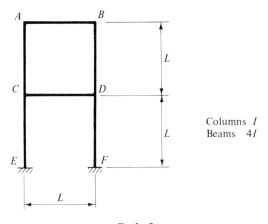

Columns I
Beams $4I$

Prob. 9

10. Analyze the continuous beam shown for the effects of the uniform load by the
   method of moment distribution.

11. a. and b. Repeat Prob. 1 by the moment-distribution method.

12. Repeat Prob. 2 by the moment-distribution method.

13. a. to c. Repeat Prob. 3 by the moment-distribution method.

14. a. to b. Repeat Prob. 4 by the moment-distribution method.

15. a. to c. Repeat Prob. 5 by the moment-distribution method.

16. a. to c. Repeat Prob. 6 by the moment-distribution method.

17. a. and b. Repeat Prob. 8 by the moment-distribution method.

18. a. Analyze the frame shown in part (a) of the figure by moment distribution for the effects of the uniform gravity loads shown.

    b. Analyze the frame shown in part (b) for the effects of the uniform gravity loads shown.

    c. Compare results of parts (a) and (b) of the problem, and draw conclusions as to the feasibility of isolating one floor of a multistory building frame, with far ends of adjacent columns fixed, for the purpose of analyzing the effects of floor loads on girder and column moments.

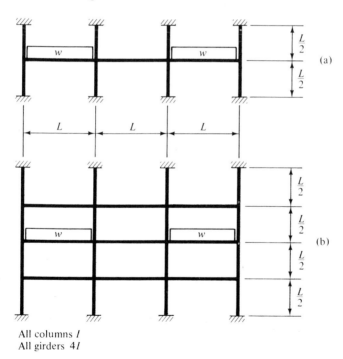

All columns I
All girders $4I$

Prob. 18

CHAPTER 15

Some Supplementary Topics

The topics in this chapter do not, strictly speaking, belong in a part of this book devoted to the displacement method. Rather, they are intended to provide a unifying guide to the different methods that have been presented, and background for some of the ideas presented in the next chapter. Accordingly, Sec. 15.1 presents a comparison between the force and displacement methods which illustrates the analogy that exists between the two approaches.

Sections 15.2 to 15.4 discuss a number of energy methods that find frequent application in structural mechanics. In deriving these approaches, stress is placed on two factors:

1. The derivations are intended to show that all the energy methods can be considered restatements of the theorem of virtual work.

2. The duality of the methods is pointed out; those based on the theorem of virtual displacements result in equilibrium equations in terms of displacements and lead to an alternative formulation of the displacement method. Those based on the theorem of virtual forces result in geometrical equations in terms of forces and lead to an alternative formulation of the force method.

It is hoped that by establishing such a unifying scheme, some light is shed on the interrelations and uses of these various procedures.

Finally in Sec. 15.5, the all-important theorem of virtual work is derived for a special case in order to clarify the approach and its basic assumptions.

15.1. Comparison of the Force and Displacement Methods

It may have been observed by the reader that there are some striking analogies in the formulation of the force and the displacement methods; this becomes particularly obvious in formal matrix expressions, such as Eqs. (10.8) and (13.12).

Indeed, the two basic approaches are analogous methods in the sense that identical mathematical operations are performed on different physical quantities. A thorough consideration of this fact might be based on the duality of the theorem of virtual work in terms of its equilibrium version, the theorem of virtual displacements, and its geometrical version, the theorem of virtual forces. Here, however, we only wish to direct attention to the unifying features of the two methods by listing, side by side, the basic operations involved in each:

Force Method	*Displacement Method*
1. Reduce the structure to static determinacy by removing r forces (or *redundants*). The r points at which forces have been removed are called *cuts*. The resulting statically determinate structure is called the primary structure. The conditions of geometry of the actual structure are violated at the r cuts. The aim of the subsequent calculation is to determine those values of the r redundants necessary to restore the specified conditions of geometry at the cuts.	1. Reduce the structure to geometric determinacy by removing k displacements (or *degrees of freedom*). The k points at which displacements have been removed are called *constraints*. The resulting geometrically determinate structure is called the primary structure. The conditions of equilibrium of the actual structure are violated at the k constraints. The aim of the subsequent calculation is to determine those values of the k displacements necessary to restore the specified conditions of equilibrium at the constraints.
2. Write the displacements $\{\Delta^\circ\}$ at cuts i of the primary structure due to the applied loads.	2. Write the forces $\{X^F\}$ at constraints i of the primary structure due to the applied loads.
3. Calculate the displacements at the cuts i of the primary structure due to unit values of the redundants at cuts j. These are the *flexibilities* F_{ij}, which can be contained in a matrix $[F]$ of order $(r \times r)$. The displacements at	3. Calculate the forces at the constraints i of the primary structure due to unit values of the displacements at constraints j. These are the *stiffnesses* $-K_{ij}$, which can be contained in a matrix $[K]$ of order $(k \times k)$. The forces

cuts i due to the actual redundant at cut j, X_j, will be $F_{ij}X_j$, and can be contained in a matrix $[F]\{X\}$.

4. The *conditions of geometry* at the r cuts are satisfied by adding the displacements due to applied loads, $\{\Delta^\circ\}$, and due to the redundants, and equating to the specified displacements, $\{\Delta\}$:

$$\{\Delta^\circ\} + [F]\{X\} = \{\Delta\}$$

5. Solve this set of equations for the r redundants $\{X\}$:

$$\{X\} = [F]^{-1}[\{\Delta\} - \{\Delta^\circ\}]$$

at constraints i due to the actual displacement at constraint j, Δ_j, will be $-K_{ij}\Delta_j$, and can be contained in a matrix $[-K]\{\Delta\}$.

4. The *conditions of equilibrium* at the k constraints are satisfied by equating the forces on the joints due to applied loads, $\{X^F\}$, due to the joint displacements, and due to the specified loads, $\{X\}$, to zero:

$$\{X^F\} - [K]\{\Delta\} + \{X\} = 0$$

5. Solve this set of equations for the k joint displacements $\{\Delta\}$:

$$\{\Delta\} = [K]^{-1}[\{X\} + \{X^F\}]$$

The comparison could be carried on to other operations, but the foregoing will suffice to illustrate the matter.

In general, it appears that we can switch from one method to the other by exchanging a few terms. Wherever, in one method, terms pertaining to statics appear, the other method will contain the corresponding geometrical terms, and vice versa. Thus, the words "force" and "displacement," "cut" and "constraint," "redundant" and "degree of freedom," "equilibrium" and "geometry," "flexibility" and "stiffness," and so on, are analogous to each other.

15.2. Energy Methods

The theorem of virtual work in the form of Eq. (5.1) is a convenient starting point for the derivation of a number of other energy methods, which are widely cited in the literature and have a variety of applications. It is important to understand that these different approaches can all be derived from the principle of virtual work; they can, in fact, be looked on simply as different ways of expressing the virtual-work principle, which can thus serve as a means of unifying the various energy concepts. Such a systematization is particularly important in view of the variety of derivations existing in the literature, which tends to obscure the relations between the different theorems.

In this section, some of the energy theorems are derived in such a fashion as to show them to be only specific applications of the theorem of virtual work. Their use is then demonstrated by means of examples.

We start in each case with the theorem of virtual work,

$$\sum X \cdot \Delta = \int_{\text{vol}} \sigma\epsilon \, dV \qquad (15.1)$$

in which the forces X and the stresses σ denote a statically compatible force system, and Δ, ϵ constitute a geometrically compatible displacement system.

Strain energy. If we consider a fiber shown in Fig. 15.1(a) of material with the stress–strain relations shown in Fig. 15.1(b) under uniaxial stress

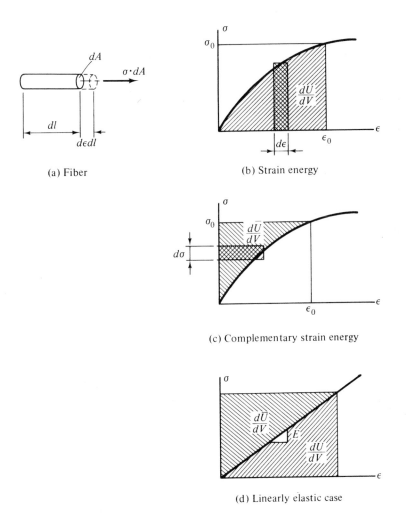

(a) Fiber   (b) Strain energy

(c) Complementary strain energy

(d) Linearly elastic case

Fig. 15.1 Strain energy and complementary strain energy.

σ, then the work done by the force ($\sigma \, dA$) moving through the deformation ($d\epsilon \cdot dl$) is

$$(\sigma \cdot dA)(d\epsilon \cdot dl) = \sigma \, d\epsilon \cdot dV$$

Under a load history that increases from ($\sigma = 0$, $\epsilon = 0$) to (σ_0, ϵ_0), the element dV is subjected to an amount of work

$$dU = \left[\int_{\epsilon=0}^{\epsilon_0} \sigma \, d\epsilon \right] dV$$

and the entire body is subjected to an amount of work

$$U = \int_V \left[\int_{\epsilon=0}^{\epsilon_0} \sigma \, d\epsilon \right] dV \tag{15.2}$$

If this work is stored without loss of energy in a deformable body, we call it the *strain energy*. The integral $\int_{\epsilon=0}^{\epsilon_0} \sigma \, d\epsilon$, which represents the strain energy per unit of volume, or *strain energy density*, is shown by the shaded area under the stress–strain curve of Fig. 15.1(b).

There is an analogous energy quantity called the *complementary strain energy density*, defined by the integral $\int_{\sigma=0}^{\sigma_0} \epsilon \, d\sigma$. It can be interpreted as the shaded complementary area over the stress–strain curve of Fig. 15.1(c). For the entire body of volume V, the complementary strain energy is written as the integral

$$\bar{U} = \int_V \left[\int_{\sigma=0}^{\sigma_0} \epsilon \, d\sigma \right] dV \tag{15.3}$$

There is no physical meaning attached to the complementary energy; it will be used in a purely formal sense in what follows.

In the case of a linearly elastic material of modulus E, Fig. 15.1(d) shows that the quantities U and $\bar{U}$ are identical. Since now $\sigma = E\epsilon$, or $\epsilon = \sigma/E$, we can integrate Eqs. (15.2) and (15.3) over the stress or strain path to obtain

$$U = \bar{U} = \int_V \frac{E\epsilon^2}{2} \, dV = \int_V \frac{\sigma^2}{2E} \, dV \tag{15.4}$$

In the case of pure bending of an elastic beam of moment of inertia I and length L under moment M, Eq. (15.4) can be further integrated over the cross section to yield

$$U = \bar{U} = \int_L \frac{EI \cdot \Phi^2}{2} \, dl = \int_L \frac{M^2}{2EI} \, dl \tag{15.5}$$

where $\Phi = M/EI$ is the curvature of the beam.

For a prismatic straight two-force member of length L and cross-sectional area A, subjected to a constant force N, Eq. (15.4) leads to

$$U = \bar{U} = \frac{AE(\Delta L)^2}{2L} = \frac{N^2 \cdot L}{2AE} \tag{15.6}$$

where ΔL is the total elongation of the member.

Similar expressions could be derived for the strain energy due to shear or torsional deformations.

15.3. Two Methods Based on the Theorem of Virtual Displacements

1. *Theorem of minimum potential energy.* In the virtual work equation, Eq. (15.1), we consider two sets of virtual displacements, Δ_1^{**} and Δ_2^{**}, which are compatible with the real specified support conditions of the structure, as shown by the two virtual beam deflections of Fig. 15.2(a); the virtual strains corresponding to the two systems are ϵ_1^{**} and ϵ_2^{**}. The real force systems are identical in each case, equal to the real external forces on the structure, X^*, and the corresponding real internal stresses σ^*.

We now write the virtual work equations for the real forces riding through each of the two virtual displacement systems:

$$\int_V \sigma^* \epsilon_1^{**} dV = \sum X^* \Delta_1^{**}$$

$$\int_V \sigma^* \epsilon_2^{**} dV = \sum X^* \Delta_2^{**}$$

Subtracting the first from the second equation, we obtain

$$\int_V \sigma^* (\epsilon_2^{**} - \epsilon_1^{**}) \, dV = \sum X^* (\Delta_2^{**} - \Delta_1^{**})$$

or, denoting the differences

$$(\epsilon_2^{**} - \epsilon_1^{**}) \equiv d\epsilon, \qquad (\Delta_2^{**} - \Delta_1^{**}) \equiv d\Delta$$

and transposing, we obtain

$$\int_V \sigma \, d\epsilon \, dV - \sum X \, d\Delta = 0 \tag{15.7}$$

We now integrate both terms of Eq. (15.7) over the deformation path traveled; and then write the differential, to obtain

$$d\left[\int_{\epsilon=0}^{\epsilon} \left(\int_V \sigma \, d\epsilon \, dV \right) - \sum \int_{\Delta=0}^{\Delta} X \, d\Delta \right] = 0 \tag{15.8}$$

We recognize by comparison with Eq. (15.2) that the first integral represents the internal strain energy, U, and the second, the work W done by the real forces riding through the virtual displacements; these forces consist of the applied loads P and the reactions R, so that we can write the second term

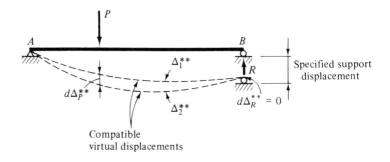

(a) Geometrically compatible virtual displacements

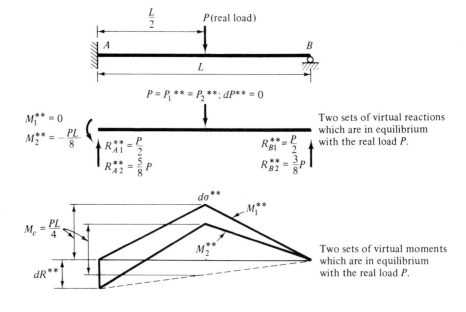

(b) Statically compatible virtual forces

Fig. 15.2 Compatible virtual displacement and force fields.

of Eq. (15.8) as

$$W = \sum \int_{\Delta=0}^{\Delta} X \, d\Delta = \int_{\Delta=0}^{\Delta} [\sum P \cdot d\Delta_P + \sum R \cdot d\Delta_R]$$

where $d\Delta_P$ are the virtual displacements of the load points and $d\Delta_P$ are the virtual displacements of the reactions. Because of the assumed compatibility of the virtual displacements with the specified support conditions [see Fig. 15.2(a)], $d\Delta_R = 0$, and the second term vanishes. The external work is thus due only to that of the applied loads P.

We now define the *potential energy* Ψ of the system as

$$\Psi = U - W$$

where U is the internal strain energy due to a geometrically compatible displacement system, and W is the external work of the applied loads. Then we write Eq. (15.8) in the form

$$d\Psi = d(U - W) = 0 \qquad (15.9)$$

We recognize this as a condition for an extremum of the function Ψ. Since it is based on the theorem of virtual displacements, it expresses satisfaction of the equilibrium equations between loads and stresses, and we can state it in words as follows:

> Of all geometrically compatible states, that one which extremizes the potential energy Ψ will also satisfy the equilibrium of the system.

It can be shown that the potential energy is unbounded in a positive sense, as shown in Fig. 15.3(a), so that the only extremum is a minimum value. For this reason the above statement, expressed by Eq. (15.9), is called the *theorem of minimum potential energy*; it provides an equilibrium condition in terms of displacements.

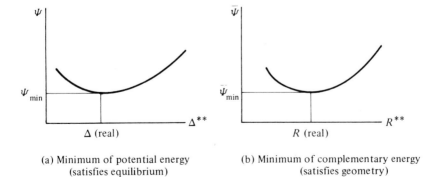

(a) Minimum of potential energy   (b) Minimum of complementary energy
(satisfies equilibrium)   (satisfies geometry)

Fig. 15.3 Minima of potential and complementary energy.

Since a linearly elastic problem has always one unique solution, it follows that once a solution satisfying both equilibrium and geometry has been found, it must be the correct one. Thus, this theorem provides a useful way of solving structural problems.

Example 15.1: Determine the displacements Δ_1 and Δ_2 of point A of the three-member truss shown in part (a) of the figure due to the loads P_1 and P_2.

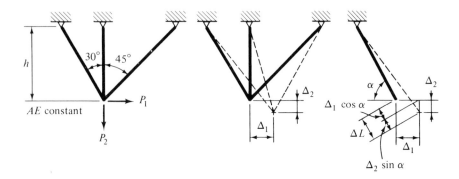

   (a) Structure   (b) Compatible displacement   (c) Member geometry

Fig. E15.1

Solution: We consider the geometrically compatible displacement system shown in part (b). To calculate the internal strain energy U of each member, we consider the typical member shown in part (c) inclined at an angle α with the horizontal; its elongation ΔL due to the end displacements Δ_1 and Δ_2 is

$$\Delta L = \Delta_1 \cos \alpha + \Delta_2 \sin \alpha$$

The strain energy in the member is then, according to Eq. (15.6),

$$U = \frac{AE}{2L}(\Delta L)^2 = \frac{AE}{2}\frac{\sin \alpha}{h}(\Delta_1 \cos \alpha + \Delta_2 \sin \alpha)^2$$

Substituting appropriate values of α for all three members, computing the internal strain energy for each, and adding, we obtain the total strain energy of the system:

$$U = \frac{AE}{2h}(.570\Delta_1^2 + .042\Delta_1\Delta_2 + 2.004\Delta_2^2)$$

The work of the applied loads is

$$W = P_1\Delta_1 + P_2\Delta_2$$

so that the total potential energy of the system becomes

$$\Psi = \frac{AE}{2h}(.570\Delta_1^2 + .042\Delta_1\Delta_2 + 2.004\Delta_2^2) - (P_1\Delta_1 + P_2\Delta_2)$$

To minimize the potential energy, we differentiate with respect to each of the unknown displacements Δ_1 and Δ_2, and equate to zero:

$$\frac{\partial\Psi}{\partial\Delta_1} = \frac{AE}{2h}(1.14\Delta_1 + .042\Delta_2) - P_1 = 0$$

$$\frac{\partial\Psi}{\partial\Delta_2} = \frac{AE}{2h}(.042\Delta_1 + 4.008\Delta_2) - P_2 = 0$$

We thus obtain a set of simultaneous equations equal to the number of unknown displacement components or degrees of freedom, from which we solve for the unknowns:

$$\Delta_1 = \frac{h}{AE}(1.752P_1 - .019P_2)$$

$$\Delta_2 = \frac{h}{AE}(-.019P_1 + .499P_2)$$

All required member forces and reactions can be computed from these displacements.

The reader should observe the striking similarities between this procedure and that of the displacement method. Indeed, the displacement method can be interpreted as an application of the theorem of minimum potential energy, and is sometimes derived in this manner.

The theorem of minimum potential energy is particularly suitable for approximate solutions by assuming approximate compatible displacements and finding the one for which the potential energy of the system is least. The *Ritz method*, which is used in Ex. 15.2, consists of assuming such a set of displacements in the form of a series with unknown coefficients, computing the corresponding potential energy, and then minimizing it with respect to the unknown coefficients, leading to a set of equations that may be solved for these coefficients.

A *Fourier series* of the form

$$\sum_n \left(a_n \sin\frac{n\pi x}{L} + b_n \cos\frac{n\pi x}{L} \right) \tag{15.10}$$

can be used to represent any function to the desired degree of accuracy by including a sufficient number of terms. A deflected shape expressed by this type of series is particularly convenient for expressing the potential energy, because it leads to integrals of the type

$$\int_{x=0}^{L} \sin^2 \frac{n\pi x}{L}\,dx = \int_{x=0}^{L} \cos^2 \frac{n\pi x}{L}\,dx \equiv \frac{L}{2} \qquad m = n$$

$$\int_{x=0}^{L} \sin \frac{m\pi x}{L} \sin \frac{n\pi x}{L}\,dx = \int_{x=0}^{L} \cos \frac{m\pi x}{L} \cos \frac{n\pi x}{L}\,dx \equiv 0, \qquad \text{for } m \neq n$$

$$(15.11)$$

This property of the Fourier series is called *orthogonality.*

The Fourier series can be used for representation of the deformed shape only when it satisfies the geometric boundary conditions of the problem, since this is a basic requirement of any displacement function used to compute the potential energy.

> ***Example 15.2:*** Using the first two terms of a Fourier representation of the deformed shape and the theorem of minimum potential energy, calculate the deflected shape of the beam under quarter-point load shown in part (a) of the figure.

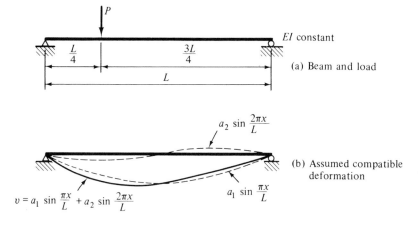

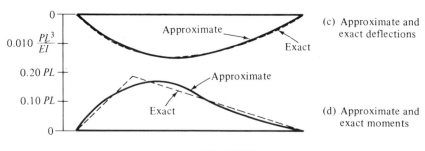

Fig. E15.2

Solution: We assume an approximate deflected shape of the form shown in part (b):

$$v = a_1 \sin \frac{\pi x}{L} + a_2 \sin \frac{2\pi x}{L}$$

We differentiate twice to obtain the approximate curvature:

$$\Phi = \frac{d^2 v}{dx^2} = -\left(\frac{\pi}{L}\right)^2 \left(a_1 \sin \frac{\pi x}{L} + 4a_2 \sin \frac{2\pi x}{L}\right)$$

Using this curvature, we compute the strain energy using Eq. (15.5) and the integrals given in Eq. (15.11):

$$U = \int_{x=0}^{L} \frac{EI}{2} \Phi^2 \, dx$$

$$= \frac{EI}{2}\left(\frac{\pi}{L}\right)^4 \int_{x=0}^{L} \left(a_1^2 \sin^2 \frac{\pi x}{L} + 8a_1 a_2 \sin \frac{\pi x}{L} \sin \frac{2\pi x}{L} + 16a_2^2 \sin^2 \frac{2\pi x}{L}\right) dx$$

$$= \frac{EI}{2}\left(\frac{\pi}{L}\right)^4 \left(a_1^2 \frac{L}{2} + 16a_2^2 \frac{L}{2}\right)$$

Because of the orthogonality of the Fourier series, the coupling terms vanish. The work of the applied load is

$$W = P \cdot v\left(x = \frac{L}{4}\right) = P\left(a_1 \sin \frac{\pi}{4} + a_2 \sin \frac{\pi}{2}\right) = P(.707 a_1 + 1.000 a_2)$$

so that the potential energy becomes

$$\Psi = (U - W) = \frac{EI}{2}\left(\frac{\pi}{L}\right)^4 \cdot L\left(\frac{1}{2}a_1^2 + 8a_2^2\right) - P(.707 a_1 + 1.000 a_2)$$

To minimize this quantity, we differentiate with respect to the unknown coefficients a_1 and a_2, and equate to zero:

$$\frac{\partial \Psi}{\partial a_1} = \frac{EI}{L^3}\frac{\pi^4}{2}a_1 - .707 P = 0$$

$$\frac{\partial \Psi}{\partial a_2} = \frac{EI}{L^3}8\pi^4 a_2 - P = 0$$

Each equation can be solved directly for the unknown coefficient:

$$a_1 = .0146\frac{PL^3}{EI}; \qquad a_2 = .0013\frac{PL^3}{EI}$$

Resubstituting these quantities into the original assumed deflection curve, we obtain

$$v = \left(.0146 \sin \frac{\pi x}{L} + .0013 \sin \frac{2\pi x}{L}\right)\frac{PL^3}{EI}$$

In particular, the approximate deflection under the load, at $x = L/4$, becomes

$$v\left(x = \frac{L}{4}\right) = .0116\frac{PL^3}{EI}$$

compared to the exact value of $.0117(PL^3/EI)$. The deflected shape is shown in part (c); part (d) shows in solid line the plot of the approximate moments derived by two successive differentiations of the approximate deflection function:

$$M = -EI\frac{d^2v}{dx^2} = PL\left(.144 \sin \frac{\pi x}{L} + .0512 \sin \frac{2\pi x}{L}\right)$$

The exact moment diagram is shown in dashed line.

We note that in Ex. 15.2 only two terms of the deflection series already yield remarkably good results, and that the approximate deflection is on the low side. Indeed, a solution obtained by minimizing the potential energy based on an approximate (but geometrically compatible) displacement function will always lead to an underestimation of the deflections, or an overestimation of the stiffness, a fact that can be explained physically by realizing that such an approximation amounts to introducing constraints on the deformed shape which prevent the full distortions of the structure. This conclusion will be useful in evaluating the results of the finite-element method discussed in Chapter 16.

Finally, a word of warning: Any approximations in the deflected shape will be amplified upon differentiation; thus, the errors in forces (such as shear or moments) obtained by differentiation of approximate displacements (such as that of Ex. 15.2) will be greater than those of the displacements. This is clearly shown in part (d) of Ex. 15.2.

2. *Castigliano's first theorem.* We write Eq. (15.9) in the form

$$dU = dW$$

where $W = P \cdot \Delta_P$. Therefore,

$$dW = P \cdot d\Delta + \Delta \cdot dP$$

Since the applied loads P are specified, $dP = 0$ and the second term vanishes, leaving

$$dU = P \cdot d\Delta$$

or

$$P = \frac{dU}{d\Delta} \tag{15.12}$$

Equation (15.12), which is called *Castigliano's first theorem*, is an equilibrium equation in terms of displacements, since it is based on the theorem of virtual displacements. Its use will be demonstrated in the following example.

Example 15.3: Compute the stiffness K_{11} of the three-member truss shown in part (a) of the figure.

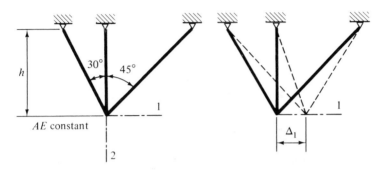

(a) Structure and nodes   (b) Compatible displacements

Fig. E15.3

Solution: We recognize the stiffness K_{11} as the force P necessary to cause a unit displacement Δ_1, all other displacements equal to zero. Accordingly, the displaced shape is as shown in part (b), which as already computed in Ex. 15.1 causes the strain energy

$$U = \frac{AE}{2h}(.570\Delta_1^2)$$

Equation (15.12) yields the force

$$P = \frac{dU}{d\Delta} = \frac{AE}{h}.570\Delta_1$$

The stiffness is the value of P corresponding to $\Delta_1 = 1$; thus

$$K_{11} = .570\frac{AE}{h}$$

15.4. Two Methods Based on the Theorem of Virtual Forces

1. *Theorem of minimum complementary energy.* In the virtual work equation, Eq. (15.1), we consider two sets of virtual forces X_1^{**} and X_2^{**}, which are

in equilibrium with the specified loads, as, for instance, shown by the force systems 1 and 2 in the beam of Fig. 15.2(b); the corresponding virtual internal stresses are σ_1^{**} and σ_2^{**}. The real displacements are identical in each case, equal to the real displacements Δ^* and the corresponding strains ϵ^*.

We now write the virtual work equations for each of the two virtual force systems riding through the real displacements:

$$\int_V \epsilon^* \sigma_1^{**} \, dV = \sum \Delta^* X_1^{**}$$

$$\int_V \epsilon^* \sigma_2^{**} \, dV = \sum \Delta^* X_2^{**}$$

Subtracting the first from the second equation, we obtain

$$\int_V \epsilon^* (\sigma_2^{**} - \sigma_1^{**}) \, dV = \sum \Delta^* (X_2^{**} - X_1^{**})$$

or, denoting the differences

$$(\sigma_2^{**} - \sigma_1^{**}) \equiv d\sigma; \qquad (X_2^{**} - X_1^{**}) \equiv dX$$

and transposing, we obtain

$$\int_V \epsilon \, d\sigma \, dV - \sum \Delta \, dX = 0 \tag{15.13}$$

We now integrate both terms of Eq. (15.13) over the loading path, and then write the differential to obtain

$$d \left[\int_{\sigma=0}^{\sigma} \left(\int_V \epsilon \, d\sigma \, dV \right) - \sum \int_{X=0}^{X} \Delta \, dX \right] = 0 \tag{15.14}$$

We recognize by comparison with Eq. (15.3) that the first integral represents the complementary strain energy, $\bar{U}$, and the second the work $\bar{W}$ done by the virtual forces riding through the real displacements; as before, we can split this external virtual work into the part done at the load points and that done at the supports, so that we can write the second term as

$$\sum \int_{X=0}^{X} \Delta \, dX = \int_{X=0}^{X} \left(\sum \Delta_P \, dP + \sum \Delta_z \, dR \right)$$

where dP and Δ_P are the difference of virtual loads, and real displacements, respectively, at the load points, and dR and Δ_R are the corresponding terms at the support points. Because the virtual load system was assumed in equilibrium with the real specified loads, the quantity $dP = 0$, and the first term vanishes. The external work is thus due only to that of the virtual reactions dR.

We now define the *complementary energy* $\bar{\Psi}$ of the system as

$$\bar{\Psi} = \bar{U} - \bar{W}$$

where $\bar{U}$ is the complementary strain energy due to a force system in equilibrium with the applied loads and $\bar{W}$ is the work of the reactions. Then we write Eq. (15.14) in the form

$$d\bar{\Psi} = d(\bar{U} - \bar{W}) = 0 \qquad (15.15)$$

We recognize this as a condition for an extremum of the function $\bar{\Psi}$. Since it is based on the theorem of virtual forces, it expresses a condition of geometry, and we can state it in words as follows:

Of all equilibrium states, that one which also extremizes the complementary energy $\bar{\Psi}$ will also satisfy the conditions of geometry of the system.

It can be shown that the complementary energy is unbounded in the positive sense, as shown in Fig. 15.3(b), so that the only extremum is of minimum value. For this reason the above statement, expressed by Eq. (15.15), is called the *theorem of minimum complementary energy*; it provides a geometrical condition in terms of forces.

By the uniqueness of solutions of linearly elastic problems, any solution satisfying both equilibrium and geometry must be the correct one. For this reason, the minimization of the complementary energy of an equilibrium system must lead to the correct answer. The following example will demonstrate this approach in solving for the reactions of a statically indeterminate structure.

Example 15.4: Determine the value of the reaction S_1 of the three-member truss of part (a) of the figure due to the applied loads P_1 and P_2, and due to a support displacement $-\Delta$ of member 3, as shown.

Solution: The complementary strain energy of the system can, according to Eq. (15.6), be written as

$$\bar{U} = \frac{1}{2} \sum \frac{L}{AE} S^2 = \frac{h}{2AE}(1.154 S_1^2 + 1.000 S_2^2 + 1.414 S_3^2)$$

where we now seek to express all member forces in terms of the desired quantity S_1. Summing forces along axes 1 and 2, respectively, of the free body of part (b), we obtain

$$\sum F_1 = 0: \quad S_2 = P_1 + P_2 - 1.366 S_1$$
$$\sum F_2 = 0: \quad S_3 = .707 S_1 - 1.414 P_1$$

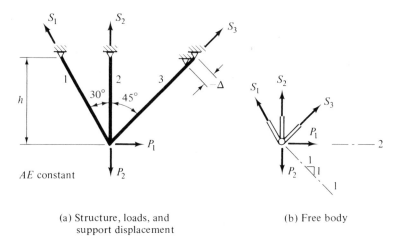

(a) Structure, loads, and
support displacement

(b) Free body

Fig. E15.4

and substituting these values we find $\bar{U}$ in terms of S_1:

$$\bar{U} = \frac{h}{2AE}[1.154S_1^2 + (P_1 + P_2 - 1.366S_1)^2 + 1.414(.707S_1 - 1.414P_1)^2]$$

The external work of the reactions riding through the specified support displacement Δ is

$$\bar{W} = -S_3\Delta = -(.707S_1 - 1.414P_1)\Delta$$

so that the complementary energy of the system is

$$\bar{\Psi} = (\bar{U} - \bar{W})$$
$$= \frac{h}{2AE}[1.154S_1^2 + (P_1 + P_2 - 1.366S_1)^2 + 1.414(.707S_1 - 1.414P_1)^2]$$
$$+ [(.707S_1 - 1.414P_1)\Delta]$$

To minimize $\bar{\Psi}$ with respect to S_1, we set

$$\frac{d\bar{\Psi}}{dS_1} = \frac{h}{AE}(3.726S_1 - 2.780P_1 - 1.366P_2) + .707\Delta = 0$$

from which

$$\underline{S_1 = .739P_1 + .362P_2 - .190\Delta\frac{AE}{h}}$$

2. *Castigliano's second theorem.* We write Eq. (15.15) in the form

$$d\bar{U} = d\bar{W}$$

where $$\bar{W} = R \cdot \Delta_R$$

Therefore,

$$d\bar{W} = R \cdot d\Delta_R + \Delta_R \cdot dR$$

Since the displacements at support points, Δ_R, are specified, $d\Delta_R = 0$ and the first term vanishes, leaving

$$d\bar{W} = \Delta \cdot dR$$

or
$$\Delta = \frac{d\bar{U}}{dR} \tag{15.16}$$

Equation (15.16), which is called Castigliano's second theorem, is a geometrical relation in terms of forces, since it is based on the theorem of virtual forces. Its use will be demonstrated in the following example.

Example 15.5: Find the reaction R of the propped-cantilever beam of part (a) of the figure.

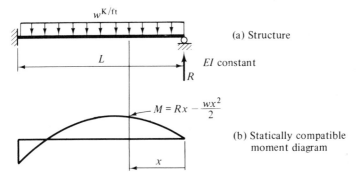

Fig. E15.5

Solution: The moment variation $M(x)$ in terms of the unknown reaction R, as shown by the moment diagram of part (b), is

$$M = Rx - \frac{wx^2}{2}$$

and the complementary strain energy is, according to Eq. (15.5),

$$\bar{U} = \frac{1}{2} \int_{x=0}^{L} \frac{M^2}{EI}\,dx = \frac{1}{2EI}\left(\frac{1}{3}R^2L^3 - \frac{1}{4}RwL^4 + \frac{1}{20}w^2L^5\right)$$

Differentiating $\bar{U}$ with respect to R, we obtain

$$\Delta_R = \frac{d\bar{U}}{dR} = \frac{1}{2EI}\left(\frac{2}{3}RL^3 - \frac{1}{4}wL^4\right)$$

from which, since the displacement Δ_R of the support point is specified zero,

$$\Delta_R = \frac{1}{2EI}\left(\frac{2}{3}RL^3 - \frac{1}{4}wL^4\right) = 0$$

Therefore,

$$R = \frac{3}{8}wL$$

With a greater number of redundants R_i, and corresponding specified displacements Δ_i, the differentiation of Eq. (15.16) would be carried out with respect to each one of the redundants, leading to a set of equations identical to the conditions of geometry, Eq. (10.1), of the force method. Thus, Castigliano's second theorem provides an alternative derivation of the force method of solving statically indeterminate problems, as shown in the following example.

Example 15.6: Find the rotational stiffnesses k_{11} and k_{21} of the prismatic beam with the nodal numbering shown in part (a) of the figure.

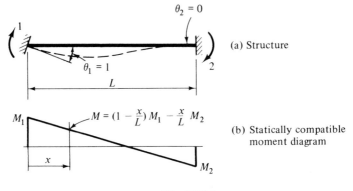

(a) Structure

(b) Statically compatible moment diagram

Fig. E15.6

Solution: We write the statically compatible moments in terms of the unknown end moments M_1 and M_2, as shown in part (b):

$$M = \left(1 - \frac{x}{L}\right)M_1 - \frac{x}{L}M_2$$

The complementary energy of the beam, according to Eq. (15.5), is

$$\bar{U} = \frac{1}{2EI}\int_0^L M^2\,dx = \frac{1}{2EI}\int_0^L \left[\left(1 - \frac{x}{L}\right)M_1 - \frac{x}{L}M_2\right]^2 dx$$

$$= \frac{1}{2EI}\left(\frac{L}{3}M_1^2 + \frac{L}{3}M_2^2 - \frac{L}{3}M_1M_2\right)$$

Differentiating partially with respect to the two redundant end moments, and substituting the specified end rotations $\theta_1 = 1$, $\theta_2 = 0$, we obtain the set of conditions of geometry:

$$\theta_1 = \frac{\partial \bar{U}}{\partial M_1} = \frac{1}{2EI}\left(\frac{2}{3}L \cdot M_1 - \frac{1}{3}L \cdot M_2 \right) = 1$$

$$\theta_2 = \frac{\partial \bar{U}}{\partial M_2} = \frac{1}{2EI}\left(-\frac{1}{3}L \cdot M_1 + \frac{2}{3}L \cdot M_2 \right) = 0$$

which can be solved for the end moments:

$$M_1 = k_{11} = \frac{4EI}{L}$$

$$M_2 = k_{21} = \frac{2EI}{L}$$

15.5. Derivation of the Theorem of Virtual Work for Beams

The theorem of virtual work can be derived in general for deformable bodies, but this involves some advanced mathematical concepts; to avoid these, we shall restrict ourselves to a derivation for plane members in pure flexure, in order to gain additional insight into the nature of the method. The technique used is suitable for the derivation of the theorem for any member subject to uniaxial stress states.

Before proceeding, we shall recall some facts of the theory of bending. By purely statical means, the relation between the bending moment M, the shear V, and the load w is given by

$$w = \frac{dV}{dx} = \frac{d^2M}{dx^2}; \qquad V = \frac{dM}{dx} \tag{15.17}$$

Furthermore, Figs. 15.4(a) and (b) show the situation in the vicinity of a concentrated force P and a cranked-in moment M_0; from equilibrium of the free bodies,

$$\sum F = 0: \qquad P = V_B^+ - V_B^- \tag{15.18a}$$

$$\sum M = 0: \qquad M_0 = M_C^+ - M_C^- \tag{15.18b}$$

By purely geometrical means, the relation between the displacement v, the slope θ, and the curvature Φ is, for small displacements,

$$\Phi = \frac{d\theta}{dx} = \frac{d^2v}{dx^2}; \qquad \theta = \frac{dv}{dx} \tag{15.19}$$

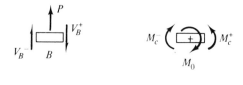

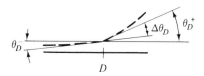

(a) Applied load (b) Applied moment

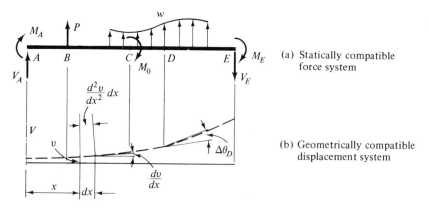

(c) Concentrated kink

Fig. 15.4 Beam discontinuities.

In the case of a kink of value $\Delta\theta_D$ in the displaced curve at point D, Fig. 15.4(c) shows that

$$\Delta\theta_D = \theta_D^+ - \theta_D^- = \left(\frac{dv}{dx}\right)_{D^+} - \left(\frac{dv}{dx}\right)_{D^-} \qquad (15.20)$$

where the superscripts $+$ and $-$ indicate quantities immediately to the right and to the left of section D.

We now consider a beam as in Fig. 15.5(a), subject to a distributed load w, a concentrated load P applied at point B, and a concentrated moment M_0 applied at point C. These forces are all in equilibrium so that Eqs. (15.17) and (15.18) apply. We call these forces a *statically compatible* or *equilibrium* system.

(a) Statically compatible force system

(b) Geometrically compatible displacement system

Fig. 15.5 Force and displacement systems.

Next, and entirely independent of the force system, we consider the beam subjected to a set of displacements that are continuous at least over finite segments so that the slope θ and the curvature Φ can be determined by the differentiations of Eq. (15.19). In this case, we shall assume a deformed shape that is continuous at all points except at D, where it contains a kink of value $\Delta\theta_D$, as shown in Fig. 15.5(b). We call these displacements a *geometrically compatible* or *kinematically compatible* system.

We now set up a quantity composed of the internal moment M of the force system, multiplied by the corresponding curvature Φ of the independent displacement system, integrated over the length of the beam. We recognize this quantity as the internal virtual work in flexure, which has been discussed in Sec. 8.1:

$$\text{IVW} = \int_L M\Phi \, dx = \int_L M\frac{d^2v}{dx^2} \, dx \tag{15.21}$$

where we must remember to break the integral at each point of discontinuity of moment or curvature, such as points B, C, and D in Fig. 15.5(a).

To prove the theorem of virtual work, we now resort to the process of integration by parts, according to the formula

$$\int_A^B f \cdot dg = (f \cdot g)\Big|_A^B - \int_A^B g \cdot df \tag{15.22}$$

Applying this expression to Eq. (15.21), we let

$$f \equiv M; \qquad df = \frac{dM}{dx} dx = V \, dx$$

$$dg \equiv \frac{d^2v}{dx^2} dx; \qquad g = \frac{dv}{dx} = \theta$$

so that Eq. (15.21) becomes

$$\int_L M \cdot \frac{d^2v}{dx^2} dx = (M \cdot \theta)\Big| - \int \frac{dv}{dx} V \, dx \tag{15.23}$$

where the first term on the right-hand side must be evaluated separately at the limits of each region of continuity, that is, from A to B, B to C, and so on.

The second term on the right-hand side of Eq. (15.23) is subjected once more to an integration by parts according to Eq. (15.22), where now

$$f \equiv V; \qquad df = \frac{dV}{dx} dx = w \, dx$$

$$dg \equiv \frac{dv}{dx} dx; \qquad g = v$$

and Eq. (15.23) becomes now

$$\int_L M \frac{d^2v}{dx^2}dx = (M \cdot \theta)\Big| -(V \cdot v)\Big| + \int_L w \cdot v \, dx \qquad (15.24)$$

We now evaluate the first term on the right-hand side of Eq. (15.24) at the limits of the four separate regions of continuity:

$$(M \cdot \theta)\Big|_A^B + (M \cdot \theta)\Big|_B^C + (M \cdot \theta)\Big|_C^D + (M \cdot \theta)\Big|_D^E$$
$$= (M_B^-\theta_B^- - M_A\theta_A) + (M_C^-\theta_C^- - M_B^+\theta_B^+) + (M_D^-\theta_D^- - M_C^+\theta_C^+)$$
$$+ (M_E\theta_E - M_D^+\theta_D^+)$$

At point B, the moments M_B^- and M_B^+ as well as the slopes θ_B^- and θ_B^+ are equal, so that the first and fourth terms cancel. At point C, $\theta_C^- = \theta_C^+ = \theta_C$ because of the continuity of the deformed shape; but because of the cranked-in moment, $M_0 = M_C^+ - M_C^-$, according to Eq. (15.18b), so that

$$M_C^-\theta_C^- - M_C^+\theta_C^+ = -M_0\theta_C$$

which we recognize as the work done by the cranked-in moment M_0 rotating through the slope θ_C of the independent deformed shape.

At point D, equilibrium demands that $M_D^+ = M_D^- = M_D$, but due to the kink in the deformed shape, $\Delta\theta_D = \theta_D^+ - \theta_D^-$, so that the fifth and eighth terms become

$$M_D^-\theta_D^- - M_D^+\theta_D^+ = -M_D \Delta\theta_D$$

The first term on the right-hand side of Eq. (15.24) now becomes

$$-M_A\theta_A - M_0\theta_C + M_D \Delta\theta_D + M_E\theta_E$$

Similar reasoning involving Eq. (15.18a) reduces the second term on the right-hand side of Eq. (15.24) upon evaluation at the limits to

$$P \cdot v_B + V_A \cdot v_A - V_E \cdot v_E$$

Substituting these two expressions into Eq. (15.24), we obtain

$$\int_L M \frac{d^2v}{dx^2} dx = -M_A\theta_A - M_0\theta_C - M_D\Delta\theta_D + M_E\theta_E + Pv_B + V_Av_A - V_E\dot{v}_E$$
$$+ \int_L w \, v \, dx \qquad (15.25)$$

We now identify the various terms as belonging either to the internal or to the external part of the virtual work, and place the former on the left-hand, the latter on the right-hand side of the equation; thus, M_A, M_E, V_A, and V_E are, respectively, the moments and shears of the equilibrium system acting on the ends of the beam segment, so that the terms $-M_A\theta_A$, $M_E\theta_E$, $V_A v_A$, and $-V_E v_E$ contribute to the external virtual work. Similarly, the term $\int_L w\,v\,dx$ denotes the external virtual work due to the applied distributed load, and $P v_B$ that due to the load P. All these external virtual work terms remain on the right-hand side. The term $-M_D \cdot \Delta\theta_D$ is the internal work done by the internal moment M_D riding through the kink of value $\Delta\theta_D$, and thus should be listed with appropriate change of sign, on the left, or internal-virtual-work, side of the equation.

Having thus classified the various work quantities and shown the equality of those defined external to those defined internal, we deduce that for this case

Internal Virtual Work = External Virtual Work

We can now make the following observations regarding the theorem of virtual work derived in this manner:

1. There is no relation required between the force system and the displacement system; the theorem is thus not limited to any specific force–displacement relation.

2. The force system must be an equilibrium system so that Eqs. (15.17) and (15.18) can be used.

3. The geometrical system need not satisfy any physical constraint conditions, and may have finite discontinuities.

4. No real energy quantities were used in the derivation; only the terms "internal virtual work" and "external virtual work" were established by definition.

We have thus, for this particular case, proved the theorem of virtual work, which was first stated in Chapter 5. The following example extends the basic idea to a different situation.

Example 15.7: Establish, by integration by parts, the validity of the theorem of virtual work for the torsion member of part (a) of the figure subjected to a distributed applied torque $t(x)$ and to a concentrated torque T_B.

Solution: We consider the displacement field consisting of the twisted rod shown in part (b), with the continuously varying angle of twist $\theta(x)$. The relation between the torque $T(x)$ at any point in the rod and the applied distributed torque $t(x)$ is, by the statics of the free body of part (c), $t = dT/dx$; similarly, torsional equilibrium of the free body of point B is $T_B = T_B^+ - T_B^-$.

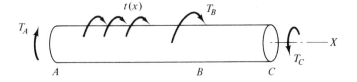

(a) Statically compatible
force system

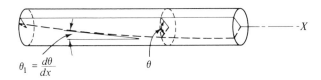

(b) Geometrically compatible
displacement system

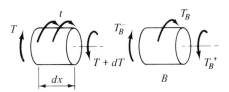

(c) Free bodies

Fig. E15.7

The geometry of the twisted bar relates the unit angle of twist, θ_1, to the total angle of twist by $\theta_1 = d\theta/dx$. The internal virtual work can now be written as

$$\text{IVW} = \int_L T \cdot \theta_1 \, dx = \int_L T \cdot \frac{d\theta}{dx} \, dx$$

An integration by parts (letting $T \equiv f$, $\frac{d\theta}{dx} \, dx \equiv dg$) leads to

$$\text{IVW} = (T \cdot \theta) \Big|_A^B + (T \cdot \theta) \Big|_B^C - \int_L \frac{dT}{dx} \theta \, dx$$

$$= (T_B^- \theta_B^- - T_A \theta_A) + (T_C \theta_C - T_B^+ \theta_B^+) - \int_L t \cdot \theta \, dx$$

$$= -T_A \theta_A - T_B \theta_B + T_C \theta_C - \int_L t \cdot \theta \, dx = \text{EVW}$$

where the first three terms give, with due respect to signs, the virtual work of the external torques T_A, T_B, and T_C, and the last term represents the virtual external work done by the distributed applied torque $t(x)$.

This concludes the proof for this case.

In generalizing the proof to other types of structural members, we consider generalized internal forces Q_{int}, from which corresponding external forces Q_{ext} can always, by purely statical means, be obtained by n successive differentiations:

$$Q_{ext} = \frac{d^n Q_{int}}{dx^n} \qquad (15.26)$$

This type of relation is, for instance, exemplified by Eq. (15.17).

Similarly, we consider generalized external displacements q_{ext}, from which corresponding internal distortions q_{int} can always, by purely geometrical means, be obtained by n successive differentiations:

$$q_{int} = \frac{d^n q_{ext}}{dx^n} \qquad (15.27)$$

This type of relation is, for instance, exemplified by Eq. (15.19).

For corresponding forces and displacements, the theorem of virtual work would be written as

$$\int_L Q_{ext} q_{ext}\, dx = \int_L Q_{int} q_{int}\, dx$$

or, in view of Eqs. (15.26) and (15.27), as

$$\int_L \frac{d^n Q_{int}}{dx^n} q_{ext}\, dx = \int Q_{int} \frac{d^n q_{ext}}{dx^n}\, dx$$

Both sides of this equation lend themselves to integrations by parts, and the identity of both sides of the equation can be proved in this manner, as was done in the specific cases discussed in this section.

For solid bodies under more general states of stress, the derivation of the theorem of virtual work depends upon the divergence theorem, which is a generalization of the method of integration by parts used previously. Just as the integration by parts serves to relate internal to external forces, so the divergence theorem serves to relate a function defined in the interior of the body (the internal stresses) to a function defined on the surface of the body (the applied forces). For the general derivation, the reader is referred to books on continuum mechanics.

PROBLEMS

1. Derive an expression for the strain energy of a beam of length L, cross-sectional area A, and shear modulus G, due to the shear deformations occurring in consequence of a shear force $V(x)$ along the beam. Assume the shear stresses are uniformly distributed across the beam section.

2. Repeat Prob. 1 for the case of a beam of rectangular cross section of width b and depth d, in which, according to basic beam theory, the shear stresses τ vary parabolically with the distance y from the neutral axis:

$$\tau = \frac{3}{2}\frac{V}{A}\left[1 - \left(\frac{2y}{d}\right)^2\right]$$

3. Derive an expression for the strain energy of an element of length dl of a circular bar of radius R and shear modulus G, due to an applied torque T.

4. Compute the strain energy of the cantilever beam of length L and rectangular cross section $b \times d$
   a. Due to bending only.
   b. Due to bending and shear; consider the parabolic distribution of shear stresses as in Prob. 2.
   c. Plot the ratio of the strain energy due to shear to the total strain energy with varying beam proportion d/L.

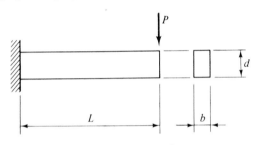

Prob. 4

5. Compute the strain energy due to bending and torsion of the semicircular cantilever bow girder under the concentrated load P. The girder is of thin walled pipe, of $G = \frac{1}{2}E$.

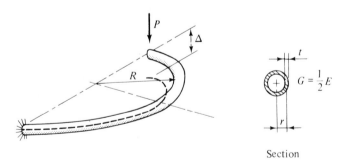

Section

Prob. 5

6. a. Compute the strain energy of an elastic spring of modulus k due to an applied force S.
   b. Apply the theorem of minimum potential energy to determine the deflection Δ of point B of the idealized structure shown, assuming that bar AB is

perfectly rigid. From the result, compute the spring forces, and draw the moment diagram for bar AB.

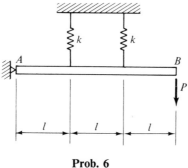

Prob. 6

7. By minimizing the potential energy of the system of a rigid bar on two springs, determine the deflection of point B and the slope of the bar.

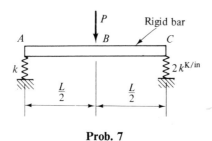

Prob. 7

8. By assuming a deflection expressible as a three-term Fourier sine series

$$V = a_1 \sin \frac{\pi x}{L} + a_2 \sin \frac{2\pi x}{L} + a_3 \sin \frac{3\pi x}{L}$$

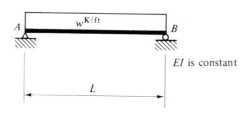

Prob. 8

determine the potential energy of the uniformly loaded beam of length L, considering only bending distortions.

9. a. Compute the strain energies $\begin{Bmatrix} U_1 \\ U_2 \end{Bmatrix}$ of the spring system with the nodal numbering shown, associated with the displacements $\begin{Bmatrix} \Delta_1 \\ \Delta_2 \end{Bmatrix}$ applied one

at a time. From these, calculate the total strain energy of the system.

b. By appropriate differentiation of the strain energy according to Castigliano's first theorem, compute the stiffness matrix of the system

$$[K] = \begin{bmatrix} K_{11} & K_{12} \\ K_{21} & K_{22} \end{bmatrix}.$$

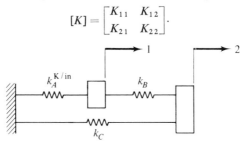

Prob. 9

10. a. Compute the spring rotations and moments due to nodal displacements Δ_1 and Δ_2 applied one at a time. From these, calculate the total strain energy of the discretized beam shown as a function of Δ_1 and Δ_2.

    b. Compute the stiffness matrix $[K] = \begin{bmatrix} K_{11} & K_{12} \\ K_{21} & K_{22} \end{bmatrix}$ by application of Castigliano's first theorem.

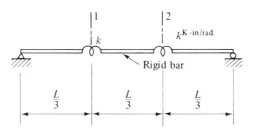

Prob. 10

11. Compute the strain energy due to nodal displacements of the cantilever beam shown, and compute the stiffness matrix $[K] = \begin{bmatrix} K_{11} & K_{12} \\ K_{21} & K_{22} \end{bmatrix}$ by appropriate application of Castigliano's first theorem.

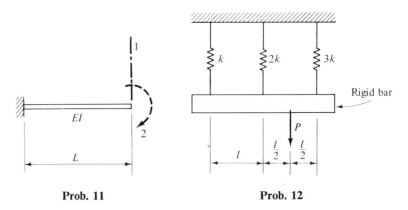

Prob. 11   **Prob. 12**

12. For the three-spring assembly shown, compute the complementary energy due to load *P*. By minimization of the complementary energy, determine all spring forces.

13. Analyze the beam shown by minimizing the complementary energy of the system due to
a. The linearly varying load shown.
b. A support settlement Δ of point *B*.

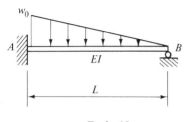

Prob. 13   **Prob. 14**

14. Determine the fixed-end moments of the beam under linearly varying load by minimizing the complementary energy of the system.

15. By applying the theorem of minimum complementary energy, analyze the propped semicircular cantilever bow girder due to
a. Its dead load, of amount *w* kips/ft.
b. A support settlement of amount Δ at point *B*.
Consider bending and torsional strain energy.

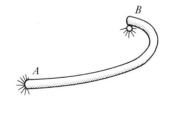

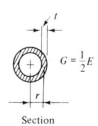

Section

Prob. 15

16. Analyze the propped cantilever beam under linearly varying loading by use of Castigliano's second theorem.

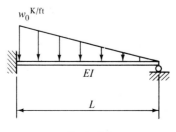

Prob. 16

17. Repeat Prob. 13, using this time Castigliano's second theorem.

18. Repeat Prob. 15, using this time Castigliano's second theorem.

19. By application of Castigliano's second theorem, determine the flexibility matrix $[F] = \begin{bmatrix} F_{11} & F_{12} & F_{13} \\ F_{21} & F_{22} & F_{23} \\ F_{31} & F_{32} & F_{33} \end{bmatrix}$ of the prismatic cantilever beam with the nodes shown.

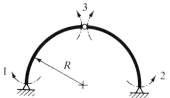

Prob. 19

20. Determine the flexibility matrix $[F]$, of order (3×3), for the semicircular three-hinged arch by use of the second theorem of Castigliano.

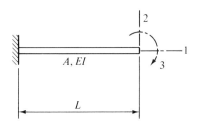

EI is constant

Prob. 20

21. Determine the (6×6) flexibility matrix $[F]$ for the semicircular cantilever bow girder by applying Castigliano's second theorem.

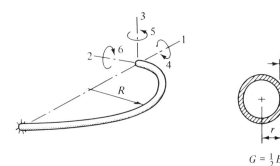

$G = \frac{1}{2}E$

Prob. 21

CHAPTER 16

The Finite-Element Method

The finite-element method, as presented here, is an extension of the matrix displacement method, as already covered in Chapter 13, applicable to the analysis of any solid body. This generalization depends on discretization of the body into finite elements and on the ability to formulate stiffness matrices for these elements.

The basis of the method is covered in Sec. 16.1, and a general method for the determination of stiffness matrices is explained in Sec. 16.2 in terms of a simple beam element, and extended in Secs. 16.3 and 16.4 to an element in a state of plane stress, to exemplify the further use of the procedure. Sections 16.5 to 16.8 present additional aspects of the finite-element method; it will be seen that in the main these are just restatements of ideas already covered in Chapter 13. Section 16.9, finally, discusses the formulation of stiffness matrices applicable to a wider class of structures, thus pointing the way toward a unified method of stress analysis, which should be pursued in advanced texts.

16.1. Basic Approach

The finite-element method is a generalization of the previously covered matrix methods that can be applied to analyze any solid body. The basic idea is shown in Figs. 16.1(a) and (b): The body to be analyzed is modeled as an assembly of finite elements interconnected at specified nodal points.

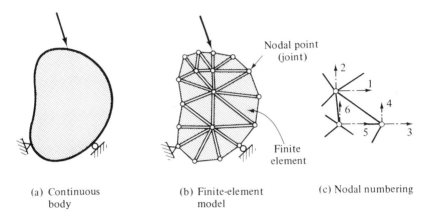

(a) Continuous   (b) Finite-element   (c) Nodal numbering
    body   model

Fig. 16.1 Finite-element model.

In the case of the plane-stress body shown in Fig. 16.1, a decomposition into triangular elements is convenient because of the ease with which such elements can simulate irregular boundaries. They also permit variation of the element size, with the smallest elements in regions of stress concentration and larger elements in areas of more regular stress variation.

In a force-method approach, it will be necessary to decompose the idealized structure into a statically determinate primary system to determine the displacements due to the loads and unit redundants (the flexibilities) of the primary system, and to compute those values of the redundants necessary to satisfy the geometry of the system.

In a displacement-method approach, the individual elements are constrained against displacements of their nodes to form a geometrically determinate primary system; the forces due to loads and unit nodal displacements (the stiffnesses) of the primary system are determined, and those values of the displacements necessary to satisfy the equilibrium of the system are computed.

In either case, a sufficiently fine decomposition of the structure into finite elements requires keeping track of such a great number of quantities that an adequate bookkeeping scheme is essential. Such a scheme is provided by the matrix method, to be used in conjunction with appropriate computing equipment.

In comparing the application of the force and the displacement methods to problems of this type, it appears that the latter has distinct advantages. In the force method, considerable judgment is necessary to reduce the structure to its statically determinate primary form, and the establishment of the [b] matrix expressing the equilibrium equations is by no means straightforward. On the other hand, the kinematically determinate primary structure required by the displacement method lends itself to a convenient automatic treatment, so that for practical applications of the finite-element method the displacement method is preferred. Further treatment in this chapter will therefore deal only with the displacement finite-element method.

An essential requirement of such a numerical method is that of convergence; that is, the solution should come closer and closer to the correct one for the prototype structure with increasingly finer subdivision into finite elements. Conditions under which this convergence is achieved will be discussed in Sec. 16.8.

As the specified loads are applied to the model of Fig. 16.1(b), the boundaries of adjacent elements between the nodal points would tend to open up or to overlap. It is therefore to be expected that the analysis of such a finite-element model would lead to a less stiff solution, with larger deformations, than the exact solution of the prototype body. To avoid such discontinuities of the deformed body, it is necessary to model the element behavior in such a fashion that the common boundaries of adjacent elements will deform together; such elements are called *compatible elements*. This matter will be discussed further in Sec. 16.8.

In the stiffness method of analysis, the displacements of the nodal points (or joints) are the primary unknowns. To label these unknowns clearly, it is necessary to introduce a node for each displacement component, as shown in Fig. 16.1(c).

According to the discussion of Sec. 13.4, there will also be a force quantity called *nodal force* associated with each of these nodes; at each node, either the displacement or the force will be specified, the other quantity being unknown. In assigning a numbering system to the nodes, those with unknown displacement should be numbered first.

Let us now assume that it is possible to determine structural stiffness values, K_{ij}, defined as in Sec. 13.3; that is, K_{ij} is the force along node i associated with a unit displacement of node j, all other nodal displacements equal to zero. The determination of the structure stiffness matrix $[K]$ consisting of elements K_{ij} will be covered in Secs. 16.2 to 16.8.

The structure stiffness matrix $[K]$ relates the nodal forces and displacements according to Eq. (13.12), in which the nodal numbering, as mentioned above, has been so arranged that the unknown displacements, Δ_α, are on top, and the unknown (reactive) nodal forces, X_β, at the bottom:

$$\begin{bmatrix} K_{\alpha\alpha} & K_{\alpha\beta} \\ \hline K_{\beta\alpha} & K_{\beta\beta} \end{bmatrix} \begin{bmatrix} \Delta_\alpha \\ \hline \Delta_\beta \end{bmatrix} = \begin{bmatrix} X_\alpha \\ \hline X_\beta \end{bmatrix} \qquad (13.12)$$

The solution for the unknown nodal displacements and the unknown nodal forces takes place according to the two-step scheme outlined in Sec. 13.4, leading to Eqs. (13.13) and (13.14):

$$[\Delta_\alpha] = [K_{\alpha\alpha}]^{-1}[[X_\alpha] - [K_{\alpha\beta}][\Delta_\beta]] \qquad (13.13)$$

$$[X_\beta] = [K_{\beta\alpha}][\Delta_\alpha] + [K_{\beta\beta}][\Delta_\beta] \qquad (13.14)$$

The solution of Eq. (13.13) involves the inversion of a matrix of order equal to the number of unknown displacements. For realistic problems, the number of unknown displacements can range from the order of tens to the order of thousands. It is apparent that problems of this type can be tackled only with the help of adequate computing equipment.

Once the reactions and displacements have been determined, it remains to solve for the internal stresses and strains. Since this involves use of the element stiffness matrices, its discussion will be postponed until Sec. 16.6.

It should be clear from the above outline that the only difference between

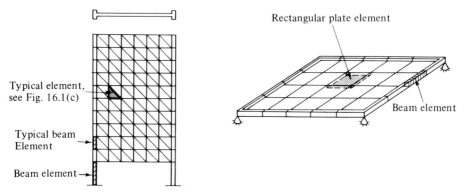

Typical element,
see Fig. 16.1(c)

Typical beam
Element

Beam element

Rectangular plate element

Beam element

(a) Deep beam structure

120 triangular elements
22 beam elements
182 nodes

(b) Plate with edge beam

20 rectangular plate elements
18 beam elements
120 nodes

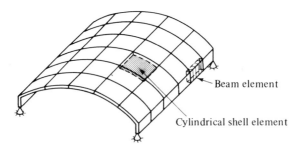

Beam element

Cylindrical shell element

(c) Cylindrical shell with edge beams
30 shell elements
12 beam elements
210 nodes

Fig. 16.2 Typical finite-element models.

the matrix displacement method of framed structures, as outlined in Sec. 13.4, and the finite-element method is the choice of the stiffness matrix; this enables different types of elements to be incorporated into the analysis in a unified fashion, and contributes greatly to the usefulness of the approach.

It also follows that the general outline of operations and the comments at the end of Sec. 13.5 apply to the finite-element displacement method. The remainder of this chapter will be devoted mainly to a method for determining the stiffness matrices of general elements.

A crucial aspect is the subdivision of the structure to be analyzed into the finite-element model. Figure 16.2 shows some typical discretizations; Fig. 16.2(a) illustrates a finite-element model of a wall on column supports, Fig. 16.2(b) a plate framed between edge beams, and Fig. 16.2(c) a cylindrical shell with edge beams. We note that each of these structures consists of different types of elements. One great asset of the method is its ability to handle any combinations' of different element types in a unified fashion.

Figure 16.3 shows an application of the finite-element method to the analysis of a modern airplane structure. The airplane shown in Fig. 16.3(a) indicates the difficulty of appropriate modeling; here, as a first step, the total structure is decomposed into the substructures shown in Fig. 16.3(b). Each substructure is in turn idealized by finer elements, as shown for a portion of the wing in Fig. 16.3(c). In aerospace structures, the dynamic behavior is all-important; in this case, one of the possible modes of vibration as determined from a dynamic finite-element analysis is shown in Fig. 16.3(d). Given such displaced shapes, we have learned in Sec. 13.4 how to determine the resulting internal forces required for design.

16.2. Element Stiffness Matrices by Assumed Displacement Function

For finite-element analysis of elastic structures, stiffness matrices should be determined in accordance with the laws of the theory of elasticity. Since an exact analysis of elastic elements, such as the triangular plane-stress element of Fig. 16.7, is a forbidding task, an approximate method is used that is completely general in its basic formulation and whose results will converge to the exact solution with decreasing element size; accordingly, the larger the number of elements, the more exact will be the solution.

This method is based on the theorem of virtual displacements, Eq. (9.2). We consider real external nodal element forces that arise due to unit nodal displacements, $[\Delta^*] = [I]$, applied one at a time; we recognize these forces as the desired element stiffnesses, $[k]$. The corresponding real stresses in the element may vary with location, and will thus be denoted by $[\sigma(x)^*]$.

The virtual nodal displacements will, likewise, be assumed of unit value, applied one at a time; thus the virtual $[\Delta^{**}] = [I]$; the corresponding com-

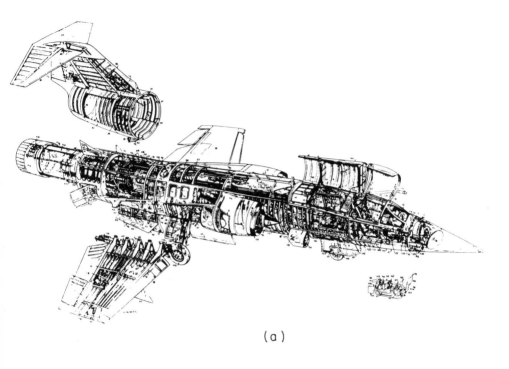

(a)

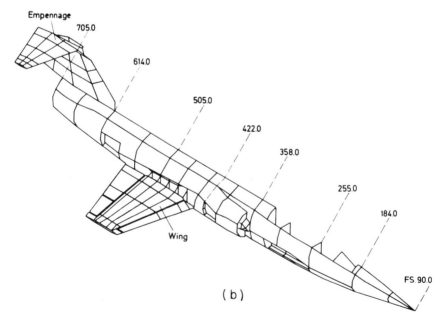

(b)

Fig. 16.3 Finite-element analysis of airplane.

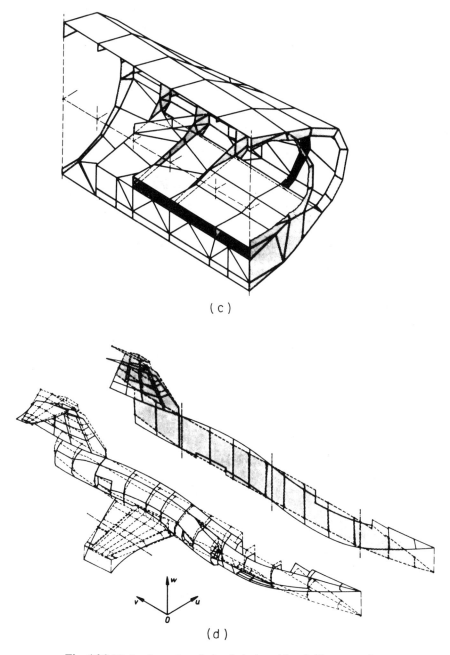

(c)

(d)

Fig. 16.3 Finite-element analysis of airplane (*Cont.*) (Courtesy of Institute of Statics and Dynamics of Air- and Space-Structures, University of Stuttgart, Germany).

patible internal strains, which may also vary over the element, will be denoted by $[\epsilon(x)^{**}]$.

The theorem of virtual displacement will then appear as

$$[I^{**}]^T[k^*] = \int_{\text{vol}} [\epsilon(x)^{**}]^T[\sigma(x)^*]\, dV \qquad (16.1)$$

	statically compatible forces	(due to real unit nodal displacements)
	geometrically compatible deformations	(due to virtual unit nodal displacements)

The unit matrix $[I]^T$ can be deleted.

The approximate nature of the method arises from the way the virtual strains and the real stresses are calculated: a *displacement function* $\{u\}$ is assumed, which must satisfy certain requirements to be discussed in Sec. 16.8 but is otherwise arbitrary. From the displacement function, the nodal displacements, the strains, and the stresses are computed, using the laws of elasticity, to be discussed in Sec. 16.3, and inserted in Eq. (16.1). The accuracy of the resulting stiffness matrix depends on the accuracy of the assumed displacement function.

To familiarize ourselves with the procedure, we shall use it to recompute the stiffness matrix for the prismatic beam element with the nodal numbering shown in Fig. 16.4(a). For this case, the right-hand side of Eq. (16.1) becomes, upon integration over the beam cross section, as discussed in Sec. 8.1 (that is,

$$\Phi = \epsilon/y, \quad M = \int_A \sigma \cdot y\, dA),$$

$$[k] = \int_L [\Phi(x)^{**}]^T[M(x)^*]\, dx \qquad (16.2)$$

We could now proceed using these symbols, but since we shall use this example to derive generally valid formulas to be used for a variety of elements, we shall adhere to the notation of Eq. (16.1); that is, the internal distortions

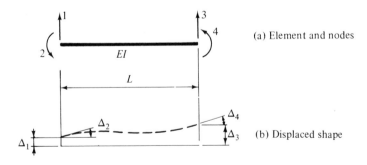

(a) Element and nodes

(b) Displaced shape

Fig. 16.4 Beam element.

(in this case, the curvatures $[\Phi(x)]$) will be denoted by $[\epsilon(x)]$, and the internal forces (in this case, the moments $[M(x)]$) will be denoted by $[\sigma(x)]$, so that the virtual work equation becomes

$$[k] = \int_L [\epsilon(x)^{**}]^T [\sigma(x)^*] \, dx \tag{16.3}$$

We proceed step by step in determining the integral:

1. To describe the deformed element shape shown in Fig. 16.4(b), we assume a four-term displacement function

$$v = \alpha_1 + \alpha_2 x + \alpha_3 x^2 + \alpha_4 x^3 \tag{16.4}$$

where the α's are as yet unknown coefficients.

Because the nodal displacements also include end rotations, we need the slope, which we obtain by differentiation:

$$\theta = \frac{dv}{dx} = \alpha_2 + 2\alpha_3 x + 3\alpha_4 x^2 \tag{16.5}$$

We can write these internal displacements, which will be designated by $\{u(x)\}$, as functions of the coefficients $\{\alpha\}$ in matrix form:

$$\begin{Bmatrix} v \\ \dfrac{dv}{dx} \end{Bmatrix} = \begin{bmatrix} 1 & x & x^2 & x^3 \\ 0 & 1 & 2x & 3x^2 \end{bmatrix} \begin{Bmatrix} \alpha_1 \\ \alpha_2 \\ \alpha_3 \\ \alpha_4 \end{Bmatrix} \tag{16.6a}$$

or

$$\{u(x)\} = [N(x)]\{\alpha\} \tag{16.6b}$$

The matrix $[N(x)]$ is a function of position along the beam.

2. We calculate the nodal displacements $\{\Delta\}$ in terms of the coefficients $\{\alpha\}$ for any one loading condition by inserting appropriate coordinates of the element nodes (that is, the beam ends) into Eq. (16.6a) (note that Δ_1 and Δ_3 are transverse end displacements, and Δ_2 and Δ_4 are rotations of the left and right ends, respectively; therefore, $x_1 = x_2 = 0$, $x_3 = x_4 = L$):

$$\begin{Bmatrix} \Delta_1 \\ \Delta_2 \\ \Delta_3 \\ \Delta_4 \end{Bmatrix} = \begin{bmatrix} 1 & 0 & 0 & 0 \\ 0 & 1 & 0 & 0 \\ 1 & L & L^2 & L^3 \\ 0 & 1 & 2L & 3L^2 \end{bmatrix} \begin{Bmatrix} \alpha_1 \\ \alpha_2 \\ \alpha_3 \\ \alpha_4 \end{Bmatrix} \tag{16.7a}$$

or

$$\{\Delta\} = [A]\{\alpha\} \tag{16.7b}$$

The matrix $[A]$ depends only on the joint coordinates.

We express the matrix $\{\alpha\}$ in terms of the joint displacements $\{\Delta\}$ by inverting Eq. (16.7b):

$$\{\alpha\} = [A]^{-1}\{\Delta\} \tag{16.8}$$

So far, we have expressed the coefficients $\{\alpha\}$ in Eq. (16.8) for a single set of nodal displacements $\{\Delta\}$. For the determination of the complete stiffness matrix, the coefficients are due to four virtual unit nodal displacements, applied one at a time:

$$[\Delta] = \begin{bmatrix} 1 & 0 & 0 & 0 \\ 0 & 1 & 0 & 0 \\ 0 & 0 & 1 & 0 \\ 0 & 0 & 0 & 1 \end{bmatrix} = [I]$$

and we therefore obtain an $[\alpha]$ matrix of order (4×4), with each column containing the deflection coefficients for one loading condition,

$$[\alpha] = [A]^{-1}[I] = [A^{-1}] \tag{16.9}$$

where, for our case, the inversion of the $[A]$ matrix of Eq. (16.7a) leads to

$$[A]^{-1} = \begin{bmatrix} 1 & 0 & 0 & 0 \\ 0 & 1 & 0 & 0 \\ -\dfrac{3}{L^2} & -\dfrac{2}{L} & \dfrac{3}{L^2} & -\dfrac{1}{L} \\ \dfrac{2}{L^3} & \dfrac{1}{L^2} & -\dfrac{2}{L^3} & \dfrac{1}{L^2} \end{bmatrix} \tag{16.10}$$

To effect the inversion, the $[A]$ matrix must be square; that is, the number of coefficients α of the displacement function $u(x)$ must match the number of element nodes.

   3. We calculate the internal distortions $[\epsilon(x)]$, which may be functions of position. In a beam in bending, these distortions are conveniently represented by the curvature Φ, which can be obtained from the displacements by two successive differentiations of Eq. (16.4):

$$[\Phi(x)] = \frac{d^2v}{dx^2} = [0 \quad 0 \quad 2 \quad 6x][\alpha] \tag{16.11a}$$

or

$$[\epsilon(x)] = [B(x)][\alpha] \tag{16.11b}$$

where the $[B(x)]$ matrix relates the internal distortions $[\epsilon(x)]$ to the coefficients $[\alpha]$. Replacing $[\alpha]$ in Eq. (16.11b) by Eq. (16.9), we obtain the internal distortions due to the four unit nodal displacements:

$$[\epsilon(x)] = [B(x)][A]^{-1} \tag{16.12}$$

or, for further use in the virtual work equation,

$$[\epsilon(x)]^T = [A^{-1}]^T[B(x)]^T \tag{16.13}$$

4. The internal forces $[\sigma(x)]$, which may be functions of position, can be represented by the moments $[M(x)]$ in the beam element. (Note that these moments M^* rotating through the curvatures Φ^{**} represent the internal virtual work per unit length.) The moments are related linearly to the curvatures through the beam stiffness EI:

$$[M] = EI[\Phi]$$

or

$$[\sigma(x)] = [D][\epsilon(x)] \tag{16.14}$$

where, in this case,

$$[D] = EI \tag{16.15}$$

In general, the matrix $[D]$ is the elasticity matrix relating internal forces and distortions. Here, it happens to be just a scalar constant.

Inserting Eq. (16.12) into Eq. (16.14), we can express the moments $[\sigma(x)]$ due to the unit nodal displacements:

$$[\sigma(x)] = [D][B(x)][A^{-1}] \tag{16.16}$$

5. We are now ready to insert the transpose of the virtual curvatures $[\epsilon(x)]^T$ from Eq. (16.13) and the real moments $[\sigma(x)]$ from Eq. (16.16) into the virtual-work equation (16.3). Since $[A^{-1}]$ is not a function of position, it can be taken outside the integral:

$$[k] = [A^{-1}]^T \int_L [B(x)]^T[D][B(x)]dx \cdot [A^{-1}] \tag{16.17}$$

Equation (16.17) is written in sufficiently general form to be valid for other elements. For our case, the matrices $[A^{-1}]$, $[B(x)]$, and $[D]$ are given by Eqs. (16.10), (16.11a), and (16.15). We perform the indicated matrix operations, integrate term by term, and finally arrive at the result

$$[k] = \frac{EI}{L^3} \begin{bmatrix} 12 & & & \text{sym} \\ 6L & 4L^2 & & \\ -12 & -6L & 12 & \\ 6L & 2L^2 & -6L & 4L^2 \end{bmatrix}$$

which has already been derived previously by other means.

The result is exact within the framework of beam theory because it so happened that the assumed displacement function $\{u\}$ of Eq. (16.4) was the exact solution of the differential beam equation.

We might question at this time the purpose of this sophisticated procedure to arrive at such a simple result. To answer this, we immediately proceed to a more involved case in the following example to demonstrate the ease with which approximate stiffness matrices of more difficult elements can be computed.

Example 16.1: Using a suitable assumed displacement function, compute an approximate stiffness matrix for the beam element of flexural stiffness $EI(x)$ varying linearly from EI_0 at the left to $2EI_0$ at the right end, as shown in parts (a) and (b) of the figure.

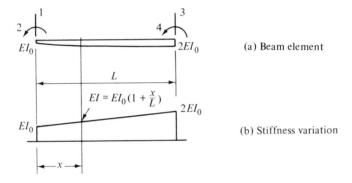

Fig. E16.1

Solution: We determine the integral of Eq. (16.3), following the same steps as before:

1. We assume a four-term displacement function identical to Eq. (16.4):

$$v = \alpha_1 + \alpha_2 x + \alpha_3 x^2 + \alpha_4 x^3$$

The function does not in this case represent an exact solution of the beam differential equation, and can therefore only yield an approximate solution. The number of terms matches that of the element nodes. The matrix $N(x)$ is identical to that of Eq. (16.6a).

2. The geometrical conditions relating the coefficients $\{\alpha\}$ to the nodal displacements $\{\Delta\}$ are identical to the preceding prismatic-beam case because the same displacement function was used. The matrices $[A]$ and $[A^{-1}]$ are therefore given by Eqs. (16.7a) and (16.10).

3. The internal curvatures are attained from the displacements by purely geometrical considerations; so, as before, the matrix $[B(x)]$ relating curvatures $\{\epsilon(x)\}$ to the coefficients $\{\alpha\}$ is given by Eq. (16.11a).

4. The moment–curvature relation, $[\sigma(x)] = [D][\epsilon(x)]$, is now different from before, since $D(x)$ now varies along the member:

$$[D(x)] = \left[EI_0\left(1 + \frac{x}{L}\right)\right] \tag{16.18}$$

5. Inserting now the matrices $[A^{-1}]$, $[B(x)]$, and $[D(x)]$ from Eqs. (16.10), (16.11a), and (16.18) into Eq. (16.17), carrying out the indicated operations, and performing the integrations term by term, we obtain the result

$$[k] = \frac{EI_0}{L^3}
\begin{bmatrix}
18 & (17.24) & & & & \text{sym} \\
8L & (7.63L) & 5L^2 & (4.82L^2) & & \\
-18 & (-17.24) & -8L & (-7.63L) & 18 & (17.24) \\
10L & (9.62L) & 3L^2 & (2.81L^2) & -10L & (-9.62L) & 7L^2 & (6.80L^2)
\end{bmatrix}$$

The quantities in parentheses are the exact values obtained by force method analysis in Example 12.6. We see that all approximate values are within 6 per cent of the exact values, in spite of the simple displacement function that had been assumed.

16.3. Some Elements of the Plane Theory of Elasticity

We now turn to the task of computing stiffness matrices for more general shapes in a state of plane stress. For this reason, we need to look at some principles of the plane theory of elasticity; we shall cover these in the usual sequence by considering equilibrium, geometry, and stress–strain relations, and supplement these with an important energy consideration.

1. *Equilibrium.* We consider an infinitesimal cartesian planar element $dx \cdot dy$, of unit thickness, as shown in Fig. 16.5. It is subjected to a general

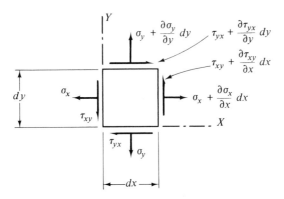

Fig. 16.5 Stresses on plane element.

state of plane stress σ_x, σ_y, $\tau_{xy} = \tau_{yx}$; all these stresses vary in intensity: for instance, the stress σ_x varies by an amount $\partial\sigma_x/\partial x$ per unit length along the X axis. Thus, the intensities of stress acting on opposite faces of the element differ by the amounts shown in Fig. 16.5.

Summing forces (that is, stress intensities times areas of element faces) along the X and Y axes, canceling terms, and simplifying, we obtain the differential equilibrium equations

$$\frac{\partial\sigma_x}{\partial x} + \frac{\partial\tau_{xy}}{\partial y} = 0$$

$$\frac{\partial\tau_{xy}}{\partial x} + \frac{\partial\sigma_y}{\partial_y} = 0$$

$$(16.19)$$

2. *Geometry.* We consider the small displacements u and v of the infinitesimal element shown in Fig. 16.6. By recalling the definitions of the linear strains ϵ_x and ϵ_y as the changes of length per unit length in the appropriate direction, and that of the shear strain γ_{xy} as the angle change between two originally normal axes X and Y, we obtain the strain–displacement relations

$$\epsilon_x = \frac{\partial u}{\partial x}; \qquad \epsilon_y = \frac{\partial v}{\partial y}; \qquad \gamma_{xy} = \frac{\partial u}{\partial y} + \frac{\partial v}{\partial x} \qquad (16.20)$$

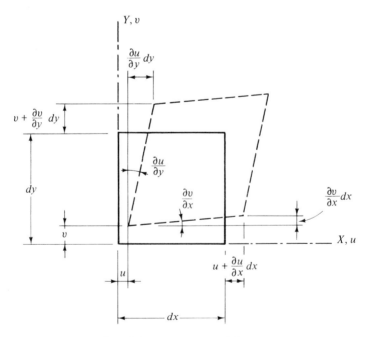

Fig. 16.6 Displacements of element.

3. *Stress–strain relations.* The relations between stress and strain in two dimensions are given by the generalized Hooke's law,

$$\begin{Bmatrix} \sigma_x \\ \sigma_y \\ \tau_{xy} \end{Bmatrix} = \frac{E}{1-\mu^2} \begin{bmatrix} 1 & \mu & 0 \\ \mu & 1 & 0 \\ 0 & 0 & \frac{1}{2}(1-\mu) \end{bmatrix} \begin{Bmatrix} \epsilon_x \\ \epsilon_y \\ \gamma_{xy} \end{Bmatrix} \qquad (16.21)$$

or

$$\{\sigma\} = [D]\{\epsilon\}$$

In Eq. (16.21), E is the modulus of elasticity and μ is Poisson's ratio of the material. The matrix $[D]$ relates the internal stresses and strains, analogous to the matrix $[D] = EI$ used in the earlier beam problem of Sec. 16.2.

4. *Theorem of minimum potential energy.* We recall from Sec. 15.3 that the minimization of the potential energy of a geometrically compatible displacement function corresponds to the satisfaction of the equilibrium conditions. If the assumed displacement function $\{u\}$ is geometrically compatible (conditions for this requirement will be discussed in Sec. 16.8), then the virtual-work equation, Eq. (16.1), can be interpreted as a process of minimization of potential energy with decreasing element size. It follows that the conditions of internal equilibrium, Eqs. (16.19), will be increasingly well satisfied as the element becomes smaller and smaller, even if the stresses corresponding to the chosen displacement function do not satisfy Eqs. (16.19) explicitly. These equations, therefore, need not be considered further in the finite-element displacement method.

16.4. Triangular Plane-Stress Element Stiffness Matrix

Whereas the beam stiffness matrices developed in Sec. 16.2 could have been found by a variety of methods, the determination of an exact stiffness matrix for the triangular plane-stress element shown in Fig. 16.7(a) would be a forbidding task. For problems of this type the approximate procedure based on an assumed displacement function takes on great importance.

We visualize this element connected to its adjacent companions at the corners, as shown in Fig. 16.1(b); since at each corner there are two components of force and displacement, we identify a total of six nodes for this element, two at each of the three vertices, and number them as shown in Fig. 16.7(a); the coordinates of these joints are shown in parentheses.

We proceed by the steps outlined in Sec. 16.2, leading to the virtual-work equation (16.17); here, the integration must be carried out over the volume,

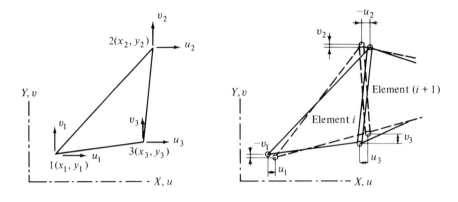

(a) Triangular element

(b) Assumed deformed shape showing
    compatibility of edge displacements

Fig. 16.7 Triangular plane element.

so this equation is generalized for a solid:

$$[k] = [A^{-1}]^T \cdot \int_{\text{vol}} [B(x)]^T [D][B(x)] dV \cdot [A^{-1}] \tag{16.22}$$

wherein we need to find the matrix $[A]$ relating the unit nodal displacements $[I]$ to the coefficients $[\alpha]$ of the assumed displacement function, the matrix $[B]$ relating the strains $[\epsilon]$ to the coefficients $[\alpha]$, and the material stiffness matrix $[D]$, already presented in Eq. (16.21). We proceed in the usual steps:

1. We assume in this case a set of linear displacement functions containing six terms to match the number of element nodes:

$$u = \alpha_1 + \alpha_2 x + \alpha_3 y$$
$$v = \alpha_4 + \alpha_5 x + \alpha_6 y \tag{16.23a}$$

or, in matrix form,

$$\begin{Bmatrix} u \\ v \end{Bmatrix} = \begin{bmatrix} 1 & x & y & 0 & 0 & 0 \\ 0 & 0 & 0 & 1 & x & y \end{bmatrix} \begin{Bmatrix} \alpha_1 \\ \cdot \\ \cdot \\ \cdot \\ \alpha_6 \end{Bmatrix} \tag{16.23b}$$

or

$$\{u(x)\} = [N(x)]\{\alpha\} \tag{16.23c}$$

2. The nodal displacements are obtained by appropriate insertion of the joint coordinates of Fig. 16.7(a) in Eq. (16.23c):

$$\begin{Bmatrix} u_1 \\ v_1 \\ u_2 \\ v_2 \\ u_3 \\ v_3 \end{Bmatrix} = \begin{bmatrix} 1 & x_1 & y_1 & 0 & 0 & 0 \\ 0 & 0 & 0 & 1 & x_1 & y_1 \\ 1 & x_2 & y_2 & 0 & 0 & 0 \\ 0 & 0 & 0 & 1 & x_2 & y_2 \\ 1 & x_3 & y_3 & 0 & 0 & 0 \\ 0 & 0 & 0 & 1 & x_3 & y_3 \end{bmatrix} \begin{Bmatrix} \alpha_1 \\ \alpha_2 \\ \alpha_3 \\ \alpha_4 \\ \alpha_5 \\ \alpha_6 \end{Bmatrix} \qquad (16.24a)$$

or

$$\{\Delta\} = [A]\{\alpha\} \qquad (16.24b)$$

or, solving for $\{\alpha\}$,

$$\{\alpha\} = [A]^{-1}\{\Delta\}$$

3. By use of the strain-displacement equations, Eqs. (16.20), we compute the element strains by differentiation of Eq. (16.23):

$$\{\epsilon\} = \begin{Bmatrix} \epsilon_x \\ \epsilon_y \\ \gamma_{xy} \end{Bmatrix} = \begin{Bmatrix} \dfrac{\partial u}{\partial x} \\[2mm] \dfrac{\partial v}{\partial y} \\[2mm] \dfrac{\partial u}{\partial y} + \dfrac{\partial y}{\partial x} \end{Bmatrix} = \begin{bmatrix} 0 & 1 & 0 & 0 & 0 & 0 \\ 0 & 0 & 0 & 0 & 0 & 1 \\ 0 & 0 & 1 & 0 & 1 & 0 \end{bmatrix} \begin{Bmatrix} \alpha_1 \\ \cdot \\ \cdot \\ \cdot \\ \alpha_6 \end{Bmatrix} \qquad (16.25a)$$

or

$$\{\epsilon\} = [B]\{\alpha\}$$

Because of the linear nature of the assumed displacement function, the strains are constant over the element in this case.

4. The element stresses are related to the element strains by the Hooke's law of Eq. (16.21), so this furnishes the [D] matrix.

With the matrices [A], [B], and [D] given by Eqs. (16.24a), (16.25a), and (16.21), we can now carry out the operations called for by Eq. (16.22). Since these matrices are in this case all independent of position, we can take them all outside the integral, leaving only the integral $\displaystyle\int_{vol} dV = \mathbf{A} \cdot t$, where $\mathbf{A}$ is the area and t the thickness of the element. Thus, the stiffness matrix becomes

$$[k] = \mathbf{A} \cdot t [A^{-1}]^T [B]^T [D][B][A^{-1}] \qquad (16.26)$$

The operations indicated by Eq. (16.26) lead to the *constant-strain triangle* stiffness matrix of Eq. (16.27) (see p. 450). This matrix is only a function of the joint coordinates, the thickness, area, and elastic constants of the element, and can thus be readily computed for each element of the structure.

$$[k] = \frac{Et}{8A(1-\mu^2)}
\begin{bmatrix}
(1-\mu)x_{32}^2 + 2y_{32}^2 & & & & & \text{sym} \\[4pt]
-(1+\mu)x_{32}y_{32} & 2x_{32}^2 + (1-\mu)y_{32}^2 & & & & \\[4pt]
-(1-\mu)x_{32}x_{31} - 2\mu y_{32}y_{31} & 2\mu x_{32}y_{31} + (1-\mu)y_{32}x_{31} & (1-\mu)x_{31}^2 + 2y_{31}^2 & & & \\[4pt]
(1-\mu)x_{32}y_{31} - 2\mu y_{32}x_{31} & -2x_{32}x_{31} - (1-\mu)y_{32}y_{31} & -(1+\mu)x_{31}y_{31} & 2x_{31}^2 + (1-\mu)y_{31}^2 & & \\[4pt]
(1-\mu)x_{32}x_{21} + 2y_{32}y_{21} & -2\mu x_{32}y_{21} - (1-\mu)y_{32}x_{21} & -(1-\mu)x_{31}x_{21} - 2\mu y_{31}y_{21} & 2\mu x_{31}y_{21} + (1-\mu)y_{31}x_{21} & (1-\mu)x_{21}^2 + 2y_{21}^2 & \\[4pt]
-(1-\mu)x_{32}y_{21} - 2\mu y_{32}x_{21} & 2x_{32}x_{21} - (1-\mu)y_{32}y_{21} & (1-\mu)x_{31}y_{21} + 2\mu y_{31}x_{21} & -2x_{31}x_{21} - (1-\mu)y_{31}y_{21} & -(1+\mu)x_{21}y_{21} & 2x_{21}^2 + (1-\mu)y_{21}^2
\end{bmatrix}$$

$$x_{ij} \equiv x_i - x_j; \quad y_{ij} \equiv y_i - y_j; \quad A \equiv \text{area of triangle} = \tfrac{1}{2}(x_{32}y_{21} - x_{21}y_{32})$$

$$(16.27)$$

16.5. Structure Stiffness Matrix

As discussed in Sec. 13.3, the structure stiffness matrix is assembled by super-position of the element stiffness matrices. The nodes have been carefully numbered, listing the free nodes first, according to a common structure numbering system within a common global set of axes, so that it is possible to insert the individual element stiffnesses k_{ij} into the appropriate slot of the structure stiffness matrix K_{ij} according to the scheme

$$K_{ij} = \sum k_{ij} \qquad (13.10)$$

where the summation extends over all elements connected to node i. This completes the assembly of the structure stiffness matrix for insertion into Eq. (13.12).

16.6. Internal Stresses and Strains

After the nodal displacements $[\Delta_\alpha]$ and the reactive forces $[X_\beta]$ have been determined by use of Eqs. (13.13) and (13.14), it remains to compute the internal strains and stresses within the structure.

To do this, we recall that Eq. (16.12) furnishes relations between internal strains and nodal displacements:

$$[\epsilon(x)] = [B(x)][A^{-1}][\Delta] \qquad (16.28)$$

and Eq. (16.16) relates the internal stresses to the nodal displacements of each element:

$$[\sigma(x)] = [D][B(x)][A^{-1}][\Delta] \qquad (16.29)$$

The nodal displacements obtained by Eq. (13.13), along with the specified boundary displacements $[\Delta_\beta]$, are thus inserted into Eqs. (16.28) and (16.29), element by element, in order to find the desired quantities.

For the linear displacement function of Eq. (16.23a), the strains and stresses are constant for any element, and the analysis will thus result in average values. It is generally sufficiently accurate to assume these values to be valid at the centroid of the element and interpolate to find the remaining values. It should, however, be realized that in the vicinity of stress concentrations, such as reentrant corners, notches, and the like, the average may be considerably less than the peak stress, and a fine subdivision of the structure may be necessary in such areas.

16.7. Local and Global Stiffness Matrices

There is one significant difference between the beam stiffness matrices of Sec. 16.2 and the triangular plane-stress element stiffness matrix of Sec. 16.4. The latter has been derived in terms of joint coordinates within a global coordinate system; thus, the symbol $[k]$ represents an element stiffness matrix in global coordinates, and the stiffnesses of all elements of the structure can be immediately superposed to form the structure stiffness matrix $[K]$.

On the other hand, the former is written in terms of local element coordinates conforming to the principal beam axes, and, according to the symbolism of Sec. 13.5, should be written $[\bar{k}]$. If the different beam elements are oriented in different directions, it is first necessary to express all element stiffness matrices in terms of a common global system by applying the transformation

$$[k] = [T]^T[\bar{k}][T] \tag{13.23}$$

derived in Sec. 13.5. Only then can these element stiffnesses be superposed to form the structure stiffness matrix.

The transformation matrix $[T]$ will always be a square matrix of the same order as the number of element nodes, as shown, for instance, by the two- and three-dimensional beam element transformation matrices of Eqs. (13.20), (13.25), and (13.28). For other types of elements, similar reasoning is applied to obtain the appropriate transformation matrix $[T]$, as shown in the following example.

Example 16.2: The rectangular plane-stress element shown in the figure, containing eight nodes, has the (8×8) stiffness matrix $[\bar{k}]$ in the local coordinate system $\bar{X}$, $\bar{Y}$ indicated. Compute the element stiffness matrix $[k]$ with respect to the global axes X, Y making an angle α with the local axes.

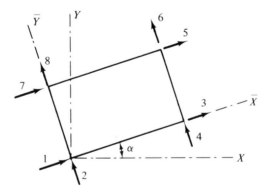

Fig. E16.2

Solution: The nodal quantities of the element consist of four plane vectors, one at each corner. According to Eq. (13.19), each vector transforms according to the law

$$\begin{Bmatrix} \bar{V}_x \\ \bar{V}_y \end{Bmatrix} = \begin{bmatrix} \cos \alpha & \sin \alpha \\ -\sin \alpha & \cos \alpha \end{bmatrix} \begin{Bmatrix} V_x \\ V_y \end{Bmatrix} \tag{a}$$

or

$$\{\bar{V}\} = [\lambda]\{V\}$$

The (8×8) transformation matrix $[T]$ transforming all four vectors simultaneously is

$$[T] = \begin{bmatrix} \lambda & 0 & 0 & 0 \\ 0 & \lambda & 0 & 0 \\ 0 & 0 & \lambda & 0 \\ 0 & 0 & 0 & \lambda \end{bmatrix} \tag{b}$$

According to Eq. (13.23), the transformation of the stiffness matrix from local to global coordinates is

$$[k] = [T]^T[\bar{k}][T]$$

where $[T]$ is given by Eqs. (b) and (a).

16.8. Convergence Criteria

We have already referred in Sec. 16.3 to the fact that the finite-element method can be interpreted as a process of establishing equilibrium by minimizing the potential energy of a geometrically compatible displacement system. For such a system, it can be proved that any approximate solution leads to an underestimation of the deformations; that is, the stiffness will be too large.

Physically, this behavior can be understood by realizing that the assumption of an inexact deformation amounts to building geometrical constraints into the displacements, which prevent full development of deformations. For instance, the linear displacement function of Eq. (16.23) for the triangular element does not allow straight lines within the element to deform into curves (which a rigorous solution would predict). The element will therefore be stiffer than it should be.

The deformed shape of any straight line of the structure, such as the transverse section of Fig. 16.8, will under this assumption be represented by a number of straight-line segments, each one passing through one element. As the number of elements is increased, as shown in Fig. 16.8(b), these straight-line segments will become more numerous and shorter, resembling more and more the curved shape of Fig. 16.8(c) predicted by rigorous theory. Thus, it

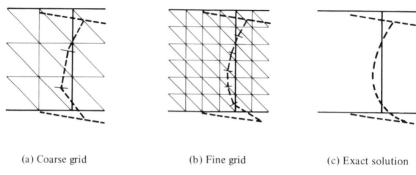

(a) Coarse grid   (b) Fine grid   (c) Exact solution

Fig. 16.8 Convergence of solution.

can be concluded that under the assumption of geometrically compatible displacements, the solution will converge to the exact one with increasing number of elements.

The conditions for convergence of the results toward exact solutions with decreasing element size are thus those required for a geometrically compatible displacement field; they are the following:

1. Continuity of the displacement function within the element: This is satisfied by the assumption of a continuous displacement field $\{u(x)\}$.

2. Interelement compatibility: At the nodes, continuity is assured by the displacement method, which assumes common displacements at common nodes. Between nodes, the common edges must deform together in order to form a continuous displacement field across the structure, as discussed in Sec. 16.1. In the linear displacement function of Eq. (16.23), this is assured because the straight-line deformed shape of two adjacent edges is completely defined by the displacements of their common end points, as shown in Fig. 16.7(b). In general, when the deformed shape of an edge is completely defined by the displacements of the nodes lying on this edge, then the displacements of two adjacent edges will be compatible.

3. The displacement function must be able to represent rigid-body displacements of the element to allow it to conform to the displacements of adjacent elements, as shown, for instance, for the beam of Fig. 16.9. In the beam displacement function, Eq. (16.4), the first, constant,

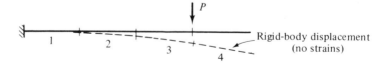

Fig. 16.9 Rigid-body displacement of element.

term permits a transverse translation, and the second, linear, term represents pure rotation of the element. In the displacement function for the triangular element, Eq. (16.23), the terms α_1 and α_4 allow translation along the X and Y axes, and the terms $\alpha_3 y$ and $\alpha_5 x$ provide for pure rotation. In general, the rigid-body displacement terms can be identified because they lead, upon appropriate differentiation, to zero strain states.

4. Constant strain states have to be accommodated. Thus, the displacement functions must contain terms that, upon appropriate differentiation, lead to constant strains. For instance, the quadratic term $\alpha_3 x^2$ in the beam displacement function, Eq. (16.4), corresponds to constant curvature $\Phi = d^2 v/dx^2 = 2\alpha_3 = $ constant. Similar terms can be identified in Eq. (16.23). Thus, an appropriate choice of displacement function is necessary to be assured of convergence to the correct solution with increasingly finer subdivision of a structure into finite elements.

16.9. Stiffness Matrices for Different Elements

The use of Eqs. (13.12) and (13.13) is entirely general; it is only necessary to insert element stiffness matrices into Eq. (13.12) that are capable of representing the response of the particular structure under investigation.

Equation (16.22) is also applicable to any linearly elastic element; however, it will be necessary in each case to identify the relevant nodal forces and corresponding internal stresses, and the nodal displacements and corresponding internal strains. Thus, in the plane-stress stiffness matrix derived in Sec. 16.4, only nodal forces were considered important, and, accordingly, Eq. (16.27) cannot accommodate cranked-in moments. (They could of course, be handled approximately as equivalent couples.) To account for such moments and the corresponding nodal rotations, "higher-order" elements should be considered.

The accuracy of the stiffness matrices can be improved by the introduction of additional nodes; for instance, nodal points could be introduced at the midpoints of the three sides of the triangular element of Fig. 16.7(a), leading to six additional nodes. This in turn would permit the addition of three quadratic terms to the displacement functions of Eq. (16.23), resulting in a linearly varying set of stresses and strains, which obviously would be better able to reproduce an actual stress variation than the constant-stress element of Sec. 16.4. Whether the improved accuracy merits the introduction of the additional nodes is a matter to be investigated from case to case.

In the following, element stiffness matrices for different structures are discussed briefly.

Quadrilateral plane-stress element. General quadrilateral elements are very versatile and widely used. As shown in Fig. 16.10, the element stiffness matrix can be obtained by superposition of two triangular stiffness matrices, or by considering the combination of four triangles as a structure and calculating the structure stiffness matrix for the exterior corners. More exact results can be obtained by a direct solution.

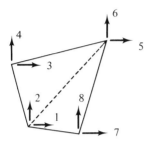

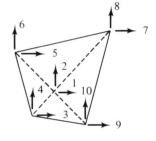

(a) Quadrilateral element by superposition of two triangles

(b) Quadrilateral element by superposition of four triangles

Fig. 16.10 Quadrilateral element.

Plane-strain element. The only difference between a plane-stress solution (as in Sec. 16.4) and a plane-strain solution, as required, for instance, for tunnel design, lies in Hooke's law, Eq. (16.21), for which should be substituted Eq. (16.30):

$$\begin{Bmatrix} \sigma_x \\ \sigma_y \\ \tau_{xy} \end{Bmatrix} = \frac{E}{(1 + \mu)(1 - 2\mu)} \begin{bmatrix} 1 - \mu & \mu & 0 \\ \mu & 1 - \mu & 0 \\ 0 & 0 & \dfrac{1 - 2\mu}{2} \end{bmatrix} \begin{Bmatrix} \epsilon_x \\ \epsilon_y \\ \gamma_{xy} \end{Bmatrix} \quad (16.30)$$

Flat-plate bending element. In plate bending, the bending moments about two axes, the corresponding rotations, and the transverse forces and corresponding deflections are taken as nodal quantities. The internal quantities are internal bending and twisting moments and the corresponding curvatures and twist. Figure 16.11 shows a rectangular plate bending element with 12 nodes, permitting a transverse deflection displacement function consisting of 12 terms.

Finite-element analysis of shells. Conceptually, it is simple to think of shell structures as being composed of flat elements oriented at appropriate angles to each other. This will require a transformation procedure such as discussed in Sec. 16.7 to rotate the nodal quantities from the local coordinate system to a common structure coordinate system.

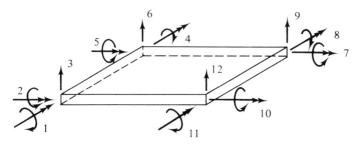

Fig. 16.11 Plate-bending element.

If only membrane action is to be considered, it is sufficient to include only the in-plane nodal quantities, as given, for instance, by Eq. (16.27). In the more general case of shell bending, the in-plane nodes and the plate-bending nodes are required, and the total element stiffness is obtained by appropriate superposition of the in-plane and the bending stiffness matrices.

A more accurate representation of the shell behavior is furnished by curved shell elements. In this case, the in-plane and the bending action cannot be decoupled as in the case of flat elements, and the formulation of the stiffness matrix must follow the rules of classical shell theory. In the general case, there will be five nodal quantities at each joint of the element, as shown in Fig. 16.12. A number of specialized shell elements for cylindrical and axisymmetric shells have been developed as well.

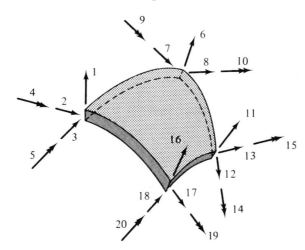

Fig. 16.12 Curvilinear-shell finite element.

Three-dimensional elastic solids. The three-dimensional analogue to the planar triangular element of Fig. 16.7(a) is the tetrahedron shown in Fig. 16.13. This element has 12 nodes and can thus accommodate a four-term

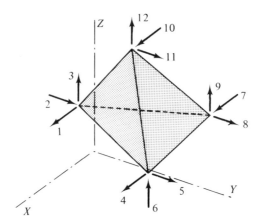

Fig. 16.13 Tetrahedral element.

linear displacement function for each of the three internal displacement components. Appropriate differentiation of these displacements will lead to constant stress and strain within the element. The simulation of realistic three-dimensional situations usually requires a larger number of nodal quantities than can be handled by available computing equipment. On the other hand, if axial symmetry is present, the formulation can be reduced to a quasi-planar case by appropriate manipulations.

We have only been able to present the bare-bones elements of the finite-element method in this short section. The literature of recent years is full of additional details of the mathematical background, programming, applications to different structures, an almost infinite variety of stiffness matrices of all degrees of sophistication and applicability. To obtain a better background and further insight in the uses of the finite-element method, a study of Zienkiewicz' work,* a very readable and interesting classic in its field, is highly recommended.

PROBLEMS

1. Compute the approximate (4 × 4) stiffness matrix for the tapering beam element, assuming the linear displacement function, Eq. (16.4). Compare results with exact values from Prob. 11.9 or 12.17, and discuss the results.

2. Compute the (4 × 4) stiffness matrix for the prismatic beam element, including the effects of bending and shear distortions. Assume the linear displacement function, Eq. (16.4), and modify the approach of Sec. 16.2 to include the internal virtual work due to shear distortion.

*O. C. Zienkiewicz, *The Finite Element Method in Engineering Science*, McGraw-Hill Book Company, London, 1971.

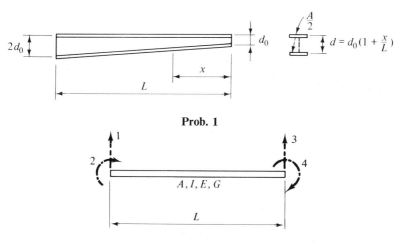

Prob. 1

Prob. 2

3. Compute the (6 × 6) stiffness matrix of the plane prismatic beam element, including the effects of bending and axial distortions, by assumption of appropriate displacement functions.

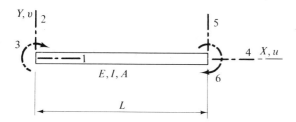

Prob. 3

4. a. Assuming an appropriate displacement function that contains a number of terms equal to the number of element nodes, compute the matrices $[A]$, $[B]$, and $[D]$ necessary to establish the (8 × 8) stiffness matrix for the rectangular elastic plane-stress element of uniform thickness.

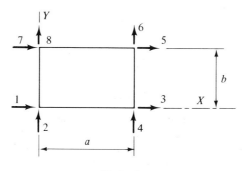

Prob. 4

b. Using the matrices established in part a, perform the necessary matrix computations to obtain the element stiffness matrix.

5. The triangular plane-stress element shown has its thickness varying linearly over its surface, with thicknesses t_1, t_2, and t_3 at the corners. Compute the stiffness matrix for the variable-thickness element, and discuss the result.

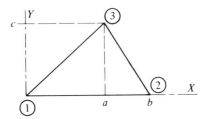

Prob. 5

6. Discuss the satisfaction of the convergence criteria of Sec. 16.8 by the displacement functions used in Probs. 4 and 5.

7. a. Write the stiffness matrix for the right triangular element shown in part (a) of the figure with respect to the X, Y axes, using a linear displacement function.

   b. By appropriate transformation, obtain the stiffness matrix for the same element oriented as shown in part (b).

   c. Check the result obtained in part b of the problem by direct application of Eq. (16.27).

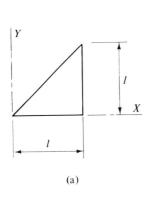

(a)

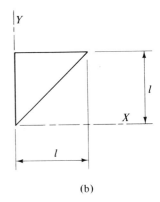

(b)

Prob. 7

8. The flat square elastic plate of constant thickness is to be analyzed for the effects of the applied concentrated load P. Due to symmetry, only one quarter of the plate needs to be considered; this is to be modeled by the triangular finite-element arrangement shown.

   a. Introduce appropriate nodal numbering, listing free nodes first. Note that

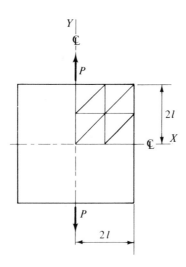

Prob. 8

along the symmetry axis Y, the displacements parallel to the X axis and the forces parallel to the Y axis are zero; similarly, along the symmetry-axis X, the displacements parallel to the Y axis and the forces parallel to the X axis must be zero.

b. Write the matrices of joint loads and joint displacements, following the numbering established in part a. Note that only half the load P is acting on the portion of the plate considered in the analysis.

c. On a sufficiently large sheet of paper, establish a tabular form for the structure stiffness matrix, insert the individual element stiffness matrices in their appropriate slots, and add the stiffness contributions to obtain the structure stiffness matrix.

d. Using subroutine SOLN (see Appendix A), solve the system of equations for the unknown displacements and forces. Plot the indicated nodal displacements, as well as the forces normal to the axis X, and discuss the results. Check the equilibrium of forces across the X axis.

9. It is required to write a computer program for the finite-element analysis of plane elastic structures by use of the constant-stress triangle discussed in Sec. 16.4. This is to be done by modification of program FRAME (see Appendix A) and its subroutines. Note that the basic organization of that program is suitable here. It is mainly the numbering and generation of the element stiffness matrix that has to be changed to suit the new problem. Use the program to solve Prob. 10.

10. The square flat elastic plate of Prob. 8 is subjected to the uniform edge loading shown; note that now it is sufficient to consider an octant of the entire structure.

a. Analyze the structure using the finite-element idealization shown, and replacing the applied load by its nodal resultants.

b. Now, subdivide the structure into right triangles of one half the previous side length, $l/2$, and analyze again.

c. Compare the results of parts a and b of the problem, and discuss the convergence of the approach.

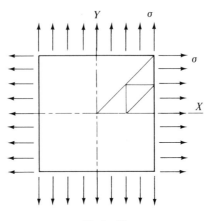

Prob. 10

APPENDIX A

Computer Programs

Main Programs:

TRUSS
FRAME

Subroutines:

LINST
LINAS
SOLN
FORCE
MATINV
MULT
TRANSM
TRMULT
MIN
MOUT

```
      PROGRAM TRUSS (INPUT,OUTPUT,TAPE5=INPUT,TAPE6=OUTPUT)
C
C  THIS PROGRAM ANALYZES THREE-DIMENSIONAL IDEAL STATICALLY DETERMINATE
C  TRUSSES.
C  FOR REQUIRED INPUT, FOLLOW COMMANDS TITLED#READ#, AND .THE ACCOMPANYING
C  FORMAT STATEMENTS.
C  THIS PROGRAM AS PRESENTLY DIMENSIONED CAN ACCOMMODATE TRUSSES OF NO
C  MORE THAN 24 JOINTS, 72 UNKNOWN FORCES ( INCLUDING MEMBER FORCES
C  AND REACTIONS ), AND 72 LOAD COMPONENTS, FOR 3 DIFFERENT LOADING
C  CONDITIONS. IF LARGER STRUCTURES ARE TO BE ACCOMMODATED, NEW
C  DIMENSION STATEMENTS MUST BE WRITTEN.
C
      DIMENSION XCRD(24),YCRD(24),ZCRD(24),LOAD(72,3)  ,FORCE(72,3)
      DIMENSION DCOS(72,72),IND(72,2)
      REAL XCRD,YCRD,ZCRD,LENGTH,DCOSX,DCOSY,DCOSZ,DCOS,LOAD,LLOAD,FORCE
      INTEGER NOJT,LC,JT,NOMEM,NOFOR,FORNO,ENDONE,ENDTWO
C
C         INPUT OF GENERAL TRUSS INFORMATION
C
C
C         DATA INPUT - ONE CARD, CONTAINING
C             NOJT - NUMBER OF JOINTS
C             NOMEM - NUMBER OF MEMBERS
C             NOREACT - NUMBER OF REACTIVE COMPONENTS
C             LC - NUMBER OF LOADING CONDITIONS
C
      READ(5,100)NOJT,NOMEM,NOREACT,LC
  100 FORMAT(4I4)
      WRITE(6,101)
  101 FORMAT(54H0NUMBER OF JOINTS,MEMBERS,REACTIONS,LOADING CONDITIONS/)
      WRITE(6,100)NOJT,NOMEM,NOREACT,LC
C
C         INPUT OF JOINT NUMBERS AND JOINT COORDINATES
C
      WRITE(6,102)
  102 FORMAT(39H0JOINT NUMBER AND X,Y,AND Z COORDINATES/)
C
C         DATA INPUT - ONE CARD FOR EACH JOINT, CONTAINING
C             JT - JOINT NUMBER
C             XCRD(I), YCRD(I),ZCRD(I) - COORDINATES OF JOINT I
C
      DO 103 I=1,NOJT
      READ(5,104)JT,XCRD(I),YCRD(I),ZCRD(I)
  104 FORMAT(1I4,3F10.2)
  103 WRITE(6,104)JT,XCRD(I),YCRD(I),ZCRD(I)
C
C         INITIALIZE DCOS MATRIX AT ZERO
C
      NOFOR=NOMEM+NOREACT
      NNJT=3*NOJT
      DO 109 I=1,NNJT
      DO 109 J=1,NOFOR
  109 DCOS(I,J)=0
C
C         INPUT OF MEMBER NUMBERS AND INCIDENCES,AND CALCULATION
```

Program TRUSS.

```
C          OF MEMBER LENGTH AND DIRECTION COSINES
C
      WRITE(6,105)
  105 FORMAT(57HOMEMBER NUMBER,NEAR AND FAR ENDS,LENGTH,DIRECTION COSINE
     1S/)
C
C          DATA INPUT - ONE CARD FOR EACH MEMBER, CONTAINING
C              FORNO - MEMBER NUMBER, BEGIN WITH 1 AND CONTINUE TO NOMEM
C              ENDONE, ENDTWO - JOINTS CONNECTED BY MEMBER ( MEMBER IN-
C                              CIDENCES )
C
      DO 106 I=1,NOMEM
      READ(5,107)FORNO    ,ENDONE,ENDTWO
  107 FORMAT(3I4)
      LENGTH=SQRT((XCRD(ENDTWO)-XCRD(ENDONE))**2+(YCRD(ENDTWO)-YCRD(ENDO
     1NE))**2+(ZCRD(ENDTWO)-ZCRD(ENDONE))**2)
      DCOSX=(XCRD(ENDTWO)-XCRD(ENDONE))/LENGTH
      DCOSY=(YCRD(ENDTWO)-YCRD(ENDONE))/LENGTH
      DCOSZ=(ZCRD(ENDTWO)-ZCRD(ENDONE))/LENGTH
      WRITE(6,108)FORNO    ,ENDONE,ENDTWO,LENGTH,DCOSX,DCOSY,DCOSZ
  108 FORMAT(3I4,1F10.2,3F10.6)
C
C          INSERTION OF DIRECTION COSINES OF MEMBER FORCES IN DCOS MATRIX
C
      DCOS(3*ENDONE-2,FORNO)=DCOSX
      DCOS(3*ENDONE-1,FORNO)=DCOSY
      DCOS(3*ENDONE  ,FORNO)=DCOSZ
      DCOS(3*ENDTWO-2,FORNO)=-DCOSX
      DCOS(3*ENDTWO-1,FORNO)=-DCOSY
  106 DCOS(3*ENDTWO  ,FORNO)=-DCOSZ
C
C          INPUT OF REACTION INFORMATION,AND INSERTION OF DIRECTION
C          COSINES OF REACTIONS IN DCOS MATRIX
C
      WRITE(6,110)
  110 FORMAT(40HOREACTION NUMBER,JOINT,DIRECTION COSINES/)
      NNMEM=NOMEM+1
C
C          DATA INPUT - ONE CARD FOR EACH REACTIVE COMPONENT, CONTAINING
C              FORNO - REACTION NUMBER, NUMBER CONSECUTIVELY WITH MEMBER
C                      NUMBERING
C              JT - NUMBER OF JOINT ON WHICH THE REACTIVE COMPONENT IS
C                   ACTING
C              DCOSX, DCOSY, DCOSZ - DIRECTION COSINES OF REACTIVE COM-
C                                   PONENT
C
      DO 112 I=1,NOREACT
      READ(5,113)FORNO    ,JT,DCOSX,DCOSY,DCOSZ
  113 FORMAT(2I4,3F10.6)
      DCOS(3*JT-2,FORNO)=DCOSX
      DCOS(3*JT-1,FORNO)=DCOSY
      DCOS(3*JT  ,FORNO)=DCOSZ
  112 WRITE(6,113)FORNO,JT,DCOSX,DCOSY,DCOSZ
C
C          WRITEOUT OF DCOS MATRIX (LEFT HAND SIDE OF FORCE EQUATIONS)
C
```

Program TRUSS (*Cont.*)

```
      WRITE(6,116)
  116 FORMAT(28H0MATRIX OF DIRECTION COSINES/)
      DO 114 I=1,NNJT
  114 WRITE(6,115)(DCOS(I,J),J=1,NOFOR)
  115 FORMAT(10F12.6)
C
C         INVERSION OF DCOS MATRIX
C
      CALL MATINV(DCOS,NOFOR,IND,72)
C
C         WRITEOUT OF INVERSE OF DCOS MATRIX
C
      WRITE(6,120)
  120 FORMAT(23H0INVERSE OF DCOS MATRIX/)
      CALL MOUT(DCOS,NNJT,NNJT,1,72,72)
C
C         INPUT OF LOADING CONDITIONS
C
      DO 128 I=1,LC
      WRITE(6,129) I
  129 FORMAT(44H0LOADS AND DIRECTIONS FOR LOADING CONDITION I2/)
C
C         DATA INPUT - FOR EACH LOADING CONDITION, ONE CARD FOR EACH
C                    LOAD COMPONENT, CONTAINING
C         JT - NUMBER OF JOINT ON WHICH LOAD IS ACTING
C         LLOAD - MAGNITUDE OF LOAD COMPONENT
C         DCOSX, DCOSY, DCOSZ - DIRECTION COSINES OF APPLIED LOAD COM-
C                            PONENTS
C
      DO 128 J=1,NOJT
      READ(5,130)JT,LLOAD,DCOSX,DCOSY,DCOSZ
  130 FORMAT(1I4,1F10.2,3F10.6)
      WRITE(6,130)JT,LLOAD,DCOSX,DCOSY,DCOSZ
C
C         CALCULATION OF RIGHT HAND SIDE (LOAD MATRIX) OF FORCE EQUATIONS
C
      LOAD(3*JT-2,I)=-LLOAD*DCOSX
      LOAD(3*JT-1,I)=-LLOAD*DCOSY
  128 LOAD(3*JT,I)=-LLOAD*DCOSZ
C
C         CALCULATION OF FORCES , (FORCE) =-( DCOS)-1*(LOAD)
C
      CALL MULT(DCOS,NOFOR,NOFOR,72,LOAD,LC,72,FORCE,72)
C
C         WRITEOUT OF FINAL MEMBER FORCES AND REACTIONS
C
      WRITE(6,125)
  125 FORMAT(28H0MEMBER FORCES AND REACTIONS/)
      CALL MOUT(FORCE,NOFOR,LC,1,72,3)
      STOP
      END
```

Program TRUSS (*Cont.*)

```
      PROGRAM FRAME(INPUT,OUTPUT,TAPE5=INPUT,TAPE6=OUTPUT)
C
C         THIS PROGRAM WILL ANALYZE TWO- OR THREE-DIMENSIONAL FRAMED
C         STRUCTURES, SUBJECT TO MEMBER OR JOINT LOADS, OR JOINT DIS-
C         PLACEMENTS. AS PRESENTLY DIMENSIONED, IT CAN HANDLE STRUCTURES
C         OF NO MORE THAN 10 JOINTS, 60 NODAL QUANTITIES, AND 3 LOAD-
C         ING CONDITIONS. IF LARGER STRUCTURES ARE TO BE ANALYZED,
C         THE PROGRAM MUST BE RE-DIMENSIONED. THE NODAL NUMBERING
C         MAY BE ARBITRARY SINCE THE PROGRAM CONTAINS AN AUTOMATIC
C         ORDERING ROUTINE. THE NODES ARE ORDERED BY MEANS OF JOINT
C         RESTRAINT NUMBERS, WHICH ARE 0 FOR FREE, 1 FOR CONSTRAINED
C         NODES.  FOR REQUIRED INPUT DATA, REFER TO EXPLANATORY COMMENTS,
C         TO COMMANDS ENTITLED *READ*, AND ACCOMPANYING FORMAT STATEMENTS
C
      DIMENSION  JT(10),XCRD(10),YCRD(10),ZCRD(10),RESTR(6),ORDER(10,6),
     1FO(60,3),U(60,3),FOSS(60,3),ENDO(20),ENDT(20),E(20),G(20),
     1AREA(20),IX(20),UU(60,3),A(30,30),
     1IY(20),IZ(20),ALPHA(20),T(12,12),SMLC(12,12),SMGC(12,12),
     1FO1(6,3),U1(6,3),SS(60,60), AML(20,12,3),AM(12,3),BM(12,3)
      REAL AML,AM,BM
      REAL XCRD,YCRD,ZCRD,FO,U,FOSS,E,G,AREA,IX,IY,IZ,ALPHA
      INTEGER JT,NOPROBS,NOLEL,NJTS,LC,NONODES,RESTR,LOAD,DEF,SELFSTR,
     1ORDER,ENDO,ENDT,NR,NF
C
C         GENERAL STRUCTURE INFORMATION
C
C
C         DATA INPUT. ONE CARD,
C            NOLEL - NUMBER OF ELEMENTS
C            NJTS - NUMBER OF JOINTS
C            LC - NUMBER OF LOADING CONDITIONS
C
      READ(5,3)NOLEL,NJTS,LC
    3 FORMAT(3I5)
      WRITE(6,40)
   40 FORMAT(H1,5X,*NUMBER OF ELEMENTS, JOINTS, LOADING CONDITIONS*/)
      WRITE(6,3) NOLEL,NJTS,LC
C
C         INITIALIZE STRUCTURE STIFFNESS, FORCE, AND DISPLACEMENT
C         MATRICES AT ZERO
C
      NONODES=6*NJTS
      DO 17 I=1,NONODES
      DO17 J=1,NONODES
   17 SS(I,J)=0.
      NF=0
      NR=0
      DO 13 I=1,NONODES
      DO 13 J=1,LC
      FO(I,J)=0.
      U(I,J)=0.
   13 FOSS(I,J)=0.
C
C         JOINT INFORMATION, ORDERING OF NODAL NUMBERING ( FREE NODES
C         FIRST ), AND SPECIFIED JOINT LOADS AND DISPLACEMENTS
```

Program FRAME.

```
C
C
C          DATA INPUT.ONE CARD FOR EACH JOINT,
C             I - NUMBER OF JOINT
C             XCRD(I),YCRD(I),ZCRD(I) - COORDINATES OF JOINT I
C             RESTR(J) - 0 FOR UNKNOWN DISPLACEMENT, 1 FOR KNOWN
C                        DISPLACEMENT OF NODE J
C             LOAD - 0 IF JOINT IS UNLOADED, 1 IF JOINT LOAD IS SPECIFIED
C             DEF - 0 IF JOINT DISPLACEMENT IS ZERO, 1 IF NON-ZERO JOINT
C                   DISPLACEMENT IS SPECIFIED
C
       DO 12  I=1,NJTS
       READ(5,4)     I,XCRD(I),YCRD(I),ZCRD(I),(RESTR(J),J=1,6),LOAD,DEF
     4 FORMAT(1I5,3F10.4,8I2)
       WRITE(6,41)I
    41 FORMAT(/6X,*X,Y,AND Z COORDINATES AND RESTRAINT NUMBERS FOR JOINT*
      1,I3,/)
       WRITE(6,42) XCRD(I),YCRD(I),ZCRD(I),(RESTR(J),J=1,6)
    42 FORMAT(1X,3F10.4,6I5)
       DO 7 J=1,6
       IF(RESTR(J).EQ.0.)GO TO 6
       ORDER(I,J)=NONODES-NR
       NR=NR+1
       GO TO 7
     6 NF=NF+1
       ORDER(I,J)=NF
     7 CONTINUE
       IF(LOAD.EQ.0.)GO TO 8
C
C          DATA INPUT FOR EACH JOINT WITH LOAD=1. ONE CARD FOR EACH
C          LOADING CONDITION,
C             FO1(J,K) - APPLIED LOAD ON JOINT NODE J FOR LOADING
C                        CONDITION K
C
       DO 10 K=1,LC
       READ(5,9)(FO1(J,K),J=1,6)
     9 FORMAT(6F10.4)
       DO 10 J=1,6
       L=ORDER(I,J)
    10 FO(L,K)=FO1(J,K)
       WRITE(6,43)I
    43 FORMAT(/6X,*SPECIFIED APPLIED LOADS IN STRUCTURE COORDINATES ON JO
      1INT*,I3,/)
       CALL MOUT(FO1,6,LC,1,6,3)
     8 IF(DEF.EQ.0.)GO TO 12
C
C          DATA INPUT FOR EACH JOINT WITH DEF=1. ONE CARD FOR EACH LOADING
C          CONDITION,
C             U1(J,K) - SPECIFIED JOINT DISPLACEMENT ALONG JOINT NODE J,
C                       FOR LOADING CONDITION K
C
       DO 5 K=1,LC
       READ(5,9)(U1(J,K),J=1,6)
       DO 5 J=1,6
       L=ORDER(I,J)
     5 U(L,K)=U1(J,K)
```

Program FRAME (*Cont.*)

```
      WRITE(6,44)I
   44 FORMAT(/6X,*SPECIFIED DISPLACEMENTS IN STRUCTURE COORDINATES OF JO
     1INT*,I3,/)
      CALL MOUT(U1,6,LC,1,6,3)
   12 CONTINUE
      N=NF
C
C         INITIALIZE MATRIX OF FIXED-END JOINT FORCES AT ZERO
C
      DO 31 I=1,NOLEL
      DO 31 J=1,12
      DO 31 K=1,LC
   31 AML(I,J,K)=0
C
C         MEMBER INFORMATION, CALCULATION OF ELEMENT STIFFNESS MATRICES,
C         ASSEMBLY INTO STRUCTURE STIFFNESS MATRIX, AND COMPUTATION
C         OF FIXED-END MEMBER FORCE MATRIX
C
C
C         DATA INPUT. ONE CARD FOR EACH MEMBER,
C             L - MEMBER NUMBER
C             ENDO(L),ENDT(L) - NUMBERS OF JOINTS CONNECTED BY MEMBER L
C             E(L), G(L), - ELASTIC AND SHEAR MODULI OF MEMBER L
C             AREA(L),IX(L),IY(L),IZ(L) - CROSS-SECTIONAL PROPERTIES
C                                       OF MEMBER L
C             ALPHA(L) - FOR GENERAL MEMBER - ANGLE BETWEEN HORIZONTAL AND
C                                       PRINCIPAL CROSS SECTION AXES
C                        FOR VERTICAL MEMBER - ANGLE FROM GLOBAL X3 TO
C                                       LOCAL X3 BAR AXIS, POSITIVE
C                                       IF TOWARD GLOBAL X1 AXIS.
C             LOAD - 0 IF MEMBER IS UNLOADED. 1 IF MEMBER LOAD IS SPECIFIED
C
      DO 14 L=1,NOLEL
      READ(5,15)L,ENDO(L),ENDT(L),E(L),G(L),AREA(L),IX(L),IY(L),IZ(L),
     1ALPHA(L),LOAD
   15 FORMAT(3I5,2F8.0,4F10.2,1F7.5,1I2)
      WRITE(6,19)L
   19 FORMAT(/6X,*MEMBER ENDS,E,G,AREA,IX,IY,IZ,ALPHA FOR ELEMENT*,I3,/)
      WRITE(6,16)ENDO(L),ENDT(L),E(L),G(L),AREA(L),IX(L),IY(L),IZ(L),
     1ALPHA(L)
   16 FORMAT(2I5,2F8.0,4F10.2,1F10.5)
      CALL LINST(L,XCRD,YCRD,ZCRD,ENDO,ENDT,E,G,AREA,IX,IY,IZ,
     1ALPHA,T,SMLC,SMGC)
      CALL LINAS(L,ENDO,ENDT,SMGC,ORDER,SS,NONODES)
      DO 32 I=1,12
      DO 32 J=1,LC
   32 AM(I,J)=0
      IF(LOAD.EQ.0.) GO TO 14
C
C         DATA INPUT FOR EACH MEMBER WITH LOAD=1. TWO CARDS FOR EACH
C         LOADING CONDITION,
C             AM(J,K) - FIXED-END MEMBER FORCE ALONG MEMBER NODE J FOR
C                       LOADING CONDITION K
C
      DO 21 K=1,LC
   21 READ(5,22)(AM(J,K),J=1,12)
```

Program FRAME (*Cont.*)

```
   22 FORMAT(6F10.2)
      WRITE(6,45)L
   45 FORMAT(/6X,*FIXED SUPPORT FORCES DUE TO LOADS ON MEMBER*,I3,*(IN M
     1EMBER COORDINATES)*/)
      CALL MOUT(AM,12,LC,1,12,3)
      DO 11 J=1,12
      DO 11 K=1,LC
   11 AM(J,K)=-AM(J,K)
      DO 23 J=1,12
      DO 23 K=1,LC
   23 AML(L,J,K)=AM(J,K)
      CALL TRANSM(T,12,12,12,AM,LC,12,BM,12)
      II=ENDO(L)
      DO 24 J=1,6
      JJ=ORDER(II,J)
      DO 24 K=1,LC
   24 FOSS(JJ,K)=FOSS(JJ,K)+BM(J,K)
      II=ENDT(L)
      DO 25 J=7,12
      JJ=ORDER(II,J-6)
      DO 25 K=1,LC
   25 FOSS(JJ,K)=FOSS(JJ,K)+BM(J,K)
   14 CONTINUE
C
C        SOLUTION OF FORCE-DEFORMATION EQUATIONS FOR UNKNOWN
C        DISPLACEMENTS AND UNKNOWN JOINT FORCES
C
      CALL SOLN(N,SS,UU,FO,U,FOSS,A,NONODES,LC,ORDER,NJTS,LL,MM)
C
C        CALCULATION OF ELEMENT FORCES IN LOCAL COORDINATES
C
      DO20 L=1,NOLEL
      CALL LINST(L,XCRD,YCRD,ZCRD,ENDO,ENDT,E,G,AREA,IX,IY,IZ,
     1ALPHA,T,SMLC,SMGC)
      CALL FORCE(L,ENDO,ENDT,ORDER,U,SMLC,T,LC,AML,AM)
   20 CONTINUE
      STOP
      END
```

Program FRAME (*Cont.*)

```
      SUBROUTINE LINST(L,XCRD,YCRD,ZCRD,END1,END2,E1,G1,AREA1,
     1IX1,IY1,IZ1,ALPHA1,T,SMLC,SMGC)
C
C     THIS SUBROUTINE GENERATES STIFFNESS MATRICES FOR THREE-
C     DIMENSIONAL PRISMATIC MEMBERS, CONSIDERING AXIAL, BENDING, AND
C     TORSIONAL DISTORTIONS, AND TRANSFORMS THEM INTO GLOBAL
C     COORDINATES
C
      DIMENSION END1(20),END2(20),AREA1(20),E1(20),G1(20),IX1(20),
     1IY1(20),IZ1(20),ALPHA1(20),T(12,12),SMLC(12,12),SMGC(12,12),
     1XCRD(10),YCRD(10),ZCRD(10)
      INTEGER ENDO,ENDT,END1,END2
      REAL AREA,IX,IY,IZ,E,G,EL,XCRD,YCRD,ZCRD,CX,CY,CZ,T,ALPHA,SMLC,
     1SMGC,AREA1,IX1,IY1,IZ1,E1,G1,ALPHA1,ELONE
      ENDO=END1(L)
      ENDT=END2(L)
      AREA=AREA1(L)
      IX=IX1(L)
      IY=IY1(L)
      IZ=IZ1(L)
      E=E1(L)
      G=G1(L)
      ALPHA=ALPHA1(L)
      EL=SQRT((XCRD(ENDT)-XCRD(ENDO))**2+(YCRD(ENDT)-YCRD(ENDO))**2
     1+(ZCRD(ENDT)-ZCRD(ENDO))**2)
      CX=(XCRD(ENDT)-XCRD(ENDO))/EL
      CY=(YCRD(ENDT)-YCRD(ENDO))/EL
      CZ=(ZCRD(ENDT)-ZCRD(ENDO))/EL
      DO1 I=1,12
      DO1 J=1,12
    1 T(I,J)=0.
      T(1,1)=CX
      T(1,2)=CY
      T(1,3)=CZ
      IF(CX+CZ)8,9,8
    8 ELONE = SQRT(CX**2+CZ**2)
      T(2,1)=(-CX*CY*COS (ALPHA)-CZ*SIN (ALPHA))/ELONE
      T(2,2)=SQRT(CX**2+CZ**2)*COS (ALPHA)
      T(2,3)=(-CY*CZ*COS (ALPHA)+CX*SIN(ALPHA))/ELONE
      T(3,1)=(CX*CY*SIN (ALPHA)-CZ*COS (ALPHA))/ELONE
      T(3,2)=-SQRT(CX**2+CZ**2)*SIN (ALPHA)
      T(3,3)=(CY*CZ*SIN (ALPHA)+CX*COS (ALPHA))/ELONE
      GO TO 10
    9 T(2,1)=-CY*COS(ALPHA)
      T(2,3)=CY*SIN(ALPHA)
      T(3,1)=SIN(ALPHA)
      T(3,3)=COS(ALPHA)
   10 DO 2 I=4,6
      DO 2 J=4,6
      JJ=J-3
      II=I-3
    2 T(I,J)=T(II,JJ)
      DO 3 I=7,9
      DO 3 J=7,9
      II=I-3
```

Subroutine LINST.

```
      JJ=J-3
    3 T(I,J)=T(II,JJ)
      DO 4 I=10,12
      DO 4 J=10,12
      II=I-3
      JJ=J-3
    4 T(I,J)=T(II,JJ)
      DO 5 I=1,12
      DO 5 J=1,12
    5 SMLC(I,J)=0.
      SMLC(1,1)=E*AREA/EL
      SMLC(1,7)=-SMLC(1,1)
      SMLC(2,2)=12*E*IZ/EL**3
      SMLC(2,6)= 6*E*IZ/EL**2
      SMLC(2,8)=-SMLC(2,2)
      SMLC(2,12)=SMLC(2,6)
      SMLC(3,3)=12*E*IY/EL**3
      SMLC(3,5)=-6*E*IY/EL**2
      SMLC(3,9)=-SMLC(3,3)
      SMLC(3,11)=SMLC(3,5)
      SMLC(4,4)=G*IX/EL
      SMLC(4,10)=-SMLC(4,4)
      SMLC(5,5)=4*E*IY/EL
      SMLC(5,9)=-SMLC(3,11)
      SMLC(5,11)=2*E*IY/EL
      SMLC(6,6)=4*E*IZ/EL
      SMLC(6,8)=-SMLC(2,6)
      SMLC(6,12)=2*E*IZ/EL
      SMLC(7,7)=SMLC(1,1)
      SMLC(8,8)=-SMLC(2,8)
      SMLC(8,12)=-SMLC(2,12)
      SMLC(9,9)=-SMLC(3,9)
      SMLC(9,11)=-SMLC(3,11)
      SMLC(10,10)=-SMLC(4,10)
      SMLC(11,11)=SMLC(5,5)
      SMLC(12,12)=SMLC(6,6)
      DO 6 I=2,12
      II=I-1
      DO 6 J=1,I1
    6 SMLC(I,J)=SMLC(J,I)
      CALL TRMULT(T,12,12,12,SMLC,12,12,SMGC,12)
      RETURN
      END
```

Subroutine LINST. (*Cont.*)

```
      SUBROUTINE LINAS(L,END1,END2,SMGC,ORDER,SS,NONODES)
C
C         THIS SUBROUTINE ASSEMBLES THE PREVIOUSLY GENERATED ELEMENT
C         STIFFNESS MATRICES INTO THE STRUCTURE STIFFNESS MATRIX
C
      DIMENSION SS(60,60),ORDER(10,6),END1(10),END2(10),SMGC(12,12)
      REAL SMGC,SS
      INTEGER L,ENDO,ENDT,ORDER,END1,END2
      ENDO=END1(L)
      ENDT=END2(L)
      DO 1 J=1,6
      DO 1 K=1,6
      JJ=ORDER(ENDO,J)
      KK=ORDER(ENDO,K)
    1 SS(JJ,KK)=SS(JJ,KK)+SMGC(J,K)
      DO 2 J=7,12
      DO 2 K=7,12
      JJ=ORDER(ENDT,J-6)
      KK=ORDER(ENDT,K-6)
    2 SS(JJ,KK)=SS(JJ,KK)+SMGC(J,K)
      DO 3 J=1,6
       DO 3 K=7,12
      JJ=ORDER(ENDO,J)
      KK=ORDER(ENDT,K-6)
    3 SS(JJ,KK)=SS(JJ,KK)+SMGC(J,K)
      DO 4 J=7,12
      DO 4 K=1,6
      JJ=ORDER(ENDT,J-6)
      KK=ORDER(ENDO,K)
    4 SS(JJ,KK)=SS(JJ,KK)+SMGC(J,K)
      RETURN
      END
```

Subroutine LINAS.

```
      SUBROUTINE SOLN(N,ST,UU,FO,U,FOSS,A,NONODES,LC,ORDER,NJTS,LL,MM)
C
C         THIS SUBROUTINE SOLVES THE SET OF FORCE-DISPLACEMENT EQUATIONS
C         FOR THE UNKNWON JOINT DISPLACEMENTS, AND FOR THE UNKNOWN RE-
C         ACTIVE FORCES
C
      DIMENSION ST(60,60),A(30,30),UU(60,3),FO(60,3),U(60,3),FOSS(60,3),
     1FF(60,3),FI(60,3),ORDER(10,6),IND(30,2)
      REAL ST,A,UU,U,FO,FOSS,FF,FI
      INTEGER N,ORDER,LC,NJTS
      NX=N+1
      M=NONODES
      DO 18 I=1,N
      DO 18 J=1,N
   18 A(I,J)=ST(I,J)
      CALL MATINV(A,N,IND,30)
      DO 31 I=1,M
      DO 31 J=1,LC
   31 UU(I,J)=0
   45 FORMAT(6F16.4)
      DO 36 I=1,N
      DO 36 K=1,LC
      UU(I,K)=0
      DO 36 J=NX,M
   36 UU(I,K)=UU(I,K)+ST(I,J)*U(J,K)
      DO 37 I=1,N
      DO 37 K=1,LC
   37 UU(I,K)=FO(I,K)+FOSS(I,K)-UU(I,K)
      CALL MULT(A,N,N,30,UU,LC,60,U,60)
      WRITE(6,49)
   49 FORMAT(/6X,*FINAL JOINT DISPLACEMENTS(IN STRUCTURE COORDINATES)*/)
      DO 1 I=1,NJTS
      DO 1 J=1,6
      II=ORDER(I,J)
    1 WRITE(6,45)(U(II,K),K=1,LC)
      DO 41 I=NX,M
      DO 41 K=1,LC
      FI(I,K)=0
      DO 41 J=1,N
   41 FI(I,K)=FI(I,K)+ST(I,J)*U(J,K)
      DO 42 I=NX,M
      DO 42 K=1,LC
      FF(I,K)=0
      DO 42 J=NX,M
   42 FF(I,K)=FF(I,K)+ST(I,J)*U(J,K)
      DO 43 I=NX,M
      DO 43 K=1,LC
   43 FO(I,K)=-FOSS(I,K)+FI(I,K)+FF(I,K)
      WRITE(6,50)
   50 FORMAT(/6X,*FINAL JOINT FORCES(IN STRUCTURE COORDINATES)*/)
      DO 2 I=1,NJTS
      DO 2 J=1,6
      II=ORDER(I,J)
    2 WRITE(6,45)(FO(II,K),K=1,LC)
      RETURN
      END
```

Subroutine SOLN.

```
      SUBROUTINE FORCE(L,END1,END2,ORDER,U,SMLC,T,LC,AML,AM)
C
C          THIS SUBROUTINE CALCULATES THE MEMBER END FORCES IN LOCAL
C          COORDINATES
C
      DIMENSION END1(20),END2(20),ORDER(10,6),U(60,3),SMLC(12,12),
     1T(12,12),D(12,3),FOSS(12,3),AML(20,12,3),AM(12,3)
      INTEGER ENDO,ENDT,ORDER,END1,END2
      REAL D,U,SMLC,T,FOSS
      ENDO=END1(L)
      ENDT=END2(L)
      DO 7 J=1,12
      DO 7 K=1,LC
    7 AM(J,K)=AML(L,J,K)
      DO 2 I=1,6
      II=ORDER(ENDO,I)
      DO 2 K=1,LC
    2 D(I,K)=U(II,K)
      DO 3 I=7,12
      II=ORDER(ENDT,I-6)
      DO 3 K=1,LC
    3 D(I,K)=U(II,K)
      CALL MULT(T,12,12,12,D,LC,12,FOSS,12)
      CALL MULT(SMLC,12,12,12,FOSS,LC,12,D,12)
      DO 4 I=1,12
      DO 4 J=1,LC
    4 SMLC(I,J) =-AM(I,J) + D(I,J)
      WRITE(6,5)L
    5 FORMAT(/6X,*FINAL MEMBER END FORCES FOR ELEMENT*,I3,* ( MEMBER COO
     )RDINATES )*/)
      CALL MOUT(SMLC,12,LC,1,12,12)
      RETURN
      END
```

Subroutine FORCE.

```
      SUBROUTINE MATINV(A,N,IND,MAX)
C
C  THIS SUBROUTINE COMPUTES
C  A = A INVERSE,
C  WHERE A ( ORDER N * N ) IS STORED IN DIMENSION OF  MAX  ROWS
C  AND MUST MATCH THE NUMBER OF ROWS FOR WHICH IT IS DIMENSIONED
C  IN MAIN PROGRAM
C
C
      DIMENSION A(MAX,MAX),IND(MAX,2)
      DO 550 I=1,N
      IND(I,1)=0
  550 CONTINUE
      II=0
  551 AMAX=-1.0
      DO 552 I=1,N
      IF (IND(I,1).NE.0) GO TO 552
      DO 554 J=1,N
      IF (IND(J,1).NE.0) GO TO 554
      TEMP=ABS(A(I,J))
      TE=TEMP-AMAX
      IF (TE.LE.0) GO TO 554
      IR=I
      JC=J
      AMAX=TEMP
  554 CONTINUE
  552 CONTINUE
      IF (AMAX.EQ.0.) GO TO 555
      IF (AMAX.LT.0.) GO TO 556
      IND(JC,1)=IR
C INTERCHANGE ROWS
      IF (IR.EQ.JC) GO TO 559
      DO 560 J=1,N
      TEMP=A(IR,J)
      A(IR,J)=A(JC,J)
      A(JC,J)=TEMP
  560 CONTINUE
      II=II+1
      IND(II,2)=JC
C INVERSION BY GAUSS-JORDAN METHOD
  559 PIVOT=A(JC,JC)
      A(JC,JC)=1.0
      PIVOT=1.0/PIVOT
      DO 563 J=1,N
      A(JC,J)=A(JC,J)*PIVOT
  563 CONTINUE
      DO 565 I=1,N
      IF (I.EQ.JC) GO TO 565
      TEMP=A(I,JC)
      A(I,JC)=0.
      DO 566 J=1,N
      A(I,J)=A(I,J)-A(JC,J)*TEMP
  566 CONTINUE
  565 CONTINUE
      GO TO 551
```

Subroutine MATINV.

```
C REARRANGE ROWS
  568 JC=IND(II,2)
      IR=IND(JC,1)
      DO 569 I=1,N
      TEMP=A(I,IR)
      A(I,IR)=A(I,JC)
      A(I,JC)=TEMP
  569 CONTINUE
      II=II-1
  556 IF (II.EQ.0) GO TO 570
      GO TO 568
  555 WRITE (6,820)
  820 FORMAT(1H1,3X,22HZERO PIVOT IN A MATRIX)
  570 RETURN
      END
```

Subroutine MATINV (*Cont.*)

```
      SUBROUTINE MULT(A,M1,N1,M1DIM,B,N2,M2DIM,C,M3DIM)
C
C       THIS SUBROUTINE COMPUTES
C       C = A * B,
C       WHERE A IS OF ORDER ( M1 * N1 ),
C             B IS OF ORDER ( N1 * N2 ),
C             C IS OF ORDER ( M1 * N2 ),
C       AND WHERE A, B, AND C ARE STORED IN DIMENSIONS OF M1DIM,
C       M2DIM, AND M3DIM ROWS, RESPECTIVELY
C
      DIMENSION A(M1DIM,1),B(M2DIM,1),C(M3DIM,1)
      DO 1 I=1,M1
      DO 1 J=1,N2
      C(I,J)=0
      DO 1 K=1,N1
    1 C(I,J)=C(I,J)+A(I,K)*B(K,J)
      RETURN
      END




      SUBROUTINE TRANSM(A,M1,N1,M1DIM,B,N2,M2DIM,C,M3DIM)
C
C       THIS SUBROUTINE COMPUTES
C       C  = A TRANSPOSE * B,
C       WHERE A IS OF ORDER ( M1 * N1 )
C       B IS OF ORDER ( M1 * N2 ),
C       C IS OF ORDER ( N1 * N2 ),
C       AND A, B, AND C ARE STORED IN DIMENSIONS OF M1DIM, M2DIM,
C       AND M3DIM ROWS, RESPECTIVELY
C
      DIMENSION A(M1DIM,1),B(M2DIM,1),C(M3DIM,1)
      DO 1 I=1,N1
      DO 1 K=1,N2
      C(I,K)=0
      DO 1 J=1,M1
    1 C(I,K)=C(I,K)+A(J,I)*B(J,K)
      RETURN
      END




      SUBROUTINE TRMULT(A,M1,N1,M1DIM,B,M2DIM,C,M3DIM,D,M4DIM)
C
C       THIS SUBROUTINE COMPUTES
C       C = A TRANSPOSE * B * A ,
C       WHERE A IS OF ORDER (M1 * N1),
C       B IS OF ORDER ( M1 * M1 ),
C       C IS OF ORDER (N1 * N1),
C       AND A, B,C, AND D ARE STORED IN DIMENSIONS OF M1DIM, M2DIM,
C       M3DIM, AND M4DIM ROWS, RESPECTIVELY
C
      DIMENSION A(M1DIM,1),B(M2DIM,1),C( M3DIM,1),D(M4DIM,1)
      CALL TRANSM(A,M1,N1,M1DIM,B,M1,M2DIM,D,M4DIM)
      CALL MULT(D,N1,M1,M4DIM,A,N1,M1DIM,C,M3DIM)
      RETURN
      END
```

Subroutines MULT, TRANSM, and TRMULT.

```
      SUBROUTINE MIN(A,NI,NJ,M,MDIM,NDIM)
C
C          THIS SUBROUTINE READS MATRIX A ( OF ORDER NI*NJ ),
C          STORED IN DIMENSION ( MDIM,NDIM ) IN MAIN PROGRAM,
C          FROM DATA CARDS OF THE FOLLOWING FORMATS,
C             IF M=0,  FORMAT OF DATA CARDS MUST BE  8F10.2
C             IF M=1,  FORMAT OF DATA CARDS MUST BE  8E10.3
C
      DIMENSION A(MDIM,NDIM)
      IF(M)30,30,35
   30 DO 40 I=1,NI
   40 READ(5,45)(A(I,J),J=1,NJ)
   45 FORMAT(8F10.2)
      GO TO 60
   35 DO 50 I=1,NI
   50 READ(5,55)(A(I,J),J=1,NJ)
   55 FORMAT(8E10.3)
   60 RETURN
      END




      SUBROUTINE MOUT(A,NI,NJ,M,MDIM,NDIM)
C
C THIS SUBROUTINE PRINTS OUT MATRIX  A  ( ORDER NI,NJ ), STORED IN
C DIMENSION ( MDIM,NDIM) IN MAIN PROGRAM.
C FOR ROW AND COLUMN HEADINGS, SET  M = 1, OTHERWISE  M = 0 .
C
      DIMENSION A(MDIM,NDIM)
      JE=0
      KEND=0
   10 JB=JE+1
      JE=JB+5
      IF(JE-NJ)25,20,20
   20 JE=NJ
      KEND=1
   25 IF(M)41,41,30
   30 WRITE(6,40)(J,J=JB,JE)
   40 FORMAT(11HOROW/COLUMN,I5,5I19)
   41 DO 50 I=1,NI
   50 WRITE(6,60)I,(A(I,J),J=JB,JE)
   60 FORMAT(I4,6E19.5)
      IF(KEND)10,10,70
   70 RETURN
      END
```

Subroutines MIN and MOUT.

Some Fundamentals of Matrix Algebra

1. Definitions

Matrix [A] is a rectangular array of m rows and n columns of numbers.

$$[A] = \begin{bmatrix} a_{11} & a_{12} & \cdots & a_{1n} \\ a_{21} & a_{22} & & \\ \cdot & & & \\ \cdots & \cdots & a_{ij} & \\ \cdot & & & \\ a_{m1} & a_{m2} & \cdots & a_{mn} \end{bmatrix}$$

The *element* a_{ij} is the number in the ith row, jth column.

The *order* of a matrix denotes the number of rows and columns; order $[A] = (m \times n)$.

Row matrix is a matrix consisting of one row ($m = 1$).

Column matrix or *vector* $\{A\}$ is a matrix consisting of one column ($n = 1$).

Square matrix is a matrix with equal numbers of rows and columns ($m = n$).

Symmetric matrix is a square matrix with $a_{ij} = a_{ji}$, that is to say, the rows and columns can be interchanged without altering the matrix.

Example:

$$[B] = \begin{bmatrix} 1 & 2 & 3 \\ 2 & 4 & 5 \\ 3 & 5 & 6 \end{bmatrix}$$

Anti-symmetric matrix is a square matrix with $a_{ij} = -a_{ji}$.

Example:

$$[C] = \begin{bmatrix} 1 & -2 & -3 \\ 2 & 4 & -5 \\ 3 & 5 & 6 \end{bmatrix}$$

Diagonal matrix is a square matrix with elements $a_{ij} \neq 0$ for $i = j$ (diagonal elements), $a_{ij} = 0$ for $i \neq j$ (off-diagonal elements).

Example:

$$[D] = \begin{bmatrix} 1 & 0 & 0 \\ 0 & 2 & 0 \\ 0 & 0 & 3 \end{bmatrix}$$

Unit or identity matrix [I] is a diagonal matrix with elements $a_{ij} = 1$ for $i = j$.

Example:

$$[I] = \begin{bmatrix} 1 & 0 & 0 \\ 0 & 1 & 0 \\ 0 & 0 & 1 \end{bmatrix}$$

Transpose $[B] = [A]^T$ *of matrix* [A] is the matrix resulting from interchange of rows and columns, that is, $b_{ij} = a_{ji}$.

Example:

$$[A] = \begin{bmatrix} 1 & 2 & 3 \\ 4 & 5 & 6 \end{bmatrix}; \qquad [B] = [A]^T = \begin{bmatrix} 1 & 4 \\ 2 & 5 \\ 3 & 6 \end{bmatrix}$$

$$(2 \times 3) \qquad\qquad\qquad (3 \times 2)$$

Transpose of a symmetric matrix is the original matrix. Transpose of a row matrix is a column matrix, and vice versa.

2. Matrix Algebra

Scalar multiplication: Factors common to all elements can be taken outside the matrix, thus,

$$n[A] = \begin{bmatrix} na_{11} & na_{12} & \cdots & na_{1n} \\ na_{21} & & & \cdot \\ \cdot & & & \cdot \\ \cdot & & & \cdot \\ \cdot & & & \\ na_{m1} & \cdots & \cdots & na_{mn} \end{bmatrix}$$

Example:

$$[A] = \begin{bmatrix} 3 & 12 \\ 6 & 15 \\ 9 & 18 \end{bmatrix} = 3 \cdot \begin{bmatrix} 1 & 4 \\ 2 & 5 \\ 3 & 6 \end{bmatrix}$$

Addition and subtraction: Matrices are added or subtracted by addition or subtraction of corresponding elements:

$$[A] + [B] = \begin{bmatrix} (a_{11} + b_{11}) & (a_{12} + b_{12}) & \cdots & (a_{1m} + b_{1m}) \\ (a_{21} + b_{21}) & \cdots & & \\ \cdot & & & \\ \cdots & & \cdots & (a_{ij} + b_{ij}) \\ \cdot & & & \\ (a_{m1} + b_{m1}) & \cdots & & (a_{mn} + b_{mn}) \end{bmatrix}$$

It follows that only matrices of like order can be added or subtracted.

Example:

$$[A] = \begin{bmatrix} 1 & 2 & 3 \\ 4 & 5 & 6 \end{bmatrix}; \quad [B] = \begin{bmatrix} -1 & -3 & -5 \\ -2 & -4 & -6 \end{bmatrix}; \quad [A] + [B] = \begin{bmatrix} 0 & -1 & -2 \\ 2 & 1 & 0 \end{bmatrix}$$

Matrix addition is commutative and associative:

$$[A] + [B] = [B] + [A]$$
$$[A] + [[B] + [C]] = [[A] + [B]] + [C]$$

Matrix Multiplication:

$$\underset{(m \times n)}{[C]} = \underset{(m \times l)}{[A]} \cdot \underset{(l \times n)}{[B]} \quad \text{if } C_{ij} = \sum_{\alpha=1}^{l} a_{i\alpha} b_{\alpha j}$$

This means that the element C_{ij} is obtained by multiplying the first element of row i of $[A]$ by the first element of column j of $[B]$, adding to it the product of the second elements of row i of $[A]$ and column j of $[B]$, respectively, and so on to the end of row and column. The matrices $[A]$ and $[B]$ must be conformable, that is, the number of columns of $[A]$ must be equal to the number of rows of $[B]$.

Example:

$$[A] = \begin{bmatrix} 1 & 2 \\ 3 & 4 \end{bmatrix}; \quad [B] = \begin{bmatrix} 1 & 2 & 3 \\ 4 & 5 & 6 \end{bmatrix}.$$

$$[C] = \begin{bmatrix} 1 & 2 \\ 3 & 4 \end{bmatrix}\begin{bmatrix} 1 & 2 & 3 \\ 4 & 5 & 6 \end{bmatrix} = \begin{bmatrix} 9 & 12 & 15 \\ 19 & 26 & 33 \end{bmatrix}.$$

For instance, $C_{21} = 3 \times 1 + 4 \times 4 = 19$.

Matrix multiplication is not commutative: $[A][B] \neq [B][A]$. For $[A][B]$, $[B]$ is *premultiplied* by $[A]$, and $[A]$ is *postmultiplied* by $[B]$. Associative and distributive laws are valid:

$$[A][[B] + [C]] = [A][B] + [A][C]$$
$$[A][[B][C]] = [[A][B]][C]$$

Special Case: Multiplication by unit or identity matrix $[I]$:

$$[I][A] = [A][I] = [A]$$

Transpose of matrix products:

$$[[A][B]]^T = [B]^T[A]^T$$
$$[[A][B][C]]^T = [C]^T[B]^T[A]^T$$

A matrix product is transposed by transposing the individual matrices, and multiplying them in reversed order.

Example:

$$\left[\begin{bmatrix} 1 & 2 \\ 3 & 4 \end{bmatrix}\begin{bmatrix} 6 & 5 & 4 \\ 3 & 2 & 1 \end{bmatrix}\right]^T = \begin{bmatrix} 6 & 3 \\ 5 & 2 \\ 4 & 1 \end{bmatrix}\begin{bmatrix} 1 & 3 \\ 2 & 4 \end{bmatrix} = \begin{bmatrix} 12 & 30 \\ 14 & 23 \\ 6 & 16 \end{bmatrix}$$

$$(2 \times 2) \quad (2 \times 3) \qquad (3 \times 2)(2 \times 2) \qquad (3 \times 2)$$

Partitioning of matrices:

Matrices can be partitioned into *sub-matrices*; each sub-matrix can be treated as an element, subject to all rules of matrix algebra:

$$[A] = \begin{bmatrix} a_{11} & a_{12} & a_{13} \\ a_{21} & a_{22} & a_{23} \\ a_{31} & a_{32} & a_{33} \end{bmatrix} = \begin{bmatrix} [A_{11}] & [A_{12}] \\ [A_{21}] & [A_{22}] \end{bmatrix}$$

in which

$$[A_{11}] = [a_{11} \quad a_{12}]; \qquad [A_{12}] = [a_{13}]$$

$$[A_{21}] = \begin{bmatrix} a_{21} & a_{22} \\ a_{31} & a_{32} \end{bmatrix}; \qquad [A_{22}] = \begin{bmatrix} a_{23} \\ a_{33} \end{bmatrix}$$

3. Simultaneous Equations and the Matrix Inverse

A set of simultaneous equations

$$a_{11}x_1 + a_{12}x_2 + a_{13}x_3 = b_1$$
$$a_{21}x_1 + a_{22}x_2 + a_{23}x_3 = b_2$$
$$a_{31}x_1 + a_{32}x_2 + a_{33}x_3 = b_3$$

can be expressed in matrix form as

$$\begin{bmatrix} a_{11} & a_{12} & a_{13} \\ a_{21} & a_{22} & a_{23} \\ a_{31} & a_{32} & a_{33} \end{bmatrix} \begin{bmatrix} x_1 \\ x_2 \\ x_3 \end{bmatrix} = \begin{bmatrix} b_1 \\ b_2 \\ b_3 \end{bmatrix}$$

or

$$[A][X] = [B]$$

in which $[A]$ is a square matrix of known coefficients, $[X]$ is the matrix of unknowns, and $[B]$ is a known matrix.

To solve for the unknowns $[X]$, we premultiply both sides of the matrix equation by a matrix $[A]^{-1}$, called the *inverse* of $[A]$, defined such that

$$[A]^{-1}[A] = [I]$$

Thus,

$$[A]^{-1}[A][X] = [A]^{-1}[B]$$

or

$$[I][X] = [X] = [A]^{-1}[B]$$

The solution $[X]$ is thus found by premultiplying $[B]$ by $[A]^{-1}$. It remains to determine the inverse $[A]^{-1}$.

We shall present a method for inverting a square matrix (the operation is undefined for nonsquare matrices) using determinants; we assume that the notion of a determinant is known to the reader.

We define the following additional concept:

Cofactor A_{ij} of a determinant $|A|$, of order $(m \times m)$:

The cofactor A_{ij} of a determinant $|A|$ is the determinant of the matrix $[A_{ij}]$, obtained by deleting row i and column j of the matrix $[A]$; its sign is given by the formula

$$A_{ij} = (-1)^{i+j} |A_{ij}|$$

that is, when the sum of the struck row and column numbers is even, a plus sign is affixed, when odd, a minus sign.

Adjoint of matrix $[A]$:

The adjoint of matrix $[A]$ is the transpose of the matrix whose elements are the cofactors A_{ij} of the determinant $|A|$; it can therefore by written as $[A_{ij}]^T$.

We now give the rule for obtaining the inverse $[A]^{-1}$ of matrix $[A]$:

$$[A]^{-1} = \frac{[A_{ij}]^T}{|A|}$$

Example: Find the inverse of the matrix

$$[A] = \begin{bmatrix} 3 & 2 & 0 \\ -1 & -2 & 4 \\ 2 & -1 & -3 \end{bmatrix}$$

Solution:

$$|A| = 3(6+4) - 2(3-8) = 40$$

$$[A_{ij}] = \begin{bmatrix} (6+4) & -(3-8) & (1+4) \\ -(-6-0) & (-9-0) & -(-3-4) \\ (8-0) & -(12-0) & (-6+2) \end{bmatrix}$$

$$= \begin{bmatrix} 10 & 5 & 5 \\ 6 & -9 & 7 \\ 8 & -12 & -4 \end{bmatrix}$$

$$[A]^{-1} = \frac{[A_{ij}]^T}{|A|} = \frac{1}{40} \begin{bmatrix} 10 & 6 & 8 \\ 5 & -9 & -12 \\ 5 & 7 & -4 \end{bmatrix}$$

To check, we can verify that $[A]^{-1}[A] = [I]$.

The inversion of matrices of larger order follows the same procedure.

Some Fundamentals of Vector Analysis

In vector analysis, it is essential to distinguish between vector and scalar quantities. The former are designated here by bold-face symbols, such as **A**, whereas the latter are given by the regular weight letters, such as the component a_i; however, to emphasize the difference between a vector, such as **A**, and its magnitude, we shall at times write this by its absolute value symbol $|\mathbf{A}|$.

Vectors

A vector is a quantity of properties defined by magnitude and direction. It is conveniently specified within a system of coordinate axes, as shown in Fig. C.1. Here, the vector **A** is defined by its components a_1, a_2, a_3 along the axes X_1, X_2, X_3:

$$\mathbf{A} = a_1\mathbf{i} + a_2\mathbf{j} + a_3\mathbf{k}$$

where **i**, **j**, **k** are unit vectors along the coordinate axes. The magnitude of **A** is given by

$$|\mathbf{A}| = \sqrt{a_1^2 + a_2^2 + a_3^2}$$

The direction of **A** can be specified in terms of the *direction cosines*, that is,

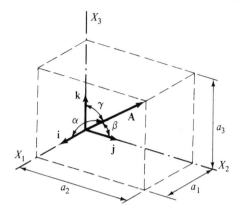

Fig. C.1 Vector.

the cosines of the angles α, β, γ, between it and the coordinate axes:

$$l_1 = \cos \alpha = \frac{a_1}{|\mathbf{A}|}$$

$$l_2 = \cos \beta = \frac{a_2}{|\mathbf{A}|}$$

$$l_3 = \cos \gamma = \frac{a_3}{|\mathbf{A}|}$$

If the axes form an orthogonal set,

$$l_1^2 + l_2^2 + l_3^2 = 1$$

Transformation of Vectors

We consider a vector $\mathbf{A}$, defined by its components a_1, a_2, a_3 within an orthogonal set of axes X_1, X_2, X_3, as shown in Fig. C.2. We now introduce a new set of orthogonal axes $\bar{X}_1$, $\bar{X}_2$, $\bar{X}_3$, and seek the components $\bar{a}_1, \bar{a}_2, \bar{a}_3$ of vector $\mathbf{A}$ along these new axes; they are given by the transformation

$$\begin{Bmatrix} \bar{a}_1 \\ \bar{a}_2 \\ \bar{a}_3 \end{Bmatrix} = \begin{bmatrix} l_{11} & l_{12} & l_{13} \\ l_{21} & l_{22} & l_{23} \\ l_{31} & l_{32} & l_{33} \end{bmatrix} \begin{Bmatrix} a_1 \\ a_2 \\ a_3 \end{Bmatrix}$$

or

$$\{\bar{a}\} = [\lambda]\{a\}$$

in which the elements of the *transformation matrix* $[\lambda]$, l_{ij}, are the direction cosines of the angles between the new axis $\bar{X}_i$ and the old axis X_j.

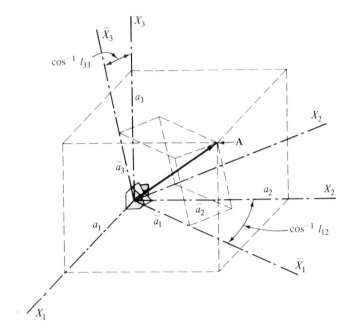

Fig. C.2 Vector transformation.

For orthogonal sets of axes X and $\bar{X}$,

$$[\lambda]^{-1} = [\lambda]^T$$

so that the inverse relation between the vector components is

$$\{a\} = [\lambda]^T\{\bar{a}\}$$

Vectors are also defined as quantities which obey this transformation.

Addition and Subtraction of Vectors

Vectors are added or subtracted by adding or subtracting their components; for instance, in Fig. C.3,

$$\mathbf{C} = \mathbf{A} + \mathbf{B} = (a_1 + b_1)\mathbf{i} + (a_2 + b_2)\mathbf{j} + (a_3 + b_3)\mathbf{k}$$

where a_i and b_i are the components of $\mathbf{A}$ and $\mathbf{B}$, respectively. The vectors $\mathbf{A}$ and $\mathbf{B}$ can also be added, without reference to the coordinate axis system, by the parallelogram law.

Vector addition is associative: $\mathbf{A} + (\mathbf{B} + \mathbf{C}) = (\mathbf{A} + \mathbf{B}) + \mathbf{C}$, as well as commutative: $\mathbf{A} + \mathbf{B} = \mathbf{B} + \mathbf{A}$, that is, the sequence of addition is immaterial.

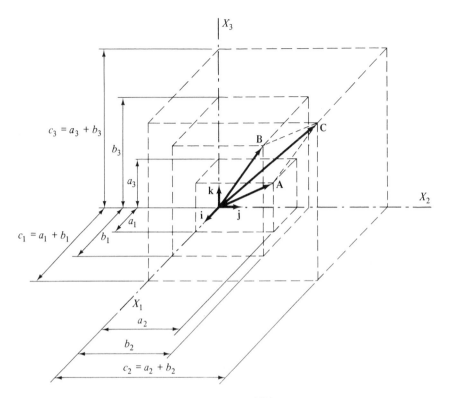

Fig. C.3 Vector addition.

Scalar or Dot Multiplication of Vectors

The product C of a dot multiplication of two vectors **A** and **B** is a scalar representing the magnitude of the projection of one vector upon the other:

$$C = \mathbf{A} \cdot \mathbf{B} = |\mathbf{A}| \cdot |\mathbf{B}| \cdot \cos \alpha,$$

in which α is the angle between **A** and **B**, as shown in Fig. C.4. In terms of the components of **A** and **B**,

$$C = a_1 b_1 + a_2 b_2 + a_3 b_3$$

If **A** and **B** are perpendicular to each other, $\mathbf{A} \cdot \mathbf{B} = 0$; if **A** and **B** are colinear, $\mathbf{A} \cdot \mathbf{B} = |\mathbf{A}| \cdot |\mathbf{B}|$.
The distributive and commutative laws hold for dot multiplication:

$$\mathbf{A} \cdot (\mathbf{B} + \mathbf{C}) = \mathbf{A} \cdot \mathbf{B} + \mathbf{A} \cdot \mathbf{C}; \qquad \mathbf{A} \cdot \mathbf{B} = \mathbf{B} \cdot \mathbf{A}$$

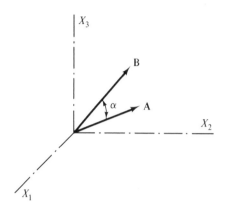

Fig. C.4 Dot multiplication.

Vector or Cross Multiplication of Vectors

The product of a cross multiplication of two vectors **A** and **B** is a vector **C** of direction normal to the plane containing **A** and **B**, of positive sense given by the right-hand rule, turning from **A** to **B**, and of magnitude

$$|\mathbf{C}| = |\mathbf{A}||\mathbf{B}|\sin\alpha$$

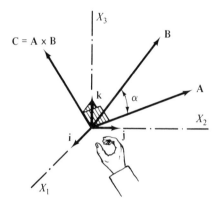

Fig. C.5 Cross multiplication.

in which α is the angle between A and B, as shown in Fig. C.5. In terms of the components of **A** and **B**, the vector **C** is given by the determinant

$$\mathbf{C} = \mathbf{A} \times \mathbf{B} = \begin{vmatrix} \mathbf{i} & \mathbf{j} & \mathbf{k} \\ a_1 & a_2 & a_3 \\ b_1 & b_2 & b_3 \end{vmatrix}$$

$$= (a_2 b_3 - a_3 b_2)\mathbf{i} - (a_1 b_3 - a_3 b_1)\mathbf{j} + (a_1 b_2 - a_2 b_1)\mathbf{k}$$

A useful special case is the following: if $\mathbf{i}$, $\mathbf{j}$, $\mathbf{k}$ are unit vectors along a right-handed orthogonal set of axes X_1, X_2, X_3, then

$$\mathbf{k} = \mathbf{i} \times \mathbf{j}$$

Cross multiplication is non-commutative: $\mathbf{B} \times \mathbf{A} = -(\mathbf{A} \times \mathbf{B})$, and non-associative: $(\mathbf{A} \times \mathbf{B}) \times \mathbf{C} \neq \mathbf{A} \times (\mathbf{B} \times \mathbf{C})$; it is, however, distributive: $\mathbf{C} \times (\mathbf{A} + \mathbf{B}) = \mathbf{C} \times \mathbf{A} + \mathbf{C} \times \mathbf{B}$.

The Triple Scalar Product

The product D of a triple scalar multiplication is a scalar defined by the operation

$$D = (\mathbf{A} \times \mathbf{B}) \cdot \mathbf{C} = \begin{vmatrix} a_1 & a_2 & a_3 \\ b_1 & b_2 & b_3 \\ c_1 & c_2 & c_3 \end{vmatrix}$$

in which the elements within the determinant are the components of the vectors $\mathbf{A}$, $\mathbf{B}$, and $\mathbf{C}$, respectively.

Index